# TOPICS IN INTERSTELLAR MATTER

W0071869

# ASTROPHYSICS AND
# SPACE SCIENCE LIBRARY

A SERIES OF BOOKS ON THE RECENT DEVELOPMENTS

OF SPACE SCIENCE AND OF GENERAL GEOPHYSICS AND ASTROPHYSICS

PUBLISHED IN CONNECTION WITH THE JOURNAL

SPACE SCIENCE REVIEWS

VOLUME 70
PROCEEDINGS

# TOPICS IN
# INTERSTELLAR MATTER

INVITED REVIEWS GIVEN FOR COMMISSION 34 (INTERSTELLAR MATTER)
OF THE INTERNATIONAL ASTRONOMICAL UNION, AT THE
SIXTEENTH GENERAL ASSEMBLY OF IAU, GRENOBLE, AUGUST 1976

*Edited by*

### HUGO VAN WOERDEN

*Kapteyn Astronomical Institute,*
*University of Groningen, Groningen, Netherlands*

President, IAU Commission 34, 1973/76

*Co-Sponsors of sessions:*

IAU Commissions 28 (Galaxies), 33 (Structure and
Dynamics of the Galactic System), 40 (Radio Astronomy)
and 44 (Space Research)

## D. REIDEL PUBLISHING COMPANY

DORDRECHT-HOLLAND/BOSTON-U.S.A.

Library of Congress Cataloging in Publication Data

International Astronomical Union.
    Topics in interstellar matter.

    (Astrophysics and space science library ; v. 70)
    Bibliography: p.
    Includes index.
    1. Interstellar matter–Congresses. I. Woerden, Hugo van. II. Title. III. Series.
QB 790.157  1977        523.1'12        77–21839
ISBN-13: 978-94-010-1256-0        e-ISBN-13: 978-94-010-1254-6
DOI: 10.1007/ 978-94-010-1254-6

Published by D. Reidel Publishing Company,
P.O. Box 17, Dordrecht, Holland

Sold and distributed in the U.S.A., Canada and Mexico
By D. Reidel Publishing Company, Inc.
Lincoln Building, 160 Old Derby Street, Hingham,
Mass. 02043, U.S.A.

TABLE OF CONTENTS

FOREWORD

Interstellar matter is one of the most active fields of research
in present-day astronomy. Observational information spans the full
electromagnetic spectrum from gamma rays through rocket-ultraviolet,
optical, infrared and millimeter to long radio waves. Results of research
in physical chemistry find as much application as mathematical methods.
Interstellar matter plays a leading role in studies of our Galaxy and
of external galaxies, and contributes increasingly to stellar astronomy.

At the 16th General Assembly of the International Astronomical Union,
held in August 1976 in Grenoble, France, the many new developments in
this vast field were surveyed in a number of sessions of Commission 34
(Interstellar Matter), mostly jointly with other Commissions of the
Union. Separate sessions were devoted to: The hot interstellar gas phase,
Interaction of stars and interstellar medium, Interstellar molecules and
dust, The large-scale distribution of interstellar matter in the Galaxy,
and Interstellar matter in external galaxies. Twenty-four invited review
papers were presented and discussed in these sessions.

The quality and success of these topical reviews made it seem
desirable to make them available to a wider audience. Professor Edith
Müller, the new General Secretary of the IAU, enthusiastically supported
the idea. Most importantly, the reviewers - who had originally been pro-
mised that an oral paper was the only requirement - agreed to prepare
written versions. I am grateful to Mrs. Müller, to the authors, and to
Reidel Publishing for their collaboration in the preparation of this
book.

I further wish to thank the Organizing Committee of Commission 34
(Drs. J.E. Baldwin, F.D. Kahn, G.S. Khromov, B. Lynds, T.K. Menon,
D.C. Morton, M. Peimbert and B.J. Robinson) and especially its Vice-
President, Dr. George B. Field, for their share in the preparation of
the meetings. Equally, I am grateful to the (Vice-)Presidents of
Commissions 28, 33, 40 and 44: E.B. Holmberg, L. Perek and F.J. Kerr,
N. Parijskij and A.D. Code for their cooperation. Special thanks go to
Dr. E.B. Jenkins, who organized the Hot-Gas session and edited its Pro-
ceedings. Important suggestions also came from several commission mem-
bers, especially Dr. W.B. Burton.

The sessions were prepared while I was on leave from Groningen at
the Division of Radiophysics, CSIRO, Sydney in 1975/76. I wish to thank
Drs. J.P. Wild and B.J. Robinson for their hospitality, and the office
staff for their efficient help. Mrs. Joan Jones, in particular, through
her tireless efforts at most improbable hours, ensured the timely preparation of the meetings.

Finally, and above all, I wish to thank the Groningen secretaries, Joke Nunnink, Ina Cameron and Roelie Olde, for their dedicated help. They retyped almost half of the manuscripts, and made them really "camera-ready". The final shape of the book is very largely their making.

Hugo van Woerden
Editor

Chapter 1

# THE HOT INTERSTELLAR GAS PHASE

Papers presented in a joint session of IAU Commissions 44
(Space Research) and 34 (Interstellar Matter), Grenoble,
25 August 1976.

Session Chairman:  George B. Field

Chapter Editor:    Edward B. Jenkins

# INTRODUCTION

E.B. Jenkins

One measure of progress in research on interstellar matter is our expanded awareness of the broad variations in the physical state of material dispersed throughout our galaxy.  As they become recognized, the various regimes of density, temperature and ionization are often characterized as discrete "phases," although in some circumstances the boundaries between such phases may not be as well defined as we once thought.  At one end of the spectrum of conditions are the compact gas clouds, rich in dust grains and complex molecules, having temperatures below a few tens of degrees K and densities in excess of $10^{-19}$ g cm$^{-3}$. If we exclude suprathermal particles (cosmic rays), we may identify the other extreme as collisionally ionized gas with a temperature on the order of $10^5$ to $10^7$ K and a mean density of around $10^{-27}$ g cm$^{-3}$.

Our perspective on the existence of a very hot phase of interstellar material, similar to that in the corona of our sun, may be traced back to a proposal by Spitzer in 1956 (Astrophys. J. 124, 20), who outlined indirect evidence for the presence of this gas and showed that such material is relatively immune to dissipation by conductive or radiative cooling.  After a period of dormancy, the viewpoint that coronal gas may be an important constituent of interstellar material was reactivated by a recent surge of theoretical inquiry on origin and maintenance of the gas, coupled with the advent of new observational data which directly confirm its existence in space.  The four articles in this section of the volume, two of them observational and two theoretical, outline some recent conclusions on the properties of the coronal gas and its evolution.

# OBSERVATIONS OF O VI

Edward B. Jenkins
Princeton University Observatory

ABSTRACT

A useful spectroscopic tracer for a hot phase of interstellar gas is the O VI ion, which reaches its maximum concentration in collisional ionization between $10^5$ and $10^6$ K. Presently, over 70 stars have been observed for O VI absorption by the Copernicus satellite. Nearly all of the stars show broad, weak lines, but no evidence favoring a circumstellar origin for the gas can be found. An overall average for $n_e$ of the hot gas in the galactic plane is of order $10^{-3}$ cm$^{-3}$. The relative volume in space occupied by the hot gas regions (and hence their internal density) is uncertain, but a filling factor in the range 0.02 to 0.2 seems most plausible. Fluctuations in radial velocities and column densities suggest there are roughly 6 regions per kpc, each with N(O VI) $\approx 10^{13}$ cm$^{-2}$. The observed rms dispersion of radial velocities for these regions is 26 km s$^{-1}$.

## 1. INTRODUCTION

A prime objective for the ultraviolet spectrometer aboard the Copernicus satellite has been to exploit a spectral region containing resonance lines of interstellar atoms in a variety of ionization stages. As expected, the stellar spectra yielded a rich array of narrow atomic lines, and also many from $H_2$, from which we have distilled information on the abundances, temperatures, excitation and distribution of the dominant, low-temperature phase of interstellar gas (Spitzer and Jenkins, 1975). Somewhat unanticipated, however, was the discovery of broad, shallow absorptions by O VI (Rogerson, York et al., 1973). This lithium-like ion has a strong resonance doublet at 1032 and 1038 Å which is observed in absorption toward a large fraction of the O and B stars studied with the satellite.

We can easily dismiss the notion that the O VI absorption features are attributable to cosmic-ray or X-ray ionization of cool gas clouds. The profile widths are always significantly broader than those from

*Hugo van Woerden (ed.), Topics in Interstellar Matter, 5-16. All Rights Reserved.*
*Copyright © 1977 by D. Reidel Publishing Company, Dordrecht-Holland.*

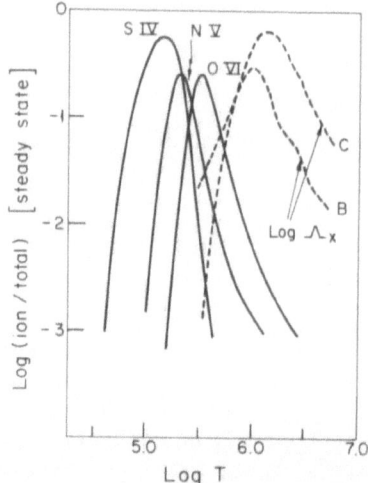

Fig. 1. Relative concentration, as a function of temperature, of coronal ions whose absorption lines can be observed by the <u>Copernicus</u> satellite. These curves, from Shapiro and Moore (1976), are for a plasma whose temperature is not changing with time; the curves exhibit a significant shoulder on the left side if the gas is allowed to cool radiatively since the recombination rate is slower than the cooling rate. For comparison, the X-ray emissivity of the plasma in the lowest energy bands of the Wisconsin experiment is shown by the dashed curves (see Cox 1977).

neutral or weakly ionized atoms, and the minimum velocity dispersion for O VI components is consistent with the doppler broadening expected for a low-density plasma whose temperature gives a maximum fraction of the five-times ionized species of oxygen.

The first survey of O VI absorption over many different directions was carried out by Jenkins and Meloy (1974), who reviewed the spectra recorded in a general survey of interstellar lines during the satellite's first year in orbit. A more detailed examination of the spectra of a few stars was performed by York (1974), who looked for absorptions by N V and S IV, in addition to measuring those of O VI. The near absence of N V and S IV indicated most of the gas is at a temperature above about $3 \times 10^5$ K (York, 1976) -- see Fig. 1. Encouraged by the success of these early observations, Jenkins (1977a) carried out an observing program dedicated to obtaining higher-quality data on O VI toward a well-chosen sample of stars observable by the satellite. The results of this most recent effort, together with conclusions from the earlier work, will be summarized here.

2.  CIRCUMSTELLAR INTERPRETATION

An important phase in the analysis of any lines seen in a stellar

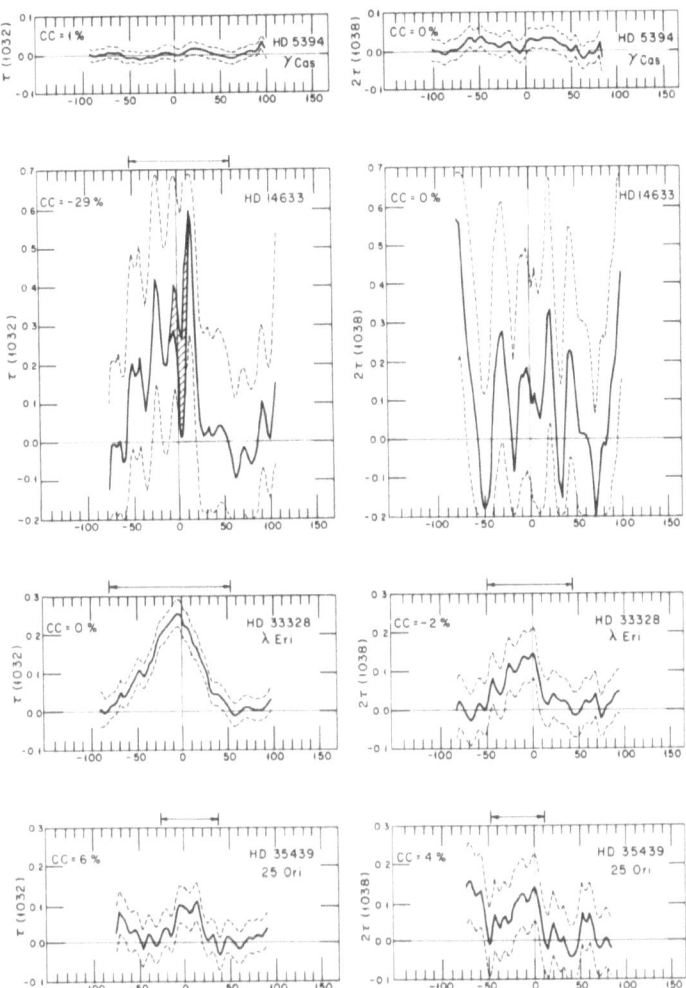

Fig. 2. Typical optical depth tracings for the 1032 and 1038 Å O VI absorptions toward several stars. Velocity scales (abscissae) are in km s$^{-1}$. Dashed lines on either side of the main traces indicate the envelope of a ± 2σ error due to photon count statistics. In many cases, systematic errors in defining the stellar continuum level are more important. The plots here show the variability in data quality from one star to the next, ranging from poor (HD14633) to excellent (HD36695). An upper limit of $10^{12}$ cm$^{-2}$ for the column density toward one star, γ Cas, indicates the sun is not immersed in a coronal gas region with a temperature near the O VI peak shown in Fig. 1.

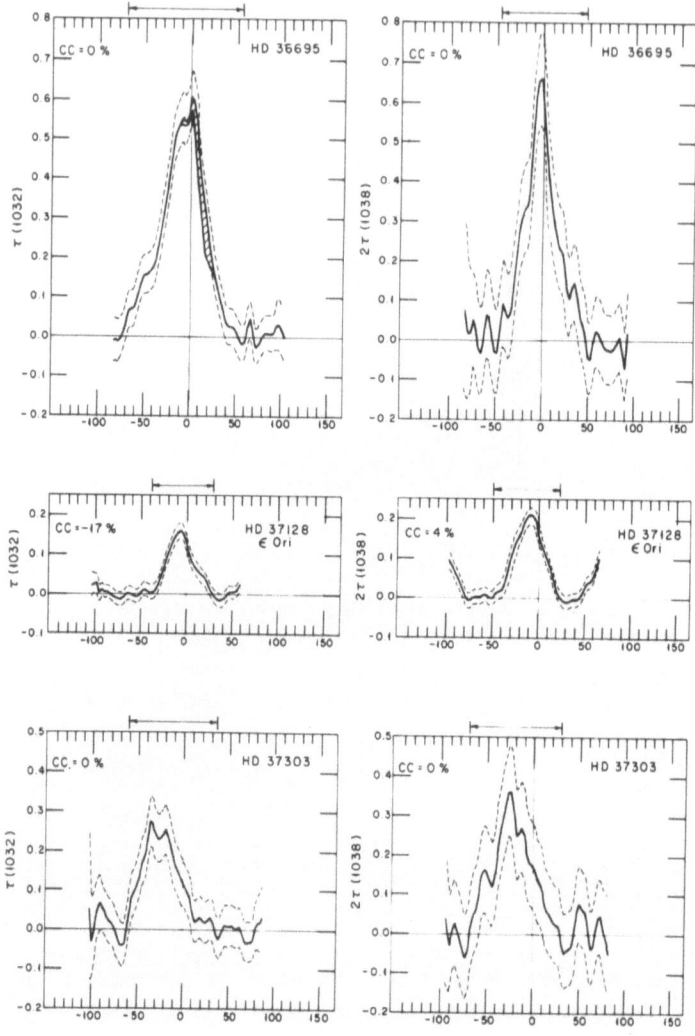

Fig. 2 (continued)

spectrum is to differentiate those originating within the star's photosphere, or a circumstellar shell, from absorptions which are truly interstellar. For stars which are rotating rapidly, O VI profiles are
narrower than normal photospheric lines, although Rogerson and Lamers
(1975) have identified extremely broad stellar O VI components associated with weak mass—loss activity in the B0 V star $\tau$ Sco, and strong
P Cygni profiles are seen in more luminous stars (Morton 1976). One
approach for identifying a circumstellar origin for absorptions of the
type shown in Fig. 2 is to rely on a statistical comparison of the data with various attributes of the stars observed.

Fig. 3 indicates that there is no obvious dynamical relationship
between the O VI gas and the stars, since the points are scattered

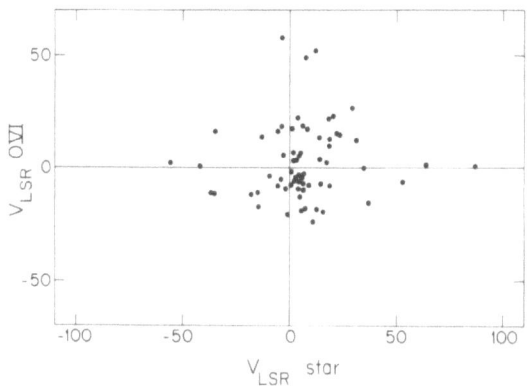

Fig. 3.   Velocity centroids of the O VI profiles versus stellar radial velocities.   All velocities are corrected to the
local standard of rest.

along a horizontal line instead of an incline with a slope of unity.
While there seems to be no overall preponderance of negative velocities,
such as one might expect if gas were being ejected from the stars, we
must acknowledge that an absolute velocity reference for all of the
points in the diagram is poorly determined because of a fairly significant uncertainty in the laboratory wavelength of the O VI transitions.
Nonetheless, it is still possible the gas may be associated with a star
without sharing its velocity; an envelope of hot material may have been
produced by the interaction of the star with ambient interstellar gas.

Castor et al. (1975) suggested that as material ejected from early-
type stars collides with the surrounding gas, a shock is established
and a circumstellar shell of O VI should be observed (see McCray (1977)
for a more up-to-date theoretical treatment).   Even though their theory
suggests the O VI column density varies weakly with the energy deposition rate of the stellar wind, it is still instructive to examine whether or not the amount of O VI seen correlates with mass loss activity
which exhibits striking variations over different spectral types.

The bolometric magnitude $M_b$ of a star is a good index for the
strength of its P Cygni profiles which show the mass loss (Snow and
Morton 1976).   An interpretation of the relationship between $M_b$ and the
O VI column density N(O VI) is, however, made more complicated by a
good correlation between $M_b$ and the star's distance r -- a result of
selecting target stars in the survey to a certain limiting apparent
brightness in the ultraviolet.   We can disentangle the effects of
distance and magnitude by examining the three-way regression of log
N(O VI), $M_b$ and log r.   The analysis shows that at a fixed $M_b$ values
of log N(O VI) are well correlated with log r (partial correlation coefficient $\rho = 0.51$), while there is practically no correlation of log
N(O VI) with $M_b$ for a fixed log r ($\rho = 0.07$).   This lack of correlation, along with other null correlations examined by Jenkins (1977b),

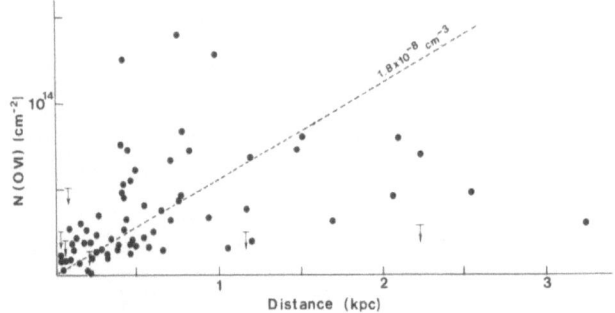

Fig. 4. O VI column density versus distance to the star.
The slope of the dashed line corresponds to $\langle n \, (\text{O VI}) \rangle \equiv$
$\Sigma N \, (\text{O VI})/\Sigma r$ for all 72 stars surveyed by Jenkins and Meloy
(1974) and Jenkins (1977a). To derive an overall average
for the electron density $\langle n_e \rangle$, one should multiply $\langle n \, (\text{O VI}) \rangle$
by a factor which ranges from $1 \times 10^4$ to $5 \times 10^4$, depending
upon the temperature distribution of the material

offers no encouragement for the O VI being primarily attributable to
the effects of mass loss from the star under observation.

## 3. INTERSTELLAR INTERPRETATION

There is good reason to believe the O VI gas regions may be distributed through much of the interstellar space in our galaxy. The production and maintenance of these regions can be the consequence of the
late stages of evolution of supernova remnants, and a general network
of coronal gas may be established with a plausible value for the supernova birth rate (Cox and Smith 1974, Smith 1977, McKee 1977).

The observed relationship of N(O VI) with distance is very irregular, even though a meaningful correlation does exist. Large fluctuations about mean densities for various other interstellar species are
not uncommon however. In effect, we may interpret the variability in
N(O VI)/r as resulting from the random placement of discrete coronal
gas domains, and we are viewing statistical fluctuations in the numbers
of these regions over various paths. The analysis of these fluctuations
by Jenkins (1977b) indicates that roughly 6 regions per kpc, each having
a column density of about $10^{13}$ cm$^{-2}$, are responsible for about 75% of
the gas, while a separate population of thicker, but more sparsely distributed regions makes up the remaining 25%.

An independent confirmation of the discrete domains is provided
by the statistics of profiles' velocity centroids and widths. The
data are consistent with viewing the superposition of components, each
with a mean N(O VI) of $10^{13}$ cm$^{-2}$ which takes on some particular random
velocity. The dispersion of radial velocities corresponds to 26 km s$^{-1}$
after a compensation for the thermal doppler broadening inside each re-

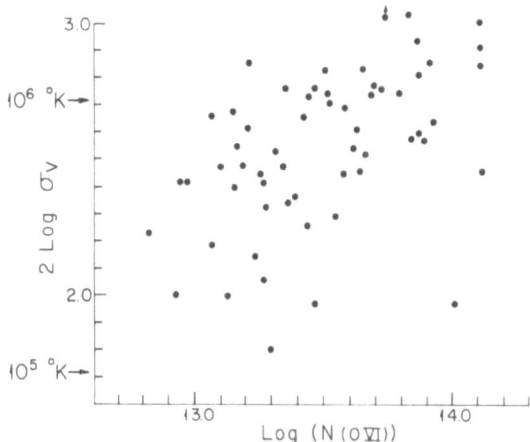

Fig. 5. Log of each profile's velocity width squared plotted against the log of the respective column density. The temperature indicators show the corresponding velocity variance from thermal doppler broadening.

gion. Fig. 5 shows the plausibility of this interpretation: the apparent velocity width increases by approximately the factor $[(n-1)/n]^{\frac{1}{2}}$, where n is the number of components.

A special effort was made by Jenkins (1977a) to observe stars at some distance from the galactic plane, even though such stars suitable for observing are hard to find. Although the scatter of points in Fig. 6 makes the general behavior hard to discern, the suggested trend of high z points below the diagonal is consistent with a density distribution which falls off as $n_0 \exp(-z/h)$, where $n_0 = 2.8 \times 10^{-8}$ $cm^{-3}$ and h = 300pc. It is unclear, however, whether this dropoff represents a decrease in the total amount of coronal gas or whether the gas is tending to avoid the temperatures which favor O VI in collisional equilibrium. The pervasiveness of hot gas in the galactic halo regions is an attractive theoretical concept (Spitzer 1956, Field 1975).

We have yet to define another important attribute of the distribution of coronal gas, namely the relative volume of the galaxy occupied by the hot regions. This average fraction, which we denote as a filling factor f, is a difficult quantity to determine from the observations. One approach is to assume the O VI gas is in approximate pressure equilibrium with ordinary interstellar gas, whose pressure p/k is estimated to be between $10^3$ and $10^4$ $cm^{-3}K$. Any gross pressure imbalance would probably result in velocity dispersions which are higher than observed.

The top of Fig. 7 illustrates how a line of sight of length L might be intersected by the coronal gas regions discussed earlier.

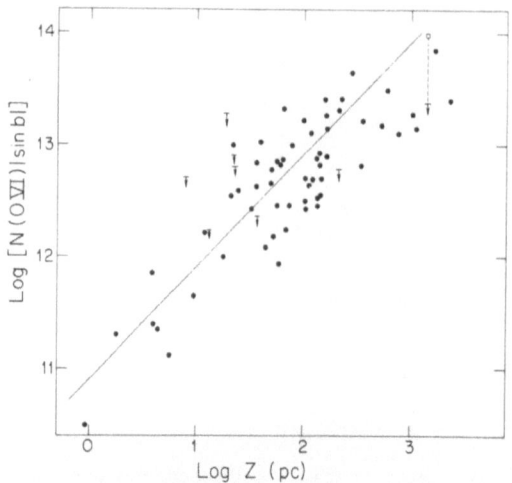

Fig. 6.  An illustration of the behavior of O VI column densities for various distances z above the galactic plane.  The
ordinate is the logarithm of the effective column density
perpendicular to the plane, averaged along a slanted path
with a galactic latitude of b.  One high z star (upper right)
gives an upper limit at 1037 Å which markedly disagrees with
a measurement at 1032 Å.  The diagonal line shows the expected
relation if O VI were distributed uniformly with different
z (i.e., if the scale height were infinite).

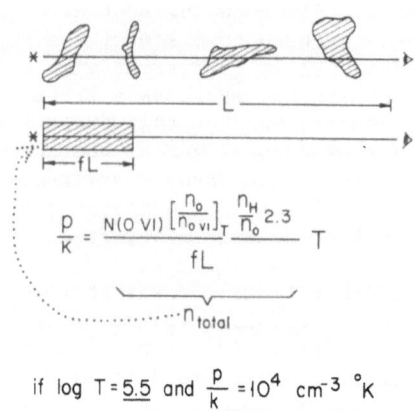

$$\frac{p}{k} = \frac{N(O\ VI)\left[\frac{n_0}{n_{0\ VI}}\right]_T \frac{n_H}{n_0} 2.3}{fL}\ T$$

$n_{total}$

if $\log T = \underline{5.5}$ and $\frac{p}{k} = 10^4\ cm^{-3}\ ^\circ K$

f = 0.014

Fig. 7.  Schematic representation of clouds filling a line of
sight.  The internal pressure p/k is inversely proportional
to the volume filling factor f.

This picture is equivalent to viewing through a single chunk of material

of total length fL.  For a constant density of all particles $n_{total}$
and temperature T inside the chunk, the pressure is given by
the formula shown in the figure, where $[n_O/n_{O\ VI}]_T$ is the inverse of
the curve shown in Fig. 1, $n_H/n_O$ is the abundance of hydrogen to oxygen
(cosmic ratio = 1470), and the factor 2.3 accounts for the electrons
from hydrogen and helium plus the helium nuclei (assuming the number
ratio of helium to hydrogen is 10%).  Values for f as large as unity
are allowed if log T differs appreciably from 5.5 where O VI has its
maximum concentration.  A firm lower limit for f of 0.014 follows by
assuming the highest possible concentration of O VI in the hot regions
and a maximum allowable pressure of $10^4$ $cm^{-3}K$.

A conditional upper limit for f was derived by Jenkins (1977b) who
found no anticorrelation in the relative excesses or deficiencies of
N(O VI) and the stars' color excesses E(B-V).  If f were greater than
0.20 and relatively homogeneous O VI gas regions forced aside or destroyed the normal gas clouds containing the dust, a noticeable anti-
correlation of the departures for N(O VI) and E(B-V) from their average
values would have been observed.  Smith's (1977) models for a hot gas
tunnel network suggest the presence of a slightly larger f, in the
range 0.2 to 0.4, while Jones' (1975) interpretation of the apparent
absence of large supernova remnants in our galaxy suggests $f \approx 0.8$.

Measurements of soft X-ray emission can, in principle, also help
us to define f, since the emissivity of the gas is proportional to
$\langle n_e^2 \rangle$ while the O VI measurements define $\langle n_e \rangle$.  Unfortunately, an evaluation of $f = \langle n_e \rangle^2/\langle n_e^2 \rangle$ is confused by the large difference in the sensitivity to temperature for the two types of observation (see Fig. 1)
and the absorption of X-rays by intervening normal interstellar material (however, see Cox (1977) for ways of overcoming these complications).  A general conclusion from the comparison of O VI and X-rays,
first discussed by Shapiro and Field (1976), is that small f and large
p/k values are needed if indeed the two observations sample the same
gas.  While the large ratio of X-ray flux to O VI density could be
reconciled by having log T as high as 5.9, this value is inconsistent
with most of the O VI profile widths (see Fig. 5).  Even if the ions
in the hot gas volumes are allowed to assume reasonable distributions
over temperature, one is almost forced to conclude the X-rays are coming
from thin, overpressured regions possibly within high-velocity shocks
having negligible O VI column densities, while the O VI observations
represent quiescent coronal gas with a very low X-ray emissivity.

If we change our outlook from general properties of the distribution of coronal gas to specific locations in the sky, we see the
situation illustrated in Fig. 8.  There seem  to be almost no coherent
variations over different regions.  Furthermore, with one exception
there appears to be no obvious correlation with the locations of unusual
phenomena.  For instance, the star HD 149881 (at $\ell = 31°$, $b = 36°$) is
located well behind the North Polar Spur -- a prominent X-ray and radio
continuum source which is probably an old supernova remnant.  Yet this
star displays a smaller than average density of O VI.  On the other

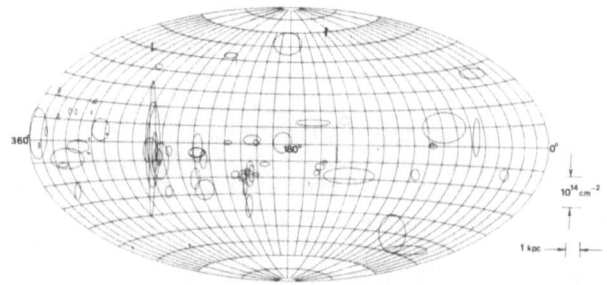

Fig. 8. Locations of O VI measurements in galactic coordinates. Each observation is shown by an ellipse whose height is proportional to N(O VI) and horizontal diameter is proportional to the star's distance. (Hence, the average density of O VI along the line of sight is depicted by the ellipse's eccentricity). Dotted ellipses represent upper limits for N(O VI).

hand, a group of stars studied by Jenkins, et al. (1976a,b) which are near the Vela supernova remnant in projection (see Fig. 9) show abnormally large N(O VI), as is evidenced by the tall, thin ellipses near $\ell = 270°$ in Fig. 8.

———————

This research was supported by contract NAS 5-1810 from the U.S. National Aeronautics and Space Administration. While the discussion here essentially duplicates the subjects covered in the Grenoble meeting, many of the conclusions have been expanded or better defined quantitatively as the result of subsequent research, and these improvements have been incorporated into this paper. The author is grateful to the Lick Observatory for the hospitality extended during his sabbatical leave when much of this work was carried out.

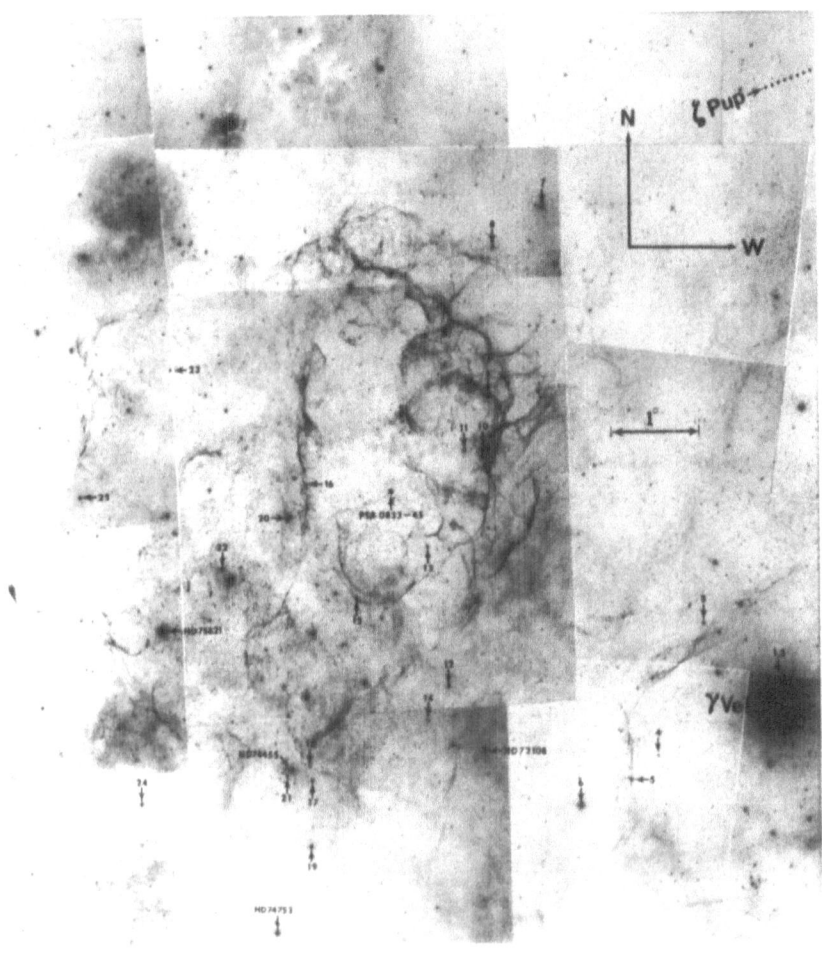

Fig. 9.  Four stars (those labeled with HD numbers) observed
by Jenkins, et al. (1976b) near the Vela supernova remnant.
These stars show abnormally large N(O VI), as well as interstellar gas with very large radial velocities.  Stars identified by small numbers have been observed for interstellar
lines in the visible (see references in Jenkins et al.
(1976b) ).  (Reproduced by courtesy of The Astrophysical
Journal, University of Chicago Press, publisher. © 1976.
American Astronomical Society.  All Rights Reserved).

REFERENCES

Castor, J., McCray, R. and Weaver, R.: 1975, Astrophys. J. Letters 200, L107.

Cox, D.P.: 1977, review article in this volume.

Cox, D.P. and Smith, B.W.: 1974, Astrophys. J. Letters 189, L105.

Field, G.B.: 1975, Astrohys. Space Sci. 38, 167.

Jenkins, E.B.: 1977a, Paper I, in preparation.

_____: 1977b, Paper II, in preparation.

Jenkins, E.B. and Meloy, D.A.: 1974, Astrophys. J. Letters 193, L121.

Jenkins, E.B., Silk, J. and Wallerstein, G.: 1976a, Astrophys. J. Letters 209, L87.

_____: 1976b, Astrophys. J. Suppl. 32, in press.

Jones, E.M.: 1975, Astrophys. J. 201, 377.

McCray, R.: 1977, review article in this volume.

McKee, C.F.: 1977, review article in this volume.

Morton, D.C.: 1976, Astrophys. J. 203, 386.

Rogerson, J.B. and Lamers, H.J.G.L.M.: 1975, Nature 256, 190.

Rogerson, J.B., York, D.G., Drake, J.F., Jenkins, E.B., Morton, D.C. and Spitzer, L.: 1973, Astrophys. J. Letters 181, L110.

Shapiro, P.R. and Field, G.B.: 1976, Astrophys. J. 205, 762.

Shapiro, P.R. and Moore, R.T.: 1976, Astrophys. J. 207, 460.

Smith, B.W.: 1977, Astrophys. J. 211, in press.

Snow, T.P. and Morton, D.C.: 1976, Astrophys. J. Suppl. 32, 429.

Spitzer, L.: 1956, Astrophys. J. 124, 20.

Spitzer, L. and Jenkins, E.B.: 1975, Ann. Rev. Astron. Astrophys. 13, 133.

York, D.G.: 1974, Astrophys. J. Letters 193, L127.

_____: 1976, preprint.

# OBSERVATIONS OF THE SOFT X-RAY BACKGROUND

Donald P. Cox
Department of Physics
University of Wisconsin-Madison

When soft X-rays were first detected with photon energies $\sim 1/4$ KeV, it was unclear whether their origin was galactic or extragalactic. Then the fact that the intensity did not go to zero in the galactic plane where the absorption optical depth is very large demonstrated that they were at least partly galactic. In the last few years, several experiments have attempted to measure the X-ray shadows of the Small Magellanic Cloud, M31, and the Large Magellanic Cloud. All have produced null results. The origin of the bulk of this radiation is convincingly galactic.

The intensity and spectral information from these investigations indicate that the characteristic energy is a fraction of a keV and that the volume emissivity and total soft X-ray luminosity are about $\varepsilon \sim 6 \times 10^{-28}$ erg cm$^{-3}$ s$^{-1}$, $\mathcal{L} \sim 3 \times 10^{39}$ erg s$^{-1}$.

The obvious next step was to find the source of this radiation. Investigations of the brightness fluctuations have shown that if the soft X-rays are produced by a collection of point sources, then these sources must be more numerous than stars. In addition, a number of nonthermal mechanisms have been ruled out because of their unacceptable side effects.

Finally, one possible origin was found which did not seem to contradict other known observations. If the interstellar medium had a component with a temperature of order 1 million degrees and rms electron density of order $4 \times 10^{-3}$ cm$^{-3}$, the soft X-ray background could be generated comfortably by this plasma, primarily by collisional excitation followed by line emission.

This scheme seems straightforward enough; there even appeared to be mechanisms available for generating and maintaining hot gas. But there are fairly serious implications for the interstellar pressure. If this hot component fills a fraction f of the interstellar volume, then the resulting pressure is roughly $P/k = 2nT \sim (8000/f^{1/2})$ cm$^{-3}$ K, $P \sim (10^{-12}/f^{1/2})$ dyne cm$^{-2}$. At the time that this idea originated, it

18

D. P. COX

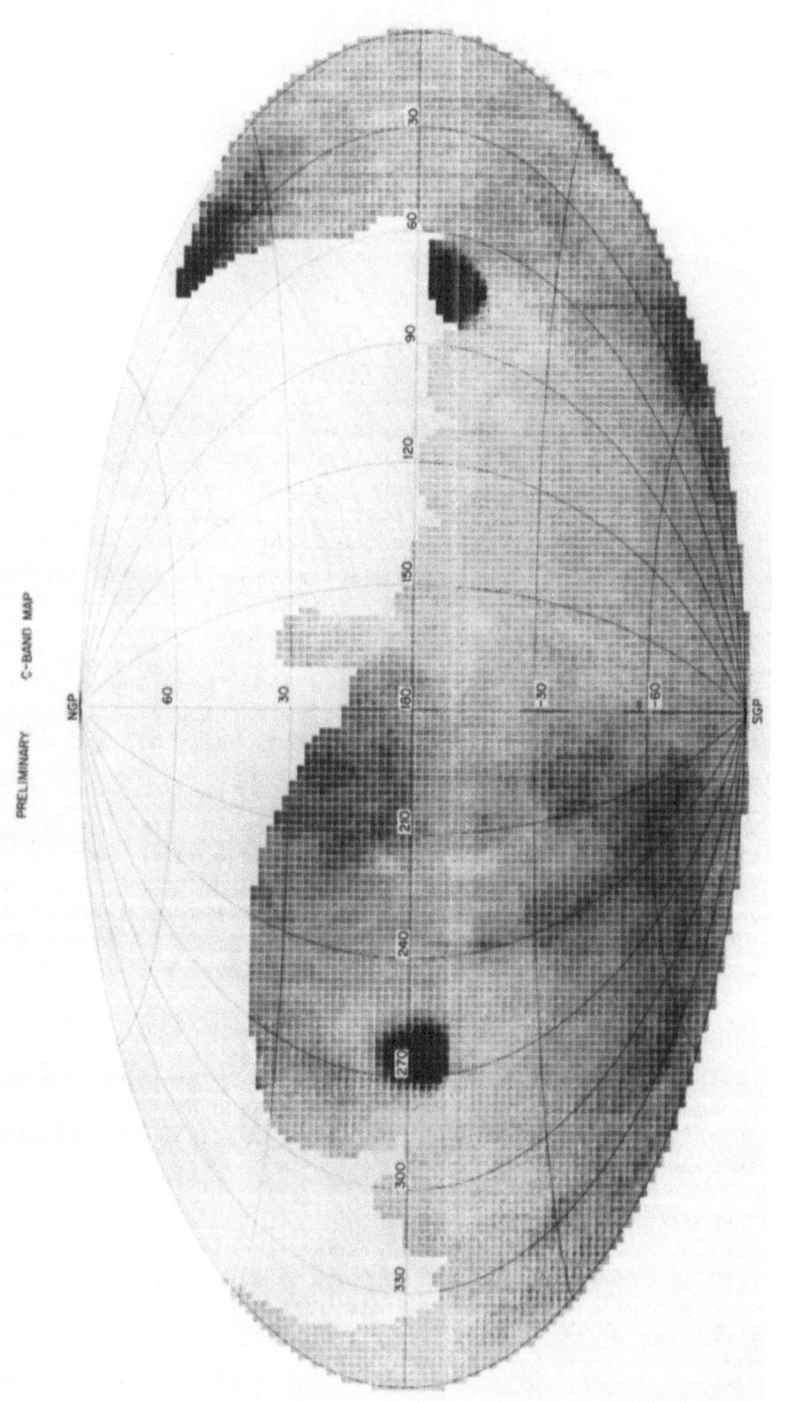

Figure 1.   Soft X-ray surface brightness.   Carbon band.

was widely held that the interstellar medium thermal pressure had $P/k$
$\sim 2000$ cm$^{-3}$ K.  A similar pressure would result from a 3 microgauss
magnetic field.  Hence it was clear that unless the hot medium occupied
a substantial fraction of the interstellar volume and the interstellar
pressure were somewhat higher than earlier believed, this scheme was in
serious jeopardy from its discrepant pressure.

These then are the developments in soft X-ray astronomy which led
to the concept of a pervasive hot phase of the interstellar medium.
These ideas, however, immediately met and intertwined with those de-
riving from O VI measurements at Princeton, leading to a rather stormy
marriage of not quite orthogonal observations; both seeming to want
some very hot interstellar material, but debating its details.  We
turn now to more recent results.

Soft X-ray astronomy is a difficult but rapidly developing field.
In the near future we can expect very significant improvements in
spectral and spacial resolution as well as in sky coverage and counting
statistics.  For the present, however, the instruments available have
spectral resolution of $\Delta$ E over E of about 1 at the relevant energies.
In addition, they have about the same spacial resolution as a light
meter looking through a toilet paper roll.  The instruments, which are
flown on rockets to get above the opaque atmosphere, collect roughly 4
minutes of useful data per flight on 1 to 2 flights per year.  Each
flight can survey about 1/8th of the sky with modest statistics, but
because of the overlap necessary to internormalize the results, about
10 flights are needed to map the entire sky.  Hence this mapping procedure requires at least 5 years of wholly successful flights to collect
the necessary 40 minutes of observations.

The present state of this effort is shown in Figure 1, a mosaic of
6 relatively homogeneous quality Wisconsin Group flights.  It is made
with data from what is called the C-band, corresponding roughly to
X-rays in the energy range 150 to 284 eV.  The galactic anti-center is
in the middle of the map; about 60% of the sky is shown with the remaining sections scheduled for future flights.  Two other such maps at
slightly lower and higher energies are also in preparation.

A number of groups besides Wisconsin have mapped portions of the
soft X-ray sky.  In regions of overlap with this map, there is satisfactory agreement.  It is clear that the features of the map are real
and reproducible.  None are artifacts.  Unfortunately, the data from
these other flights are not available in a form which can easily be
included here.

There are several things on this map which deserve attention.
1)  Two extremely bright sources are seen close to the galactic
plane:  the Cygnus Loop at $\ell \sim 75°$, and the Vela Supernova Remnant at
$\ell \sim 260°$.  Each appears only slightly larger than the resolution to
which the map is to be believed.
2)  A bright finger extends down out of the north along longitude

about 25°. This ridge is generally described as being associated with the North Polar Spur of radio-continuum maps, although the two are not actually coincident.

3) The overall surface brightness range is roughly a factor of 4, excluding the two supernova remnants.

4) If a flight were made, surveying from plane to pole at constant longitude, the variations found would not characterize the southern hemisphere as a whole. The east-west variation is just as substantial as the north-south variation.

5) Several correlations exist between bright X-ray regions and neutral hydrogen features on the Heiles survey. For example:

A. The bright region near $\ell \sim 200°$, $b \sim -45°$, known as the Eridanus Hot Spot, is surrounded by an actively expanding loop of HI.

B. The region near $\ell \sim 210°$, $b \sim +20°$, the Monoceros Hot Spot, is in a region of generally low hydrogen column density.

6) In southern regions outside the original Heiles survey, $N_H$ data are available from McGee, reviewed by Daltabuit and Meyer [1]. There is a striking comparison. The $N_H \stackrel{<}{\sim} 5 \times 10^{20}$ cm$^{-2}$ contour very clearly delineates the extended southern X-ray brightened region. Similarly, the $N_H \sim 1 \times 10^{20}$ cm$^{-2}$ contours are generally in the vicinity of the especially bright X-ray ridge boundaries.

Although bright X-ray emission clearly follows low $N_H$, the correlation thus far remains only qualitative and tantalizing.

There are two possible causes for the anti-correlation between X-ray brightness and neutral hydrogen column density. It could be that the high $N_H$ regions are dimmer only because of the greater absorption of X-rays in these directions. Alternatively, actual displacement of hot gas by neutral material and vice versa may be more significant.

This question is exciting in part because there is a straightforward test to decide which case is more true.

If absorption is the more important, it is strongest at low photon energies and we expect the dimmer regions to have harder spectra (interstellar blueing). If, on the other hand, displacement is more important, then the spectrum might either remain unchanged or perhaps (if the neutral hydrogen displacement were caused by a recent active event) even be harder in the brighter regions.

At present, in the data from 3 separate rocket flights, each taken as a whole, the brighter patches have spectra at least as hard as the less bright regions. Hence displacement appears to be at least as important as absorption in bringing about this correlation. This, by the way, seems to provide independent evidence that in the general vicinity of the sun, both the soft X-ray emitting gas and the neutral hydrogen occupy substantial volume fractions of the galactic disk. If one or the other did not, these displacement arguments would fail.

As a final remark on spacial correlations, I have been authorized to say that if one considers the O VI stars further than 100 pc and with latitudes more than 10° from the plane, the O VI column densities increase, on average, with increasing C-band surface brightness.

We now turn to a discussion of the rather limited but still intriguing spectral information. You realize of course from my earlier remarks that one cannot precisely go from an observed pulse height distribution to an incident spectrum. It is instead necessary to guess what might be an appropriate incident spectrum (for example, the emission from a hot plasma of astrophysical abundances) and to convolve the spectrum with all the instrumental transmissions, efficiencies, and redistribution functions to predict a pulse height distribution. If the number of parameters of the incident spectrum is limited (for example to plasma temperature and column density of intervening gas), this procedure can be repeated for a suitable range of these parameters and the calculated pulse height distributions compared with that observed.

For an example of an analysis relying directly on pulse height distribution matching, I would like to refer you to a recent paper by Levine, Rappaport, Halpern, and Walter of MIT [2]. This group is one of several which have attempted to push their sensitivity to extremely low energies. They report their range to be 90-280 eV and find a reasonable fit to their pulse height distribution for a thermal model with T $\sim$ 700,000 K and rms electron density $n_{rms} \sim 7 \times 10^{-3}$ cm$^{-3}$. These depend somewhat on the amount of intervening material assumed. In addition, these parameters are fit only to this very low energy range. Data from somewhat higher energies would not fit this model, confirming earlier analyses by our group and others that no single temperature spectrum can be made to fit the full range of soft X-ray data.

I would now like to describe briefly an extremely efficient and illuminating analysis by band fractions developed recently in the thesis of Paul Burstein of our group.

The transmission functions for the two thin windows of the proportional counters used by the Wisconsin Group have a high transparency in a low-energy band, a sharp cutoff at an energy which depends on the window coating, a deep minimum at energies just above this cutoff, and an increasing transparency above $\sim$ 400 eV. Because of this property of the windows, the pulse heights can be grouped into bands which correspond meaningfully to initial photon energy.

The B-band (boron) was defined to contain pulse heights in the range 60 to 400 eV for the boron coated window, corresponding to h$\nu$ < 188 eV. The C-band (carbon-coating) was for the same pulse height range in the kimfol window, corresponding to h$\nu$ < 284 eV. The M-band is the sum of pulse heights in the range 400 to 850 eV for both windows. Higher energy pulse heights are also available and are used to put upper limits on temperatures of allowed models.

For any period of observations, there is some total counting rate

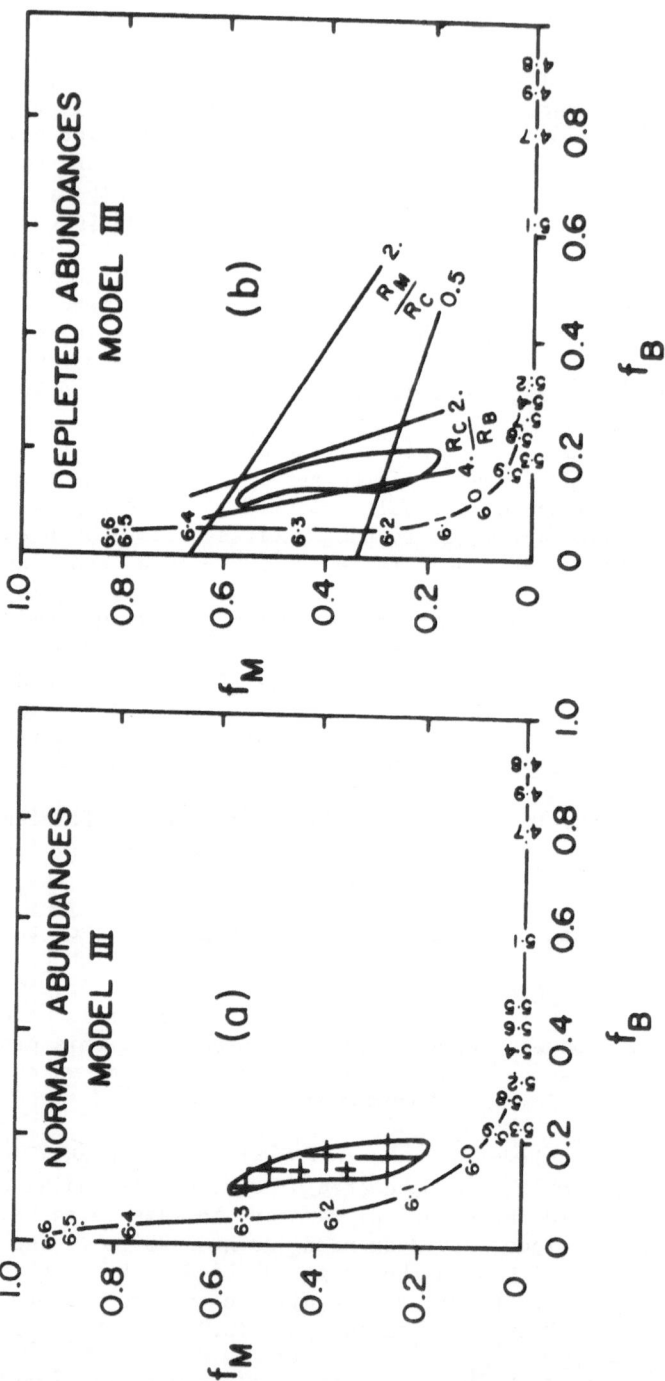

Figure 2. Band fraction plots. Absorbing gas intermixed to $N_H = \infty$.

for the two counters; this rate consists of fractions $f_B$, $f_C$, and $f_M$ (note that $\Sigma f = 1$) in the B, C, and M bands respectively.

Any particular set of band fractions can be displayed as a point in a 2 dimensional plot. Rather than do this logarithmically as in an HR diagram, we use a scheme analogous to chromaticity diagrams in optics. Still, what we are doing is 3-color photometry.

Consider the left half of Figure 2. All possible combinations of band fractions lie within a triangle joining $f_M = 1$, $f_B = 1$ and $f_M = f_B = 0$ ($f_C = 1$). Hence, the data for any particular direction in the sky can be plotted as a point within this triangle. In addition, however, band fractions can be calculated for all conceivable families of models and also plotted. One can then tell at a glance what ranges of parameters can hope to fit the data.

In this study, the flight data all fall in the irregularly shaped contour in the figure. (The more representative locations are in the bottom half of the contour.) The line shown with numbers ranging from 4.7 to 6.6 is the calculated locus of thermal models with normal astrophysical abundances, intermixed with absorbing material out to large optical depth. The numbers indicate $\log_{10}T$. A variety of other models have also been considered. The one on the right for example has depleted abundances.

You see in this figure that high-temperature models have high M-band fraction, low-temperature models have high B-band fractions, and that the model locus skirts between the data regime and the origin for intermediate temperatures. This is true for all reasonable thermal models tried by Burstein. None entered the region of observed data. We find again, and with finality, that no single temperature model is an adequate representation.

But this is only the beginning of the usefulness of the band fraction diagram. Its real strength is that it provides an immediate and straightforward scheme for evaluating 2-temperature models as well. It is easy to show that if we wish to consider two regions along a line of sight, one with a high temperature, $T_H$, and the other with lower temperature, $T_L$, that the combined band fractions must lie on the straight line joining $T_H$ and $T_L$ on the band fraction diagram. This means that if we want to match a particular observation and we know what models we wish to use, a particular choice of $T_H$ leads immediately to a specified value of $T_L$. We also see at a glance the relative proportions of the total count rate which must be due to each of the 2 components. (These are inversely proportional to the distance from the data point.)

Suppose that the models on the left side of Figure 2 are appropriate for each of the 2 temperatures. We immediately notice several things:
    1.  $T_H \gtrsim 10^{6.3} = 2 \times 10^6$ K in order to fit almost any of the data

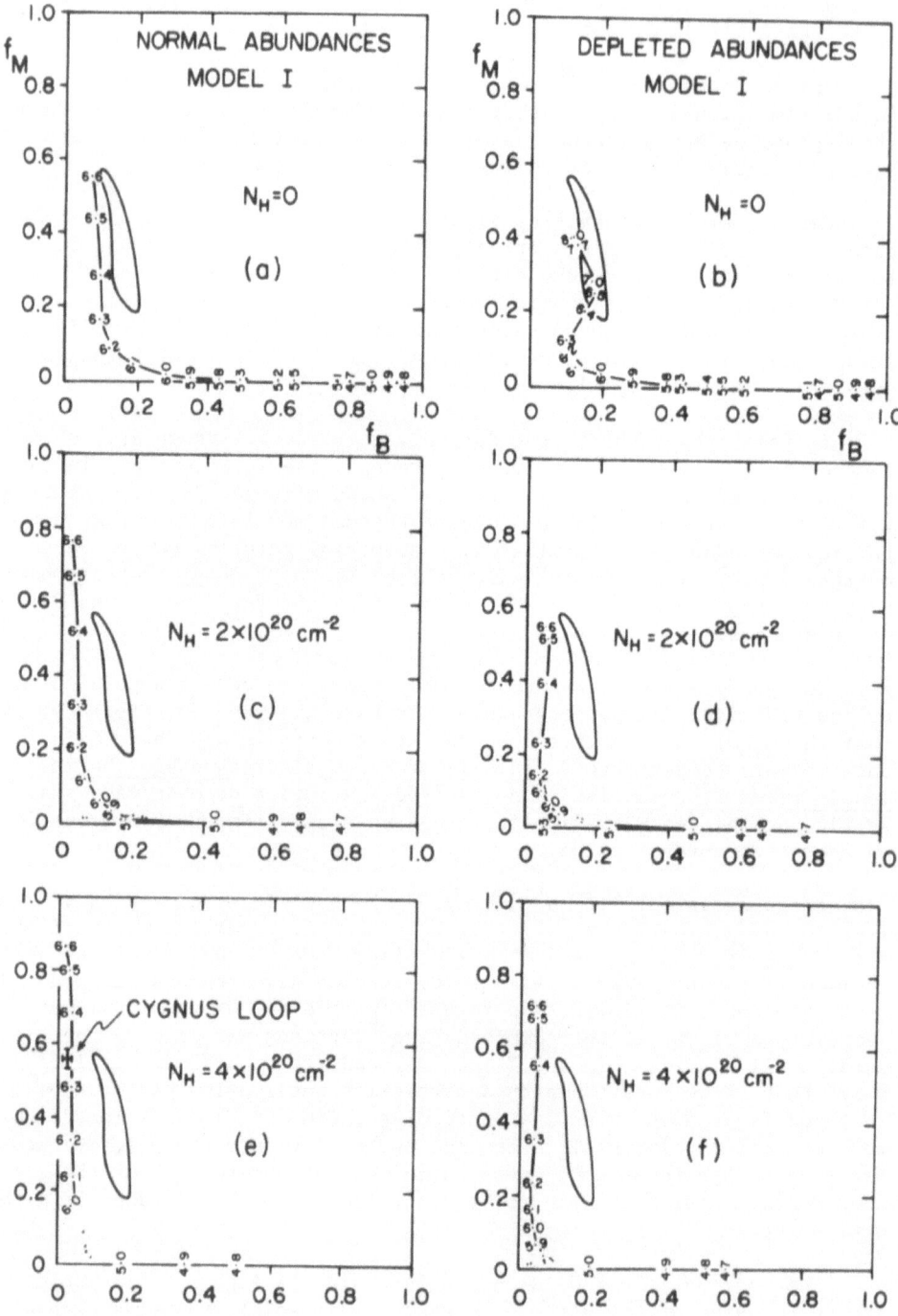

Figure 3. Band fraction plots. Intervening gas of column density $N_H$.

points.
2.  $T_L \sim 10^{5.2}$ to $10^{5.9}$ = 1.6 to $6 \times 10^5$ K.
3.  Roughly equal fractions of the total count rate are ascribed
to the two components.
4.  Adding any material of intermediate temperature forces us to
raise $T_H$ or lower $T_L$ or both.
In Figure 3, thermal spectra with normal or depleted abundances are
passed through various column densities of intervening material.  (The
points which enter the data region in 3b are so high in temperature that
they are excluded by the higher energy data.)  As you can see, the
effect of absorption is primarily to move model points to the left and
upward in the diagram.  Any model with high absorption has to be excluded because at the very low temperatures required for achieving significant B-band radiation, the X-ray emissivity is so low that tremendous
amounts of material would be required.  For the low absorption models,
the conclusions about $T_H$ and $T_L$ are about as before.

I would like to call your attention to an observation of the Cygnus
Loop made by this instrument on another flight.  The observed spectrum
falls exactly on the model line in the lower left diagram, the temperature and $N_H$ agreeing quite well with the analyses of other observers.
Although the spectrum of the Cygnus Loop is quite soft by X-ray source
standards, it is still much harder than the diffuse background.

This material is elaborated in a paper by Burstein, Borken,
Kraushaar, and Sanders [3].  In addition, they present an exhaustive
study of 2–component models which satisfy both the band fraction
analysis and constraints placed by stellar O VI column densities.  The
X-ray brightness gives information about density squared along a character distance, the O VI about density along distance to the star.
These two results can be made to fit together nicely, but again one
finds, this time for both temperature regions, that the pressure is
forced to P/k $\sim 10^4$ or P $\sim 10^{-12}$ dyne cm$^{-2}$.  In addition, the lower
temperature component is found to be accommodated most easily if $T_L$
$\sim 10^{5.8}$ = $6.3 \times 10^5$ K.  This fits nicely with the Levine et   al. pulse
height analysis results.

I would like to acknowledge the considerable assistance of Wilt
Sanders and Bill Kraushaar in preparing this brief survey.  The reader
should also be advised that I am only a theoretician looking with some
fascination over the shoulders of the experimentalists of the group.
This discussion is the fruit of their labors.

References

[1]  Daltabuit, E. and Meyer, S.: 1972, Astr. and Ap. 20, 415.
[2]  Levine, A., Rappaport, S., Halpern, J., and Walter, F.: 1976,
     preprint submitted to Astrophys. J.
[3]  Burstein, P., Borken, R. J., Kraushaar, W. L., and Sanders, W. T.:
     1977, Astrophys. J., in press (April 1).

# NATURE, ORIGIN, AND EVOLUTION OF THE HOT GAS:   INTERSTELLAR?

Christopher F. McKee
Departments of Physics and Astronomy, University of California,
Berkeley, California 94720 USA

ABSTRACT—Supernova explosions in a cloudy interstellar medium produce large volumes of hot gas with typical density $3 \times 10^{-3}$ cm$^{-3}$ and temperature $5 \times 10^5$ K. The evolution of supernova remnants in such a medium differs significantly from the conventional picture due to evaporation of cool clouds by the hot gas inside the remnant. A steady state model of the resulting three-component medium is in reasonable agreement with observations of the interstellar pressure, the OVI density, and the diffuse soft X-ray emission.

## 1.   EXISTENCE OF A HOT COMPONENT OF THE ISM

Observations of non-stellar OVI absorption lines toward many hot stars (Jenkins and Meloy 1974; York 1974) and of diffuse soft X-ray emission in the galactic plane as well as at high galactic latitudes (Williamson $et$ $al.$ 1974) both suggest the presence of large volumes of hot gas (T $\sim 10^6$ K) in the interstellar medium (ISM). The very low HI column densities observed in some directions (Bohlin 1975; Margon $et$ $al.$ 1976) are certainly consistent with this idea. The existence of a hot component of the interstellar medium, or HIM for short, requires a major revision of the classical two phase model of the ISM (Field $et$ $al.$ 1969).

The possibility of a galactic corona at a temperature of order $10^6$ K was analyzed by Spitzer (1956). However, Cox and Smith (1974) were the first to predict that supernovae would produce large volumes of such gas in the plane of the galaxy. More detailed computer models by Smith (1977) have confirmed this result. The argument is strengthened further when the inhomogeneity of the ISM is taken into account, because supernova blast waves tend to propagate in the lowest density component (McKee and Ostriker 1975, 1977). Using the results of Chevalier (1974) one finds that a supernova remnant (SNR) expands until it reaches pressure equilibrium with the ambient ISM at a radius $R_E = 55$ $E_{51}^{0.32}$ $n_0^{-0.16}$ $\tilde{P}_{04}^{-0.20}$ pc, where $E_{51}$ is the SNR energy in units of $10^{51}$ erg, $n_0$ and $P_0$ are the ambient density and pressure, and $P_{04} = k^{-1} P_0/(10^4$ K

$cm^{-3}$). The hot bubble of gas in the SNR can be destroyed by several processes (Cox and Smith 1974); the lifetime is probably set by encroachment of the ambient gas with a timescale $t_{max} = R_E/(P_0/\rho_0)^{1/2} = 7.1 \times 10^6 E_{51}^{0.32} n_0^{0.34} \tilde{P}_{04}^{-0.70}$ yr, where $\rho_0$ is the ambient mass density.

In order to estimate the fraction of the volume of the Galaxy occupied by the SNRs, define $dQ(R)$ as the probability that a given point in the Galaxy is inside a SNR with a radius between R and R + dR. For static SNRs with $R \sim R_E$, one finds $Q(R_E) = SV_E t_{max}$, where $V_E = 4\pi R_E^3/3$ and S is the supernova rate per unit volume; the SNRs fill a fraction $1 - \exp[-Q(R_E)]$ of the Galaxy. Inserting the results for $R_E$ and $t_{max}$, one finds $Q(R_E) = 0.51 E_{51}^{1.28} S_{-13} n_0^{-0.14} \tilde{P}_{04}^{-1.30}$. Thus, if the supernova rate is of order $S_{-13} = S/(10^{-13} pc^{-3} yr^{-1}) \sim 1$ (see Tammann 1974) and the interstellar pressure is not inordinately high ($P_{04} \lesssim 1$), it follows that $Q(R_E) \sim 1$ and a large fraction of the ISM is hot, low density gas.

The conclusion that there is a hot component of the ISM with a relatively large filling factor appears inescapable if our estimates of the supernova rate S and the SNR lifetime $t_{max}$ are reasonably accurate. If the ISM were initially in the two phases envisioned by Field *et al.* (1969), a large volume of hot gas would form as the cool gas was swept up into dense shells. Thus there will be a third component in the ISM in addition to the cold ($T \sim 10^2$ K) neutral clouds and regions of warm ($T \sim 10^4$ K) gas. The existence of the HIM alters the evolution of SNRs and the SNRs in turn govern the energetics of the HIM, so a self-consistent analysis is required. Here we shall first discuss some aspects of SNR evolution and then apply these results to a steady state model of the HIM.

## 2.   EVOLUTION OF SNRS IN A THREE-COMPONENT ISM

The expansion of a supernova remnant is governed by the lowest density medium which has a large filling factor. Type II supernovae may occur primarily in the cooler components of the ISM where star formation takes place, but the blast waves will soon break out into the HIM. Type I supernovae should show no correlation with the nature of the ambient ISM.

Two viewpoints may be adopted in discussing the evolution of SNRs in the inhomogeneous ISM. If the filling factor of the HIM is small, then SNRs will be connected over a small fraction of their surface area and the resulting configuration may be regarded as a "hot network of tunnels" (Cox and Smith 1974). On the other hand, if the filling factor is large then the inverse viewpoint of clouds embedded in a hot medium is more appropriate (McKee and Ostriker 1975, 1977). Here we shall adopt the latter view, which is consistent with the large filling factors estimated above.

Because of the low density of the HIM into which the SNR expands, the hot gas in the interior of the SNR will also be of low density and evaporation of the embedded clouds by this hot gas must be considered. Since most of the mass of the ISM is contained in the clouds, evaporation of just a thin outer layer of the clouds can dramatically alter the density and thus the temperature of the hot interior gas. In the early stages of evolution of a SNR the evaporation can raise the interior density to $\sim 0.1$ cm$^{-3}$, thereby mimicking the low density component of the two phase ISM (Cowie 1976).

At a later stage of evolution ($R \gtrsim 50$ pc) the temperature drops to the point that the electrons are collision-dominated and the evaporation rate is then $2.75 \times 10^4 T_h^{5/2} a_{pc} \phi$ g s$^{-1}$ per cloud (Cowie and McKee 1977), where $T_h$ is the temperature of the hot gas, $a_{pc}$ is the cloud radius in pc, and $\phi$ is a factor less than unity which takes into account the effects of turbulence and magnetic fields. Even if the magnetic field in the cloud is much larger than that in the hot gas, the evaporation is not significantly impeded provided that the field in the cloud is well-connected with that in the hot gas and the field is not too tangled.

Thermal conduction in the hot gas ensures that the interior of the SNR is at approximately constant temperature and density (Chevalier 1975; Solinger et al. 1975). In the adiabatic phase of the expansion (i.e., before radiative losses are important) the shock velocity dR/dt is comparable to the isothermal sound speed $c_h$ of the hot interior gas. We write dR/dt = $\alpha c_h$, where $\alpha = 1.68$ in the standard similarity solution (no evaporation) and $\alpha \sim 2.5$ in the evaporative case. One can readily integrate to find R(t); when the evaporated mass dominates the swept up mass ($n_h - n_0 \gg n_0$), the result is

$$R = 0.32 \ (E_{51}/n_h)^{1/5} \ t_y^{2/5} \ \text{pc} \tag{1}$$

where $t_y$ is the time in years. This is similar in form to the standard result except that here the density $n_h$ also varies with radius:

$$n_h = 96 \ E_{51}^{2/3} \ R_{pc}^{-5/3} \ \Sigma^{-1} \ \text{cm}^{-3}, \tag{2}$$

where the parameter

$$\Sigma = \alpha a_{pc}^2 / (3 \ f_{cl} \phi) \tag{3}$$

determines the effectiveness of the evaporation. The quantity $f_{cl}$ is the filling factor of both cold and warm ($10^4$ K) clouds. Altogether, then, $R \propto t^{3/5}$ in contrast to the usual $R \propto t^{2/5}$.

The adiabatic phase terminates when the radiative losses become significant. Such losses can occur directly from the hot gas or indirectly from shock compression of the clouds ("cloud crushing"). The

former dominates for small values of $f_{cl}$ (McKee and Ostriker 1977)
whereas the latter is more important for large values of $f_{cl}$ (Cox and
Smith 1974). The direct radiative losses are increased by radiation
from the evaporative interfaces around clouds (McKee and Cowie 1977).
Including this effect and neglecting losses due to "cloud crushing" one
finds that half the SNR energy is radiated away at $R_{cool} \sim 180$ pc, at
which point a shell begins to form. The density of the hot interior is
so low that direct observation of such large remnants would be difficult.
The cool gas in the remnant should be systematically expanding outward,
however, and this might be more readily detectable.

The main effect of the evaporation is to systematically alter the
density of the interior gas as the SNR expands. Cowie (1976) has
presented some admittedly weak evidence in favor of this result: the
rms electron density inferred from X-ray observations of SNRs tends to
decrease with the radius of the remnant.


3.   THE STEADY STATE HOT INTERCLOUD MEDIUM

Each supernova injects $\sim 10^{51}$ erg into the ISM, which must eventually be radiated away. Furthermore, during the adiabatic phase, cloud
evaporation injects mass into the HIM, whereas in the subsequent radiative phase of the evolution gas in the HIM is swept up into a dense
shell. Because the time scale for a SNR to sweep over a given point
in space ($\sim 4 \times 10^5$ yr) is much less than the time scale associated
with passage through spiral arms ($\sim$ few $\times 10^7$ yr) the HIM should be in
an approximate steady state governed by energy and mass balance. In
what follows we recapitulate the treatment of McKee and Ostriker (1977).

The energy balance of the HIM is governed by several processes:

$$\text{SN } (\sim 10^{51} \text{ erg}) \rightarrow \begin{cases} \text{Direct radiation from hot gas} \\ \text{Cloud crushing} \\ \text{Heating of the galactic halo} \end{cases}$$

For the first two mechanisms, the energy dissipation and generation occur in the same volume, so that in a steady state each SNR must on
average dissipate its energy before it overlaps another adiabatic SNR.
Halo heating is important only for SNRs occurring well away from the
plane of the galaxy or if energy dissipation in the plane is inefficient
(one version of a halo heating model is discussed by Shapiro and Field
1976). Here we shall focus on the first mechanism, that of direct
radiation from the hot gas.

Since each SNR must dissipate its energy before colliding with
another one, the pressure at any point in the HIM is approximately
given by the pressure in the last SNR to hit that point. In the
adiabatic phase, corresponding to a fraction $Q < Q_{cool}$ of the HIM,
$P \propto R^{-3}$ and $Q \propto R^{14/3}$; for $S_{-13} = E_{51} = 1$, this gives

$$\tilde{P}_4 = 0.29 \; Q_{cool}^{-0.69} \; (Q_{cool}/Q)^{9/14} \; , \quad Q < Q_{cool} \; . \tag{4}$$

When shell formation occurs ($Q = Q_{cool}$) the pressure drops to about half its previous value and thereafter $P \propto R^{-5}$. The dynamics in this stage are uncertain, but one can estimate

$$\tilde{P}_4 = 0.14 \; Q_{cool}^{-0.69} \; (Q_{cool}/Q)^{0.90} \; , \quad Q > Q_{cool} \; . \tag{5}$$

Insofar as supernovae are the dominant energy source for the ISM, these expressions determine the distribution of the pressure of the ISM.

The second condition for a steady state HIM is that of mass balance:

$$\begin{array}{ccc} & \longrightarrow \text{Shock compression} \longrightarrow & \\ & & \text{cold and} \\ \text{HIM} & & \text{warm clouds} \\ & \longleftarrow \text{Evaporation} \longleftarrow & \end{array}$$

This requirement that the evaporated mass balance the mass swept up into SNR shells determines the mean density of the HIM.

A complete solution for the physical conditions in the HIM requires an evaluation of the parameter $\Sigma$ (eq. 3), which depends on the size and filling factor of the clouds. Here we simply adopt $\Sigma = 50$, corresponding to $Q_{cool} = 0.5$. The *typical* conditions in the HIM (defined as the average from $Q = 0.25$ to $Q = 0.75$) are then $\tilde{P} = 3600$ K cm$^{-3}$, $n_h = 3.5 \times 10^{-3}$ cm$^{-3}$, and $T_h = 4.5 \times 10^5$ K. The *average* conditions are significantly different because of the effect of younger SNR occupying a relatively small fraction of space; for example, the average pressure is 7900 K cm$^{-3}$ in this model.

4.    COMPARISON WITH OBSERVATION

Cox and Smith (1974) assumed a value for the interstellar pressure of about 7000 K cm$^{-3}$ and showed that the temperature of the hot gas should lie in the range $10^6$ K $\gtrsim T_h \gtrsim 3 \times 10^5$ K. They pointed out that these conditions approximate those required to account for the soft X-ray background, and they were among the first to realize that the same hot gas could produce detectable OVI absorption lines. Under the assumption that the same gas is responsible for both the X-ray emission and the OVI absorption lines, Shapiro and Field (1976) showed that rather high interstellar pressures are required ($\tilde{P} \sim 2 \times 10^4$ K cm$^{-3}$).

In the model of McKee and Ostriker (1977) the typical pressure is calculated to be $\tilde{P} = 3600$ K cm$^{-3}$, which is identical to the mean value found by Jura (1975) from an analysis of molecular hydrogen in six clouds not near HII regions, and which is comparable to the value 2000 K cm$^{-3}$ usually estimated from 21 cm observations (Field 1975). The constraint

imposed by Shapiro and Field (1976) is avoided because the X-ray emission arises in the high temperature, high pressure regions inside SNRs
prior to cooling, whereas the observed OVI arises in cooler regions.

It is remarkable that both the individual OVI line widths and the
velocity dispersion of the profiles observed along different lines of
sight are small, of order 20 km s$^{-1}$. Such a small velocity is inconsistent with the production of the OVI in shock waves, which would require a velocity in excess of 100 km s$^{-1}$ to reach temperatures of order
$3 \times 10^5$ K. Instead, it is likely that the lines arise in conductive
interfaces around interstellar clouds (the circumstellar version of this
idea has been analyzed by Castor *et al.* 1975). One finds a typical
value of the OVI density of about $7 \times 10^{-8}$ cm$^{-3}$, somewhat greater than
twice the observed value in the galactic plane, $2.8 \times 10^{-8}$ cm$^{-3}$ (Jenkins
1977). Additional OVI is undoubtedly present in the hot gas inside
adiabatic SNRs, but the observations are not sensitive to such broad
lines.

Recently, the soft X-ray background has been intensively studied
by a number of workers (Burstein *et al.* 1976; Cash *et al.* 1976; de Korte
*et al.* 1976; and Levine *et al.* 1976). An emission measure of order $10^{-2}$
cm$^{-6}$ pc at T $\sim 10^6$ K is required to explain the 0.25 keV data. Since
the emission measure of a single SNR prior to cooling is about 0.03
$(T/10^6 \text{ K})^{1.75}$ cm$^{-6}$ pc, a single SNR in any direction is adequate to explain this flux, provided that the absorption is not too great. A large
flux persists at lower energies, however, where the absorption is more
severe; for example, in the lowest energy band of the Burstein *et al.*
experiment, an optical depth of unity is reached at a column density of
only $5 \times 10^{19}$ cm$^{-2}$. They suggest that the solar system may be embedded
in a "hot tunnel". One finds that this low energy data can be accounted
for if the solar system is embedded in a SNR at T = $10^{5.8}$ K and if the
absorption mean free path is set by the distance to the nearest cold
cloud ($\sim 90$ pc). Burstein *et al.* and de Korte *et al.* also present
evidence for an additional, hotter component of the background with a
temperature T $\sim 10^{6.3}$ K. An isotropic background at such a high temperature would be difficult to explain because the filling factor for such
hot SNRs is only Q $\sim 2.5 \times 10^{-3}$. The evidence for this high temperature
is based on time independent cooling functions and simple models for
interstellar absorption, however, and it is by no means conclusive.

Thus models of the ISM based on supernova heating of large volumes
of gas to temperatures of order $5 \times 10^5$ - $10^6$ K can approximately account
for observations of both the OVI absorption lines and the soft X-ray
background. More accurate models for SNR evolution in a three component
ISM, including the effects of cloud crushing, are required. It would
also be interesting to consider the effects of deviations from the
average conditions associated with spiral density waves, clusters of
young stars, or regions of anomalously low pressure. Clearly, further
observations of the hot gas will be important in developing our understanding of the interstellar medium as a whole.

I am deeply grateful to L. L. Cowie and J. P. Ostriker, with whom I have collaborated on much of the work described in this paper. I also thank E. B. Jenkins for numerous helpful conversations. Material support was provided by NSF under grant AST 75-02181.

## REFERENCES

Bohlin, R. C.: 1975, *Astrophys. J.*, 200, 402.
Burstein, P., Borken, R. J., Kraushaar, W. L., and Sanders, W. T.: 1976, Wisconsin Astrophysics preprint 39.
Cash, W., Malina, R., and Stern, R.: 1976, *Astrophys. J. (Letters)*, 204, L7.
Castor, J., McCray, R., and Weaver, R.: 1975, *Astrophys. J. (Letters)*, 200, L107.
Chevalier, R.: 1974, *Astrophys. J.*, 188, 501.
Chevalier, R.: 1975, *Astrophys. J.*, 198, 355.
Cowie, L. L.: 1976, Ph.D. thesis, Harvard University.
Cowie, L. L., and McKee, C. F.: 1977, *Astrophys. J.*, 211, (in press).
Cox, D. P., and Smith, B. W.: 1974, *Astrophys. J. (Letters)*, 189, L105.
de Korte, P. A. J., Bleeker, J. A. M., Deerenberg, A. J. M., Hayakawa, S., Yamashita, K., and Tanaka, Y.: 1976, *Astron. and Astrophys.*, 48, 235.
Field, G. B.: 1974, in *Physique Atomique et Moleculaire et Materie Interstellaire* (Les Houches 1974) ed. R. Balian, P. Encrenaz, and J. Lequeux, North Holland, Amsterdam.
Field, G. B., Goldsmith, D. W., and Habing, H. J.: 1969, *Astrophys. J. (Letters)*, 155, 49.
Jenkins, E. B.: 1977, in preparation.
Jenkins, E. B., and Meloy, D. A.: 1974, *Astrophys. J. (Letters)*, 193, L121.
Jura, M.: 1975, *Astrophys. J.*, 197, 581.
Levine, A., Rappaport, S., Doxsey, R., and Jernigan, G.: 1976, *Astrophys. J.*, 205, 226.
Margon, B., Lampton, M., Bowyer, S., Stern, R., and Paresce, F.: 1976, *Astrophys. J. (Letters)*, 210. L79.
McKee, C. F., and Cowie, L. L.: 1977, *Astrophys. J.*, (in press).
McKee, C. F., and Ostriker, J. P.: 1975, *Bull. Amer. Astr. Soc.*, 7, 415.
McKee, C. F., and Ostriker, J. P.: 1977, *Astrophys. J.* (to be submitted).
Shapiro, P. R., and Field, G. B.: 1976, *Astrophys. J.*, 205, 762.
Smith, B. W.: 1977, *Astrophys. J.*, 211, (in press).
Solinger, A., Rappaport, S., and Buff, J.: 1975, *Astrophys. J.*, 201, 381.
Spitzer, L.: 1956, *Astrophys. J.*, 124, 20.
Tammann, G.: 1974, in *Supernovae and Supernova Remnants*, ed. C. B. Cosmovici, Reidel, Dordrecht.
Williamson, F. O., Sanders, W. T., Kraushaar, W. L., McCammon, D., Borken, R., and Bunner, A. N.: 1974, *Astrophys. J. (Letters)*, 193, L133.
York, D. G.: 1974, *Astrophys. J. (Letters)*, 193, L127.

# STRUCTURE AND EVOLUTION OF WIND-DRIVEN CIRCUMSTELLAR SHELLS

Richard McCray
Joint Institute for Laboratory Astrophysics,
University of Colorado and National Bureau of Standards,
and Department of Physics and Astrophysics,
University of Colorado, Boulder, Colorado 80309 USA

## 1. INTRODUCTION

The ultraviolet spectrometer on the Copernicus spacecraft has given us many gifts. In addition to the observations of interstellar O VI absorption described here by Jenkins, the spectrometer has shown that almost all early-type stars with bolometric magnitude $M_V \lesssim -6$ have broad stellar resonance lines of N V $\lambda 1240$ and O VI $\lambda 1035$ that indicate stellar wind with mass loss rates $\dot{M}_w \gtrsim 10^{-6}$ $M_\odot$ yr$^{-1}$ and terminal velocities $V_w \approx 1000$ to $3000$ km s$^{-1}$ (Snow and Morton 1976). It was these stellar lines rather than the interstellar lines that led us to our theory (Castor, McCray and Weaver 1975, hereafter Paper I) of "bubbles" of hot gas surrounding the early type stars.

The energy deposited by a stellar wind of mechanical power $L_w = 1/2 \dot{M}_w V_w^2 \approx 1.27 \times 10^{36} (\dot{M}_w/10^{-6})(V_w/2000)^2$ ergs s$^{-1}$ over a characteristic stellar lifetime of $t \approx 3 \times 10^6$ yr is typically $E_w \approx L_w t_* \approx 10^{50}$ ergs. This value is of the same order of magnitude as the energy deposited in the interstellar medium by a supernova blast wave (cf. Woltjer 1972); and in fact we find that the structure of the interstellar bubble is similar to that of a supernova shell. Before discussing the details of the structure, I shall first take a moment to compare the properties of supernova shells with those of bubbles. Table 1 lists a number of these properties for each case. The main qualitative difference between the two is that the entire energy of the supernova shell is injected at once, whereas the energy of the bubble is deposited gradually over the lifetime of the star. (In our idealized model of a bubble the wind luminosity is assumed to be steady over the stellar lifetime.) Simple dimensional analysis then yields the power laws listed in table 1 for the shell radii $R(t)$. The adiabatic expansion phase of a supernova remnant is terminated when the radiative energy losses in the shell become comparable to the energy of the initial blast. At this point a supernova shell enters the snowplow stage and soon becomes invisible as an optical remnant. However, as Cox and Smith (1974) have pointed out, the hot ($T \gtrsim 10^6$ K) interior should persist for times $\gtrsim 10^6$ yr.

*Hugo van Woerden (ed.), Topics in Interstellar Matter, 35-44. All Rights Reserved.*
Copyright © 1977 by D. Reidel Publishing Company, Dordrecht-Holland.

Table 1

Comparison of Supernova Remnants and Bubbles

| | Supernova Remnants | Bubbles |
|---|---|---|
| Radius law | $R(t) \propto t^{2/5}$ (during blast wave phase) | $R(t) \propto t^{3/5}$ |
| End of expansion | $\sim 3 \times 10^4$ yr (limited by radiative losses of shell) | $\sim 3 \times 10^6$ yr (limited by stellar lifetime) |
| Terminal radius | $\sim 30$ pc | $\sim 30$ pc |
| Interior temperature | $\sim 10^6$ K | $\sim 10^6$ K |
| Energy | $\sim 10^{50}$ ergs | $\sim 10^{50}$ ergs |
| Galactic birth rate | $\sim (30 \text{ yr})^{-1}$ | $\sim (200 \text{ yr})^{-1}$ |
| Galactic power | $\sim 10^{41}$ ergs s$^{-1}$ | $\sim 10^{40}$ ergs s$^{-1}$ |

In contrast, the interstellar bubble continues to expand as long as the stellar wind blows, even though radiative losses are significant throughout the bubble lifetime. The bubble expands as a "driven snowplow," and its expansion terminates only when the stellar wind ceases to blow. As a result, the radius and total energy of the bubble at its termination are comparable to those of an old supernova shell. In fact, the structures are very similar in that they both have thin expanding shells surrounding a hot (T $\gtrsim 10^6$ K) low density interior.

A final comparison I wish to make between interstellar bubbles and supernova shells concerns their relative importance as sources of hot gas in the galaxy. In my view there is little doubt that supernovae are more important in the global energetics of the interstellar gas, for the simple reason that supernovae are more common occurrences than the birth of the early-type stars that are likely to develop strong stellar winds. For example, let us assume that every star of mass greater than, say, 5 $M_\odot$ terminates as a supernova, but that only stars with initial mass greater than, say, 20 $M_\odot$ develop strong stellar winds. Then assuming that the birth rate of stars more massive than $M_*$ varies as $M_*^{-1.3}$ (cf. Tinsley 1972) we find that supernova shells are more frequent than bubbles by a factor $\sim 6$. Multiplying the respective birth rates by the energies per event then gives the galactic power due to each type of source, listed in table 1.

Therefore, it seems likely that the soft X-ray background in the galaxy is due to old supernova shells rather than to interstellar bubbles, and this view is supported by the fact that the nearest stars of sufficiently early type to blow bubbles are the Orion stars, some 460 pc distant. Although the bubbles may be soft X-ray sources, the soft X-ray background we observe must have a more local origin.

However, the bubbles may be relatively more important as a source of the interstellar O VI absorption, because of the strong selection effect inherent in the Copernicus UV observations: by observing interstellar absorption lines against bright OB stars, we are virtually assured that the line of sight intersects at least one bubble.

## 2. INTERIOR STRUCTURE

Let us consider a typical interstellar bubble of age $t = 10^6$ yr, surrounding a star of mass loss rate $\dot{M}_w = 10^{-6} M_\odot$ yr$^{-1}$, and wind velocity $V_w = 2000$ km s$^{-1}$, expanding into a medium of atomic density $n_0 = 1$ cm$^{-3}$. Its temperature and density structure is shown in figure 1. Working from the inside outwards, one sees first region (a) of hypersonic stellar wind, whose density $n_w(r) \propto r^{-2}$ according to the continuity equation. This stellar wind encounters a shock at $R_1(t)$, some 6 pc from the central star. Beyond the shock is region (b) of hot ($T \gtrsim 10^6$ K) low density shocked stellar wind that occupies most of the volume of the bubble. The gas in region (b) is an inefficient radiator, so the shock at $R_1(t)$ is not an isothermal one. However, thermal conduction carries energy away from the inner shock into region (b), so that the density discontinuity at $R_1(t)$ is greater than the value of four appropriate for an adiabatic shock, and the discontinuity is not sharp.

Region (b) is bounded by a thin shell (c) at $R_2(t) \approx 25$ pc of swept-up interstellar gas. This shell may be entirely H II, or it may also contain an outer layer of H I if the stellar radiation is

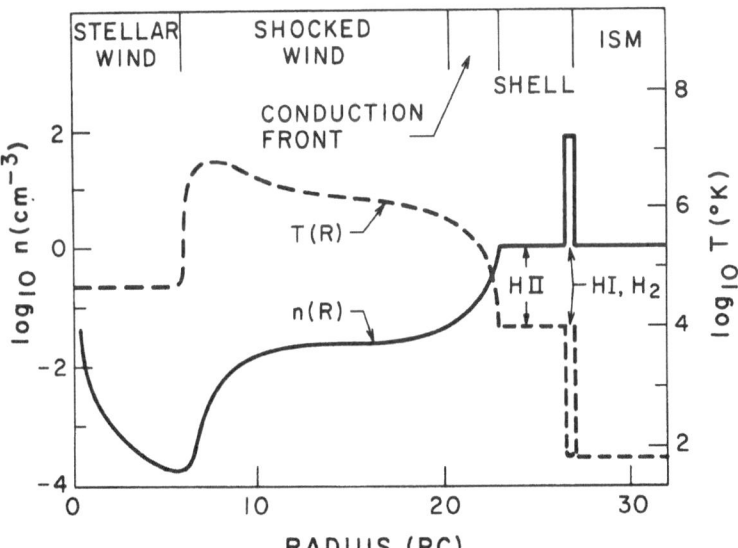

Figure 1.    Temperature and density structure of bubble model for which $L_w = 1.27 \times 10^{36}$ ergs s$^{-1}$, $n_0 = 1$ cm$^{-3}$ and $t = 10^6$ yr. ISM means ambient interstellar medium.

insufficient to photoionize all of the gas in the shell. Both options seem plausible for the typical range of stellar and interstellar para- meters. The swept-up interstellar gas in the outer shell radiates efficiently, so the outer shock is an isothermal one and the density of the gas in the shell is given by $n_s \approx n_0(\dot{R}_2/C_s)^2$, where $C_s$ is the isothermal sound speed in the shell, $C_s \approx 10$ km s$^{-1}$ for H II and $C_s \approx 1$ km s$^{-1}$ for H I, and $\dot{R}_2 \approx 13.5$ km s$^{-1}$.

The interface between region (b) of shocked stellar wind and the shell (c) of swept-up interstellar gas is particularly interesting because there the temperature spans the range $10^5$ K $< T < 10^6$ K, where highly ionized species such as O VI are the dominant ionization stages. The structure of this interface is determined by electron thermal conduction of heat from region (b) into the shell. Roughly 80 per cent of the conductive heat flux is radiated away in the inter- face by collisional excitation of UV resonance lines such as O VI $\lambda 1035$. The other 20 per cent causes evaporation of gas from the shell, which diffuses into region (b) and mixes with the shocked stellar wind. In fact, the dominant source of mass in region (b), say 20 $M_\theta$, is the interstellar gas evaporated from the shell (c) rather than the stellar wind itself. One can show that the saturation of thermal conductivity (cf. Cowie and McKee 1977) is not significant in this interface, although it is likely to be significant near the shock at $R_1$.

In Paper I we calculated the column density of O VI and other ions through the conductive interface. To our surprise and delight, the theoretical column density of O VI turned out to be right in the range of column densities observed in the survey by Jenkins and Meloy (1974)! I quote the theoretical result from Paper I:

$$N(\text{O VI}) \approx 3.4 \times 10^{16} \, X_0 n_0^{9/35} (L_w/10^{36})^{1/35} \, t_6^{8/35} \text{ cm}^{-2} \qquad (1)$$

where $X_0$ is the fractional abundance of oxygen atoms. For example, equation (1) gives $N(\text{O VI}) \approx 1.5 \times 10^{13}$ cm$^{-2}$ for typical values of stellar wind mass loss rates and terminal velocities, a cosmic abun- dance of oxygen $X_0 = 4.4 \times 10^{-4}$, $n_0 = 1$ cm$^{-3}$, and $t = 10^6$ yr. Con- sidering the extreme insensitivity of $N(\text{O VI})$ to the assumed para- meters of the system, one can hardly accuse us of adjusting our parameters to make the theory fit the observations. It also implies that the observed column densities of O VI should not be correlated very well with the properties of the central star.

In order to obtain the simple analytic description of the bubble in Paper I, we made three major approximations: first, we treated the conduction front in a plane-parallel approximation; second, we neglec- ted the radiative losses in the conduction front; and third, we assumed that the ionization of trace elements in the conduction front was given by a coronal approximation, in which a local balance between collisional ionization and radiative recombination is assumed.

Now we have constructed a much more detailed description of the structure and evolution of a bubble (Weaver et al 1977 -- hereafter

Paper II) in which each of these approximations has been removed. The results are as follows. First, going from a plane-parallel description to a spherical description reduces the column density of O VI by about 10 per cent. Removing the second approximation has a more significant effect. The effect of including radiative cooling in the energy equation is to make the temperature gradient steeper in the conduction front, and this makes the region where O VI predominates more narrow. This reduces the column density of O VI by a further factor 0.5. However, these reductions are more than compensated for by removing the third approximation of assuming a steady-state ionization balance. When we calculate the ionization structure of trace elements in the conduction front by solving time-dependent rate equations, we find that the column density of O VI is increased by a factor 3.4 over the value obtained using the steady-state assumption. This occurs because collisional ionization of O VI lags behind the temperature rise, with the result that the zone where the fractional abundance of O VI is large extends further into the hot interior of the bubble. The net result of removing all three approximations of Paper I is to increase the coefficient of equation (1) by a factor 1.5.

We have also calculated theoretical profiles of the O VI absorption lines including the effects of thermal broadening and Doppler broadening due to bulk motion of the gas. These calculations show that the optical depth $\tau_{OVI}(v)$ can be represented very well by a Gaussian whose broadening is given by an effective temperature $T_{eff} \approx 4.5 \times 10^5$ K. The velocity of the line center $v(O\ VI)$ relative to the central star is less than the outer shell velocity $\dot{R}_2$ because it is formed in gas that is flowing inward relative to the shell; typically $v(O\ VI) \approx 0.4\ \dot{R}_2$.

In Paper II we have also calculated the column densities of other highly ionized trace elements that may have observable UV absorption lines. The ratios of their column densities to those of O VI should be almost independent of the parameters of the system, and we find $\log N(X)/N(O\ VI) = -0.8, -1.2, -2.0$, and $-1.5$, respectively, for $X = C\ IV,\ N\ V,\ Si\ IV$, and S IV, assuming cosmic abundances $X_C = 3.3 \times 10^{-4}$, $X_N = 9.1 \times 10^{-5}$, $X_{Si} = 3.1 \times 10^{-5}$, and $X_S = 1.6 \times 10^{-5}$. (The approximate values quoted for Si and S in Paper I were wrong.)

The discrepancy between our theoretical value for $\log N(N\ V)/N(O\ VI) = -1.2$ and the value $-1.6$ found by York (1974) in the spectrum of $\lambda$ Sco is less than before but still exists. We have no facile explanation for this discrepancy, but we would point out that the same discrepancy should be inherent in any hot gas model for O VI, be it circumstellar bubbles or supernova tunnels, unless the gas temperature happened to be at just the right value and did not span a range -- a highly unlikely situation in our view.

The hot gas in the interior of the bubble should be a soft X-ray source. We have calculated the X-ray luminosity and spectrum of our model bubble by folding the spectral emissivity code of Shapiro

and Moore (1976) with our theoretical temperature and density structure.
The result is that the X-ray luminosity in the 44-70 Å band is
roughly $1 \times 10^{-2}$ times the total luminosity $L_b$ of the bubble, which
is expected to be comparable to the stellar wind luminosity $L_w$. (Most
of $L_b$ is emitted in the form of UV resonance lines such as O VI $\lambda 1035$.)

Therefore, a typical bubble with $L_x \approx 10^{34}$ ergs $s^{-1}$ is a very
weak soft X-ray source compared to a typical supernova shell, for
which $L_x \approx 10^{35}$ to $10^{36}$ ergs $s^{-1}$. Furthermore, the nearest stars of
sufficiently early type to blow a bubble are the Orion stars, some
460 pc away. Some soft X-ray excess in the direction of Orion has
been observed by the Wisconsin group (Williamson et al 1974) and by
the ANS group (den Boggende et al 1977), and it is reasonable that this
excess is due to bubbles around these stars.* However, the soft X-ray
source in Orion is very faint compared to the Vela supernova remnant.
Since the bubbles are intrinsically weak soft X-ray sources, and most
are so distant that the intervening interstellar gas is opaque to
soft X-rays, it is unlikely that they contribute significantly to the
galactic soft X-ray background. The suggestion by Cox and by McKee
in this volume that the background is due to an interlocking system
of old supernova remnants seems more likely.


3.  EVOLUTION OF BUBBLES

In Paper I we derived the following simple analytic formula for
the expansion of the outer shell of an idealized bubble model:

$$R_2(t) = 28 \; n_0^{-1/5} (L_w/10^{36})^{1/5} (t/10^6 \; \text{yr})^{3/5} \; \text{pc} , \qquad (2)$$

from which one may readily derive the expansion velocity

$$\dot{R}_2(t) = 16 \; n_0^{-1/5} (L_w/10^{36})^{1/5} (t/10^6 \; \text{yr})^{-2/5} \; \text{km} \; s^{-1} . \qquad (3)$$

A major approximation that was used to derive equation (2) was to
assume that the radiative losses of the interior region (b) of the
bubble are negligible, from which one can easily calculate that the
thermal energy $E_b$ of region (b) is given by $E_b = 5/11 \; L_w t$. However,
we have found that $L_b \approx L_w$, so this approximation must be removed.
In Paper II we have done so by including the radiative loss term $L_b$
in the energy equation for region (b). This has two major effects:
first, the energy and radius of the bubble are less than in Paper I;
and second, the mass evaporated from the shell into the interior is
less than that in Paper I, because a large fraction of the thermal

---

*There is substantial uncertainty in the interstellar soft X-ray
absorption toward these stars. The soft X-ray absorption is primarily due to He I and He II, and the H I column density inferred
from Lyman-$\alpha$ absorption provides only a lower limit to the column
density of interstellar gas, a good fraction of which may be H II.
We estimate that $1 \lesssim \tau(44\text{Å}) \lesssim 5$ toward the Orion stars.

energy that is conducted into the shell is radiated away instead of causing evaporation. For example, with the parameters $L_w = 1.3 \times 10^{36}$, $t = 10^6$ yr, $n_0 = 1$ cm$^{-3}$ of the model described in §2, we would find from Paper I that $R_2 = 28$ pc, $\dot{R}_2 = 16.3$ km s$^{-1}$, $M_b = 67$ $M_\Theta$, and $E_b = 1.8 \times 10^{49}$ ergs. However, the numerical calculations of Paper II give $R_2 = 25$ pc, $\dot{R}_2 = 13.5$ km s$^{-1}$, $M_b = 26$ $M_\Theta$, and $E_b = 8.1 \times 10^{48}$ ergs.

One might infer from Paper I, which shows the luminosity $L_b \propto t^{16/35}$, that $L_b$ will eventually exceed $L_w$, with the result that $E_b$ would decrease and region (b) would collapse. If so, the stellar wind would collide directly with the expanding shell, the O VI column density would be negligible, and the shell would expand according to the law $R_2(t) \propto t^{1/2}$ given by Steigman, Strittmatter and Williams (1975). However, this does not happen because of a curious thermostatic mechanism. When $L_b$ approaches $L_w$, the flow of mass into region (b) begins to decrease rapidly. This causes the density of region (b) to decrease in such a way as to suppress the further increase of $L_b$. The net effect is that the interior temperature $T_b$ remains approximately constant at a value $T_b \approx 2 \times 10^6$ K, and $L_b$ never reaches $L_w$. In the model described in §2, $L_b \approx 0.8$ $L_w$, and at greater times $L_b$ approaches the asymptotic value $L_b \approx 0.9$ $L_w$. Region (b) does not collapse. However, once $L_b$ approaches $L_w$ the expansion of the bubble obeys a law $R_2(t) \propto t^{0.55}$ that is intermediate between the law $R_2(t) \propto t^{0.5}$ of Steigman et al. (1975) and the law $R_2(t) \propto t^{0.6}$ of Paper I.

## 4.   REAL INTERSTELLAR BUBBLES

The detailed calculations of Paper II give results for the structure of our model interstellar bubble that agree fairly well with the simple formulae of Paper I, despite the fact that the major theoretical approximations of Paper I have been removed. We are now confident that our theoretical results are correct.

We cannot be so confident about the model itself, because we do not expect the real interstellar medium to have uniform density, nor the star to be at rest in that medium, nor the stellar wind to be constant in time. We have begun to consider the consequences of relaxing these basic assumptions of our idealized model.

First, consider the effect of stellar motion, and suppose that the star is moving at a constant velocity $V_*$ relative to its surrounding medium. The structure and evolution of the bubble is modified as follows. At early times, the outer shell at $R_2$ remains spherical and concentric with the initial location of the star. Its shape is not significantly affected by stellar motion until a time $t_1$ such that $R_2(t_1) \approx V_* t_1$, which one may estimate from equation (2) as $t_1 \approx 2 \times 10^6$ $n_0^{-1/2}(L_w/10^{36})^{1/2}(V_*/20$ km s$^{-1})^{-5/2}$ yr. From this time on, the inner shock at $R_1$ contacts the shell at $R_2$ at the leading edge of the bubble. The bubble becomes elongated in the direction of $V_*$, and the column density of O VI through its front end drops by more than an order of magnitude. However, the column density of O VI looking at a star moving other than toward us should be roughly as before. It is uncertain whether the stellar winds will last long enough for this situation to occur.

Relaxing the assumption that the bubble is expanding into a uniform medium raises an interesting question of hydrodynamic stability. A bubble expanding into a medium of uniform density is always decelerating, so that the gas comoving with the expanding system feels an effective outward gravity. It is therefore stably stratified because the hot low density gas in the interior of the bubble rests "on top" of the higher density gas in the shell. However, one can easily show by dimensional analysis that if the ambient density $n(r) \propto r^{-m}$ with $m > 2$, the expansion of the bubble will accelerate. It should then suffer the Rayleigh-Taylor instability and break up. A stellar wind that is increasing with time also tends to destabilize the system.

It seems reasonable to us that this instability may be a common occurrence for early type stars, because they must have been born in some high density molecular cloud and they cannot have moved far from this coeval cloud in their short lifetimes. A newly born star or OB association should begin to blow a bubble while it is inside that cloud. But when $R_2(t)$ reaches the outer boundary of the cloud where the density is decreasing abruptly, the bubble should burst and the circumstellar shell should break up into high velocity filaments. Then a new bubble of greater size should begin to form in the low density gas outside the molecular cloud. We suggest that the high velocity filaments around the Trapezium stars may be the remnant of a bubble that has recently burst.

Both our theory and the theory of supernova tunnels predict interstellar gas with temperature $T \gtrsim 10^6$ K. It seems clear to us that the galactic soft X-ray background cannot be due to bubbles. But Shapiro and Field (1976) have already argued that the same hot gas cannot be responsible for both the soft X-ray background and the O VI $\lambda 1035$ absorption lines. A distribution of gas temperatures and densities is required. We are encouraged to suggest that the dominant source of the O VI absorption lines is the bubbles, because of the selection effect and because our theory predicts column densities of O VI in the bubbles that are of the right order of magnitude and are virtually independent of the assumed parameters. However, we cannot argue strongly against the possibility that the network of supernova tunnels can contribute comparable column densities of O VI, because the column density of O VI in that theory is sensitive to assumed parameters that are uncertain, therefore adjustable. The two theories are not mutually exclusive, and it seems that the major area of contention between them is the quantitative one of what fraction of the observed O VI column densities come from each system.

How will this question be resolved? Correlations of the column densities of O VI with properties of the associated star do not appear to be a very promising way because the theory suggests that the column densities are so insensitive to the assumed parameters. Correlations of the velocities of the O VI absorption lines with the stellar lines should be a more satisfactory test, but this test is confounded at present by the uncertainty of the laboratory wavelength of O VI. However, if the velocities of the O VI features were observed to be highly correlated with the velocities of expanding circumstellar shells of H II, H I, and $H_2$ with properties predicted by our theory, we would regard such observations as strong evidence in favor of the hypothesis that most of the observed O VI column density is formed in the bubbles.

We have said little in this review about the structure of the expanding circumstellar shell surrounding the hot interior of the bubble; it is discussed further in Paper II. Here we shall only point out in conclusion that there is good evidence for expanding circumstellar shells of high pressure around early type stars, from studies of ultraviolet absorption lines of rotationally excited $H_2$ by Black and Dalgarno (1976) and by Jura (1975). Hollenbach, Chu, and McCray (1976) have shown that the observed distribution of rotational excitation of $H_2$ follows naturally from the bubble model. An excellent test of the bubble hypothesis for the origin of O VI would be to correlate the velocities of the O VI lines with those of tracers of expanding circumstellar shells. Good examples of such tracers would be the Lyman lines of $H_2(j=5)$, which select out the $H_2$ near a strong source of ultraviolet continuum, and $\lambda 1085$ of the excited fine structure levels N II$^*$, which selects H II regions of unusually high pressure such as would occur in the inner layer of H II of the circumstellar shell. Perhaps by making such observations, one can use the UV spectrometer on the Copernicus spacecraft to solve one of the most exciting mysteries it has discovered.

This work was supported by the National Science Foundation through Grant No. AST75-23590.

REFERENCES

Black, J. H., and Dalgarno, A. 1976, Astrophys. J., 203, 132.
Castor, J., McCray, R., and Weaver, R. 1975, Astrophys. J. (Letters) 200, L107 (Paper I).
Cowie, L., and McKee, C. 1977, Astrophys. J., in press.
Cox, D. P., and Smith, B. W. 1974, Astrophys. J. (Letters), 189, L108.
den Boggende, A. J. F., Mewe, R., Heise, J., Gronenschild, E. H. B. M., and Grindlay, J. 1977, Astron. Astrophys. in press.
Hollenbach, D., Chu, S.-I., and McCray, R. 1976, Astrophys. J., 208, 458.
Jenkins, E. B., and Meloy, D. A. 1974, Astrophys. J. (Letters), 193, L121.

Jura, M.   1975, Astrophys. J., 197, 575, 581.

Shapiro, P. R., and Field, G. B.   1976, Astrophys. J., 205, 762.

Shapiro, P. R., and Moore, R.   1976, Astrophys. J., 207, 460.

Snow, T., and Morton, D.   1976, Astrophys. J. Suppl., 32, 429.

Steigman, G., Strittmatter, P. A., and Williams, R. E.   1975, Astrophys. J., 198, 575.

Tinsley, B. M.   1972, Astron. and Astrophys., 20, 383.

Weaver, R., McCray, R., Castor, J., Shapiro, P. R., and Moore, R.   1977, in preparation (Paper II).

Williamson, F. O., Sanders, W. T., Kraushaar, W. L., McCammon, D., Borken, R., and Bunner, A. N.   1974, Astrophys. J. (Letters), 193, L133.

Woltjer, L.   1972, Ann. Rev. Astron. and Astrophys., 10, 129.

York, D. G.   1974, Astrophys. J. (Letters), 193, L127.

SUMMARY OF COMMENTS AND DISCUSSION

E.B. Jenkins

One participant (unidentified) questioned whether the exceptionally
low abundance of O VI toward the star γ Cas places constraints on our
interpretation of the filling factor of the hot gas. Jenkins replied
that the known irregularity in the distribution of coronal material
makes a general interpretation from a single measurement untrustworthy,
but the γ Cas measurement makes the notion that we are immersed in a
coronal region less tenable. A comment was made that the softest X-ray
emission was more isotopic than the background at higher energies; McKee
and Cox suggested this may imply we are inside a region of hot gas (perhaps too hot to produce O VI toward the very nearest stars). Jenkins
raised two possible objections to this viewpoint: one being the observation of backscattering of solar Lα emission by local neutral hydrogen,
and the other being a suggestion by Parker that the constancy of the
cosmic-ray flux with time (based on meteoritic evidence) would preclude
our being recently enveloped by a bubble of coronal material.

Field asked about the preponderance of negative velocities seen in
an early survey of O VI. Jenkins replied that the laboratory wavelengths
of the transitions are poorly known, but that a reconsideration of presently available data seemed to favour an upward revision of all the
older velocities by about 6 km s$^{-1}$. McCray emphasized that better laboratory measurements of wavelengths still need to be made, since the lack
or presence of negative velocities is crucial to deciding between circumstellar or general interstellar origins of the gas.

Radhakrishnan asked McKee what was a typical radius for a supernova
remnant when breakup occurred. McKee replied this radius should be on
the order of 170 pc. Jenkins questioned McCray on the alteration of
shell geometry by a motion of the star with respect to the surrounding
material. McCray said this problem has been considered, and he briefly
showed a diagram of the situation.

The session chairman encouraged McKee and McCray to engage in a
direct debate on their opposing viewpoints on the origin of the coronal
gas. During this confrontation McKee questioned why we do not see more
compelling observational evidence for the widespread presence of distinct shells around the hotter, more luminous stars. He suggested that
dense blobs of cool material inside the stellar wind region may quench
the formation of the shell through losses by conduction. McCray suggested
we should investigate whether or not such density inhomogeneities are
swept out by the stellar wind, or do they remain inside the cavity?

Van der Laan expressed a general caution that if we thought of the
filling factor of hot gas not being very much less than unity, we might
have trouble understanding the establishment of a "grand design" in
galactic structure.

*Hugo van Woerden (ed.), Topics in Interstellar Matter, 45. All Rights Reserved.*
*Copyright © 1977 by D. Reidel Publishing Company, Dordrecht-Holland.*

Chapter 2


INTERACTION OF STARS AND INTERSTELLAR MEDIUM


Papers presented in a session of IAU Commission 34
(Interstellar Matter), Grenoble, 28 August 1976.


Session Chairmen: Gabriel S. Khromov, Manuel Peimbert

# COMPACT HII REGIONS

P.A. Shaver
Kapteyn Astronomical Institute
University of Groningen

ABSTRACT. This paper reviews recent developments in the study of compact HII regions, particularly from the radio observational point of view. These include the association of HII regions with molecular clouds and maser sources, HII and CII regions and infrared sources in dark clouds, and aperture synthesis observations of radio recombination lines.

## INTRODUCTION

The study of compact HII regions has advanced considerably over the last few years, particularly with the explosive development of molecular line and infrared observations. Now it is possible to study directly the association of HII regions with the giant molecular complexes from which they presumably evolved. In addition it has become clear that HII regions emit the vast bulk of their radiation in the infrared, and the energy balance of HII regions has taken on new dimensions. Fortunately these developments have been paralleled by improvements in radio observations, both in sensitivity and resolution, with the advent of large synthesis telescopes; compact regions of early star formation can now be studied in detail over at least six decades of the electromagnetic spectrum. I would like to review a few of these developments, with particular emphasis on radio observations.

Fig. 1 shows a recent compilation of the electron densities and diameters of HII regions, including many new additions from observations with the Westerbork Synthesis Radio Telescope. These HII regions cover an enormous spread in these variables - fully five decades in both density and diameter. However the fact that they appear to be confined to a diagonal band running across the diagram is attributable largely to selection effects: the lower left-hand portion of the diagram is unpopulated because of inadequate sensitivity of existing radiotelescopes. On the other hand the lack of sources in the upper right-hand region is real, and reflects on the properties of the exciting stars; evidently there are no individual HII regions excited by more than one or two O4 stars (u $\sim$ 100-200 pc cm$^{-2}$), although several thousand O4 stars are required for the giant HII complexes observed in other galaxies.

*Hugo van Woerden (ed.), Topics in Interstellar Matter, 49-59. All Rights Reserved.*
*Copyright © 1977 by D. Reidel Publishing Company, Dordrecht-Holland.*

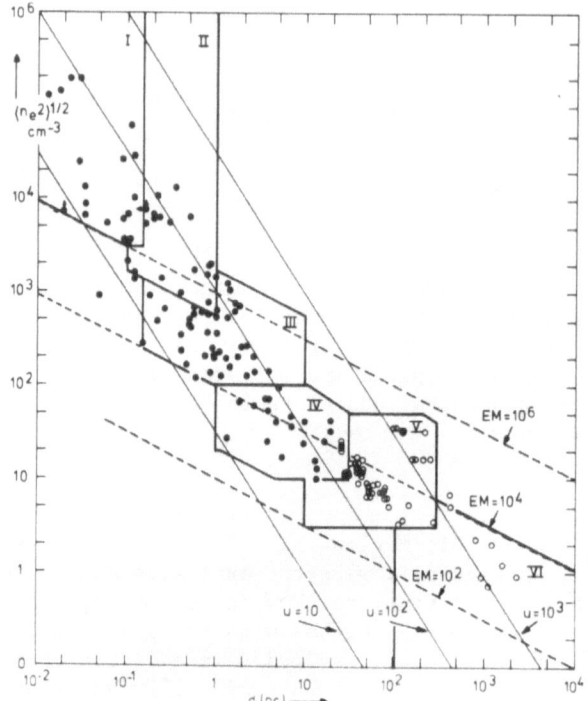

Figure 1. Electron density-diameter diagram for HII regions (Israel, 1976). Filled and open circles represent galactic and extragalactic HII regions respectively. Lines of constant emission measure and excitation parameter are marked, and the Roman numerals refer to a classification scheme outlined in Table I.

The emphasis in this paper will be placed on those HII regions with the highest densities and smallest diameters - presumably the youngest HII regions and therefore those of greatest interest in the study of star formation.


## ASSOCIATION WITH MOLECULAR CLOUDS

CO surveys (eg. Wilson et al., 1974; Dickinson et al., 1974; Blair et al., 1975) have shown that most if not all HII regions have closely associated molecular clouds - the projected distance between the CO peaks and HII regions is only 2 pc on average, compared to 6 pc for the dia-meter of the average CO concentration (Israel, 1976). The compact HII regions are generally several orders of magnitude less massive than the molecular clouds; where are they located in these clouds?

Traditionally HII regions have been idealized as homogeneous spheres, ionization bounded on all sides by the neutral cloud. Recent interest has centred on models in which the HII region is located on the surface of the molecular cloud (eg. Zuckerman, 1973; Grasdalen, 1974; Meaburn, 1975; Israel, 1976). The HII region may have been formed near the boundary of the cloud, and subsequently broken out. The ionized gas streams away from the molecular cloud (at close to the speed of sound) and is constantly replenished. Such a model resolves the paradox of the age of the Orion Nebula, according to which the dynamical age is only ∿ 20,000 years and yet the trapezium stars are on the main sequence; in the present model the HII region is constantly rejuvinated by gas from the molecular cloud, and the dynamical age is only a lower limit.

One test of this model is to compare radial velocities of the ionized and molecular gas: spectral lines from <u>visible</u> HII regions should be blue-shifted relative to the associated molecular clouds. This prediction has been verified by Liszt (1973), Wilson et al. (1974), Blair (1976), and Israel (1976). Fig. 2 shows the most comprehensive comparison to date; the visible HII regions are blue-shifted relative to the molecular clouds by $3.2 \pm 0.8$ km s$^{-1}$ on average, whereas the average radial velocity of the obscured HII regions does not differ significantly from that of the associated molecular clouds. In both cases there is a considerable scatter in the individual values of $\Delta V(HII-CO)$, which may be due to a number of factors including the detailed geometry of the HII-molecular cloud associations; however the general trend seems to clearly support the "blister" model of HII regions.

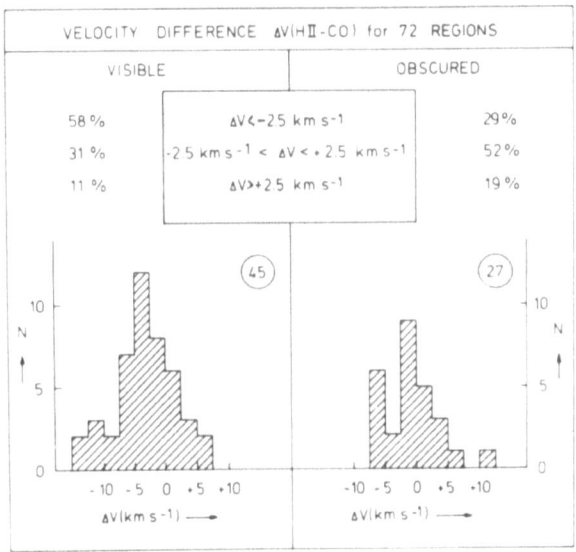

Figure 2. Comparison of the HII and CO velocities for visible and obscured HII regions shows that visible HII regions are systematically blue-shifted with respect to their associated CO clouds (Israel, 1976). The circled numbers give the total number of HII regions in each group.

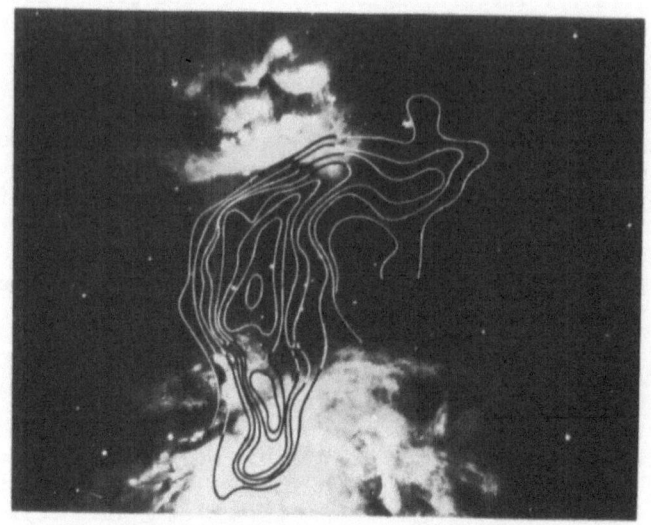

Figure 3.    $^{13}$CO peak intensity contours superimposed on a photograph of
the Orion Nebula and NGC 1977 (Kutner et al., 1976).

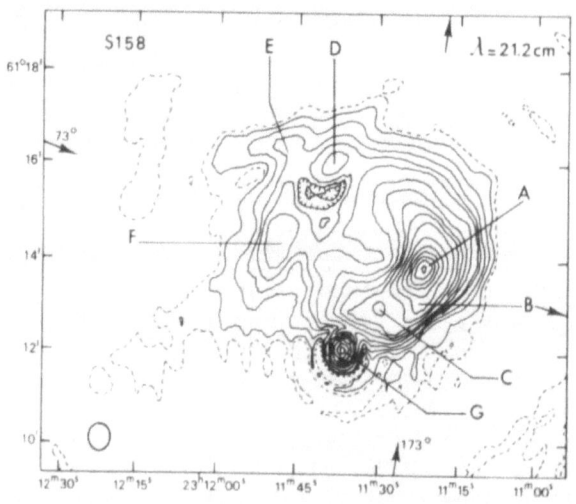

Figure 4.    λ21 cm contour map of S158(NGC 7538) (Israel et al., 1973).

Fig. 3 shows [13]CO contours superimposed on a photograph of the Orion Nebula and NGC 1977. The HII regions appear on the edges of the molecular concentration, and there is no hint of the symmetry implied in early models of the Orion Nebula. S158 (fig. 4) may be another example of the "blister" geometry: the radio contours suggest an open shell structure over the optically visible portion of the nebula, which is blue-shifted relative to the CO velocities by 3 km s$^{-1}$. The ultra-compact component G is totally obscured and coincides with OH, $H_2O$, and strong infrared sources; the CO maximum is also located somewhat to the south of S158, and these various sources may reflect an evolutionary pattern.

## COMPACT HII REGIONS AS INFRARED SOURCES

Compact HII regions emit almost all of their energy in the infrared. Fig. 5 shows the spectrum of DR-21 from $\lambda$73 cm to $\lambda$4 $\mu$m; the entire spectrum is now being filled in by a variety of techniques, and the enormous flux at infrared wavelengths is obvious. It is believed that the infrared emission is thermal radiation from dust grains heated by the absorption of stellar and/or nebular photons; the near-infrared radiation is thought to originate predominantly within the HII region itself, and most of the far-infrared from the surrounding neutral regions (Natta and Panagia (1976) and references therein). Recent observations of high angular resolution corroborate this general picture, but reveal far more complexity than it implies (eg. Gehrz et al., 1975; Werner et al., 1976; Harper et al., 1976; Becklin et al., 1976).

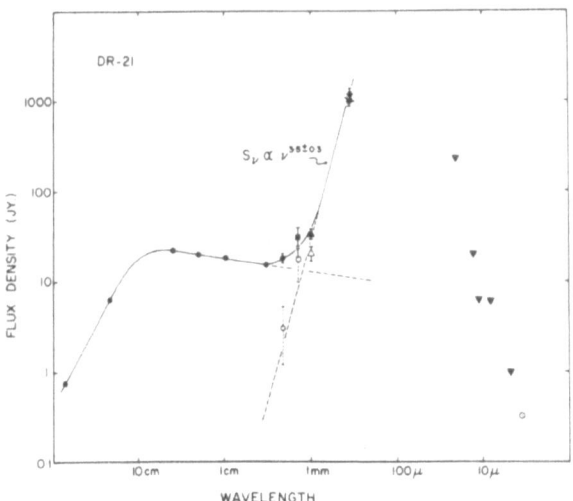

Figure 5.   Radio, submillimeter, and infrared spectrum of DR-21 (Righini et al., 1976).

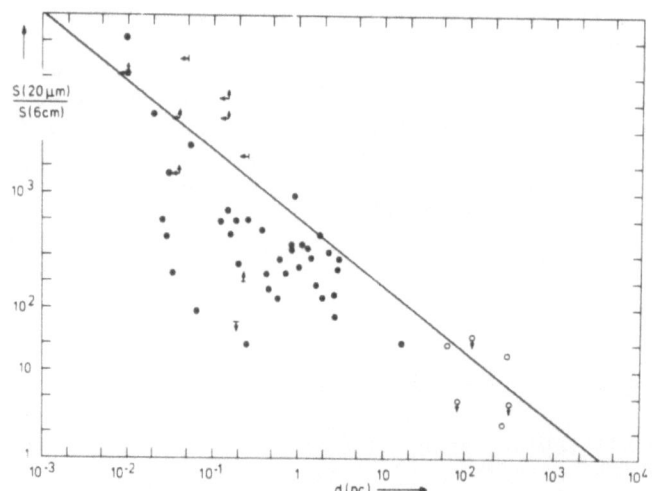

Figure 6.  Ratio of the near-infrared (λ20 μm) to radio continuum
(λ6 cm) flux densities plotted against linear diameter for HII regions
(Israel, 1976).

Ratios of the λ20 μm to λ6 cm flux densities are plotted against
HII region diameters in fig. 6. This ratio is large for many small-
diameter HII regions and comparatively small for larger HII regions.
This can be explained by a cooling of the dust or by decreasing density
(dust competes more effectively for Lyman-continuum photons at high
densities) as the HII region expands, for a given type of exciting star
(de Jong et al., 1975). The low ratios for some compact HII regions may
be due to dust depletion or to absorption by foreground dust.

The point at the upper left corner of fig. 6 is of special interest:
it represents a small HII region likely ionized by a B3 star embedded in
the Ophiuchus dark cloud. Fig. 7 shows three far-infrared sources in the
Ophiuchus dark cloud recently detected by Fazio et al. (1976); also shown
are the positions of several 2 μm sources, and CO and $H_2CO$ features.
There are four weak radio sources in this area, possibly HII regions,
which have been detected by sensitive radio interferometric methods
(Brown and Zuckerman, 1975). One of these coincides with S1, the infrared
source referred to above; the 2 μm source appears to be a heavily reddened ($A_v \sim 12$ mag) B3 star, surrounded by the compact HII region (diameter
$10'' \sim 0.01$ pc), and more extended zones of ionized carbon and sulfur and
strong far-infrared emission. This source is also near a maximum in the
CO emission and a minimum in the $H_2CO$ absorption, possible indicators of
star formation and locally intense ultraviolet radiation (Loren et al.,
1973; Myers and Ho, 1975). A total of 67  2 μm sources have been detected
in the Ophiuchus dark cloud, and most of these are thought to be embedded
F5 to B3 stars (Vrba et al., 1975). The radio observations are sensitive
to HII regions excited by stars of type B3 or earlier, with densities

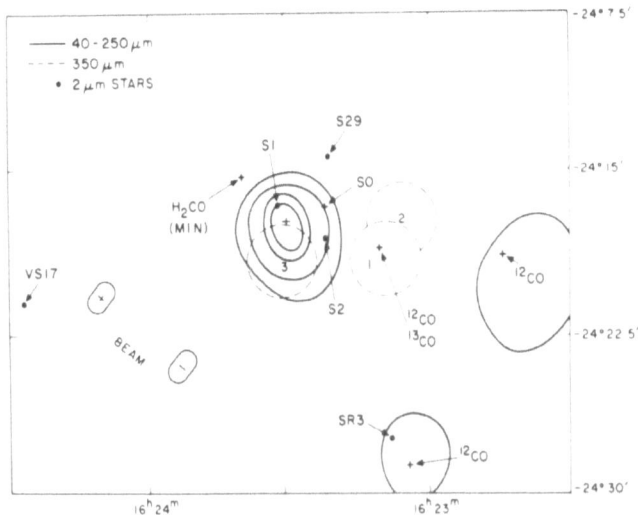

Figure 7. Map of far-infrared ($\lambda$40-250 $\mu$m) sources in the Ophiuchus dark cloud (Fazio et al., 1976). Also shown are the positions of the peak emission of $^{12}$CO, $^{13}$CO, and SO, the minimum absorption of $H_2$CO, the most intense 2 $\mu$m sources, and the 350 $\mu$m emission.

$\geqslant 10^{4.5}$ cm$^{-3}$ (Brown and Zuckerman , 1975). Such observations will be of increasing use in studying the hidden stellar content of dark clouds.

## ASSOCIATION WITH MASER SOURCES

80 percent of type I OH masers coincide with compact HII regions within a few arcseconds (Habing et al., 1974). On the other hand between 20 and 50 percent of very compact HII regions coincide with type I OH masers (Israel, 1976). It has recently been established at least in one case (W3 OH) that the coincidence is not perfect – the OH sources lie to one side of the HII region by $\sim$ 0.03 pc (Forster et al., 1977). $H_2$O masers are also frequently associated with these sources, although they are not coincident; they are found typically within $\sim$ 0.2 pc (Johnston et al., 1973; Lo et al., 1975). However OH and $H_2$O sources are not found near larger HII regions - the OH phenomenon seems to disappear when the associated HII region has expanded to $\sim$ 0.07 pc $\sim$ 15000 A.U. (Habing et al., 1974).

The very compact HII regions with which maser sources are associated are never isolated, but are found in clusters of HII regions. Low-resolution recombination line observations are therefore not necessarily reliable for comparison with the OH/$H_2$O radial velocities. Fig. 8(a) shows a $\lambda$6 cm synthesis map of the W58 region (resolution 20"); four components are indicated, including K3-50 (A) and NGC 6857 (D). The 1720 MHz OH source ON-3 coincides with component C, but the OH velocity

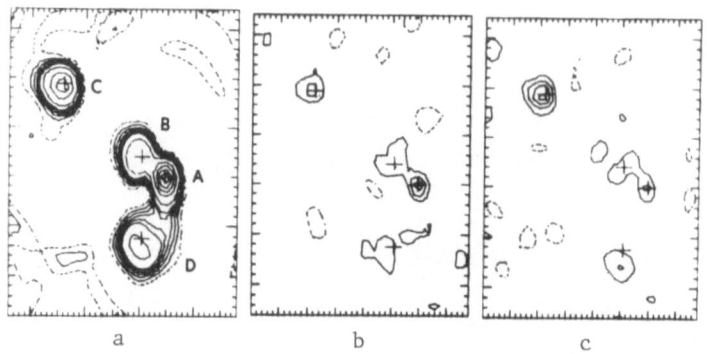

Figure 8. λ6 cm contour maps of the W58 region: (a) continuum, (b) −27 km s⁻¹, (c) −16 km s⁻¹ (van Gorkom, private communication). The maps have been smoothed to 20" resolution, and the bandwidths for (b) and (c) are 129 kHz (7.7 km s⁻¹).

(−13.5 km s⁻¹) differs from other radial velocities measured for this region (∿ −24 km s⁻¹) − HI and $H_2CO$ absorption, CO emission, and single dish measurements of recombination line emission. However the Westerbork H109α recombination line maps in figs. 8(b) and 8(c) show clearly that W58 C does have a velocity near −13.5 km s⁻¹, as distinct from the other components, a suggestion first made by Rubin and Turner (1971).

RECOMBINATION LINE STUDIES

Very high resolution recombination line observations are also use-ful in the study of the detailed physics of HII regions. One recent and perhaps surprising result is that the line-to-continuum ratios from synthesis observations have been found to be <u>lower</u> than the correspond-ing single-dish values. In fig. 8 the peak ratios are 1.7% (A) and 3% (B,C,D) compared to the single-dish value of 4.4% (Reifenstein et al., 1970), although the line widths are not markedly different. Similar results have been obtained by Wellington et al. (1976) for W3 and DR-21 at λ6 cm, and by Sullivan and Downes (1973) for W3A at λ21 cm − in the latter case a line-to-continuum ratio three times less than single-dish values. The observed ratio for K3-50 can be duplicated theoretically using the components from Harris's (1975) high resolution observations with $T_e$ = 10⁴ K, if their electron densities are increased fivefold (implying clumping within these dense components); the high densities diminish non-LTE effects, and effectively eliminate the contributions from the densest components through pressure broadening. The single-dish observations are apparently dominated not by maser effects along the lines of sight to the dense components as suggested by Brocklehurst and Seaton (1972), but rather by line emission from the much more widespread lower-density gas. In summary, high-resolution recombination line obser-vations give lower line-to-continuum ratios than do lower-resolution observations, indicating even greater electron densities in the compact

components than the rms values derived from continuum observations of
the highest resolution.

It is becoming increasingly clear how complicated the interpretation of radio recombination line observations can be, due to departures
from equilibrium, pressure broadening, density and temperature inhomogeneities, possible velocity structure, and convolution of all these by
the beam. Following the paper of Brocklehurst and Seaton (1972) which
stressed the importance of density inhomogeneities and pressure broadening, Lockman and Brown (1975) showed that temperature inhomogeneities
may also be important; recently they have illustrated (Lockman and Brown,
1976) that a combination of non-LTE effects and nebular geometry can
cause the LTE estimate of $T_e$ to vary across the nebula, even for a
perfectly homogeneous nebula.

A new field which recombination line observations have opened up in
the last few years is the study of carbon Strömgren spheres surrounding
early B-type stars embedded in dark clouds. So far they have been detected in the Ophiuchus dark cloud and the reflection nebulosities NGC 2023,
M78, and S140, and both $\alpha$ and $\beta$ transitions have been observed (Knapp
et al.(1976) and references therein). Typically $T_e \sim 20$ K, $n_e \sim 0.1-1$
$cm^{-3}$, with diameters $\sim 0.1-1$ pc. Associated near and far infrared sources
and CO maxima have been found, and ultracompact HII regions have been
detected as weak radio continuum sources in three of these cases.

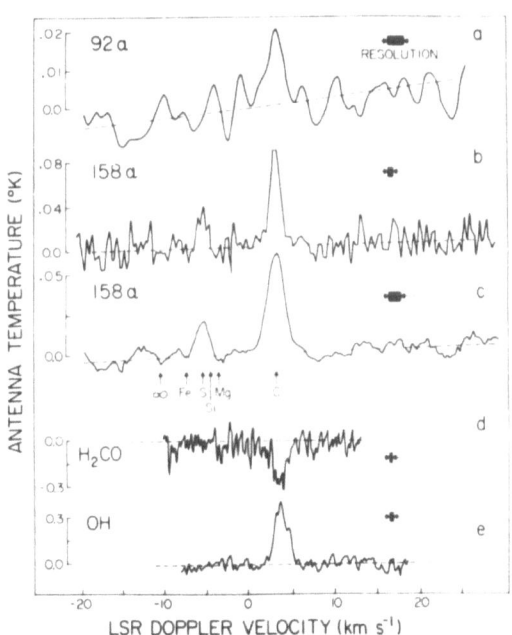

Figure 9.   Recombination and molecular line spectra at the position
$\alpha = 16^h23^m25^s$, $\delta = -24°15'50''(1950.0)$ in the Ophiuchus dark cloud
(Chaisson, 1975).

The narrowness ($\sim$ 1-2 km s$^{-1}$) of these lines permits resolution of spectral features due to different heavy elements: fig. 9 shows a sulfur recombination line from the Ophiuchus dark cloud - its intensity suggests that other heavy elements such as Mg and Si may be somewhat depleted. These recombination line observations provide a new window on star formation and the chemistry of the interstellar medium.


EVOLUTION OF HII REGIONS

The confluence of data from different sources - high resolution radio continuum observations, recombination and molecular lines, near and far infrared observations, and optical data - is beginning to reveal an evolutionary pattern for HII regions. Table I shows a tentative classification scheme for HII regions by Israel (1976); observed HII regions are classified according to their densities and sizes (see fig.1) The most compact HII regions appear to be embedded in dense molecular clouds - optically obscured and frequently relatively strong infrared sources; they are found in groups of HII regions, and they are distinguished by their unique association with the OH/H$_2$O maser phenomenon. As they expand they break out of their "cocoons" and become visible, although highly reddened at first; at this stage their morphology may be highly asymmetric, dominated by streams of ionized gas away from dense ionization fronts at the interface with the parent molecular cloud. Eventually they melt together into the giant, diffuse HII regions observed in other galaxies.


TABLE I.  Classification of HII regions.

| Class | d (pc) | $n_e$ (cm$^{-3}$) | Optical appearance | Group properties | OH/H$_2$O Masers | S(6 cm)/ S(20 μm) |
|---|---|---|---|---|---|---|
| I | <0.15 | >3000 | obscured | never isolated | + | <$10^4$ |
| II | 0.1-1.0 | >1000 | strongly reddened | never isolated | - | <$10^3$ |
| III | 0.15-10 | 100-300 | partially obscured | sometimes isolated | - | <5 x $10^2$ |
| IV | 1-30 | 10-100 | globules, bright rims | sometimes in groups | - | <$10^2$ |
| V | 10-300 | 3-50 | complex, diffuse | --- | - | -- |
| VI | >100 | <10 | diffuse | --- | - | -- |

REFERENCES

Becklin, E.E., Beckwith, S., Gatley, I., Matthews, K., Neugebauer, G.,
    Sarazin, C., Werner, M.W. 1976, Astrophys.J. 207, 770.
Blair, G.N. 1976, Unpublished PhD Thesis, University of Texas at Austin.
Blair, G.N., Peters, W.L., Vanden Bout, P.A. 1975, Astrophys.J. 200,
    L161.
Brocklehurst, M., Seaton, M.J. 1972, Monthly Notices Roy. astron. Soc.
    157, 179.
Brown, R.L., Zuckerman, B. 1975, Astrophys.J. 202, L125.
Chaisson, E.J. 1975, Astrophys.J. 197, L65.
Dickinson, D.F., Frogel, J.A., Persson, S.E. 1974, Astrophys.J. 192,347.
Fazio, G.G., Wright, E.L., Zeilik, M. II, Low, F.J. 1976, Astrophys.J.
    206, L165.
Forster, J.R., Welch, W.J., Wright, M.C.H. 1977, Astrophys.J.(in press)
Gehrz, R.D., Hackwell, J.A., Smith, J.R. 1975, Astrophys.J. 202, L33.
Grasdalen, G.L. 1974, Astrophys.J. 193, 373.
Habing, H.J., Goss, W.M., Matthews, H.E., Winnberg, A. 1974, Astron.
    Astrophys. 35, 1.
Harper, D.A., Low, F.J., Rieke, G.H., Thronson, H.A. Jr. 1976,
    Astrophys.J. 205, 136.
Harris, S. 1975, Monthly Notices Roy. astron. Soc. 170, 139.
Israel, F.P. 1976, Unpublished PhD Thesis, University of Leiden.
Israel, F.P., Habing, H.J., de Jong, T. 1973, Astron.Astrophys. 27, 143.
Johnston, K.J., Sloanaker, R.M., Bologna, J.M. 1973, Astrophys.J. 182,
    67.
Knapp, G.R., Brown, R.L., Kuiper, T.B.H., Kaler, R.K. 1976, Astrophys.J.
    204, 781.
Kutner, M.L., Evans, N.J. II, Tucker, K.D. 1976, Astrophys.J. 209, 452.
Liszt, H.T. 1973, Unpublished PhD Thesis, Princeton University.
Lo, K.Y., Burke, B.F., Haschick, A.D. 1975, Astrophys.J. 202, 81.
Lockman, F.J., Brown, R.L. 1975, Astrophys.J. 201, 134.
Lockman, F.J., Brown, R.L. 1976, Astrophys.J. 207, 436.
Loren, R.B., Vanden Bout, P.A., Davis, J.H. 1973, Astrophys.J. 185, L67.
Meaburn, J. 1975, in: HII Regions and Related Topics, Ed. T.L. Wilson
    and D. Downes, Springer, Heidelberg, p.222.
Myers, P.C., Ho, P.T.P. 1975, Astrophys.J. 202, L25.
Natta, A., Panagia, N. 1976, Astron.Astrophys. 50, 191.
Reifenstein, E.C. III, Wilson, T.L., Burke, B.F., Mezger, P.G.,
    Altenhoff, W.J. 1970, Astron.Astrophys. 4, 357.
Righini, G., Simon, M., Joyce, R.R. 1976, Astrophys.J. 207, 119.
Rubin, R.H., Turner, B.E. 1971, Astrophys.J. 165, 471.
Sullivan, W.T. III, Downes, D. 1973, Astron.Astrophys. 29, 369.
Vrba, F.J., Strom, K.M., Strom, S.E., Grasdalen, G.L. 1975, Astrophys.J.
    197, 77.
Wellington, K.J., Sullivan, W.T.III, Goss, W.M., Matthews, H.E. 1976,
    Astron.Astrophys. 47, 351.
Werner, M.W., Gatley, I., Harper, D.A., Becklin, E.E., Loewenstein, R.F.,
    Telesco, C.M., Thronson, H.A. 1976, Astrophys.J. 204, 420.
Wilson, W.J., Schwartz, P.R., Epstein, E.E., Johnson, W.A., Etcheverry,
    R.D., Mori, T.T., Berry, G.G., Dyson, H.B. 1974, Astrophys.J.191,357.
Zuckerman, B. 1973, Astrophys.J. 183, 863.

# DUST IN HII REGIONS

Syuzo Isobe
Tokyo Astronomical Observatory, University of Tokyo,
Mitaka, Tokyo

ABSTRACT

Several pieces of evidence indicate that HII regions may contain dust: 1) the continuum light scattered by dust grains (O'Dell and Hubbard, 1965),2) thermal radiation from dust grains at infrared wavelengths (Ney and Allen, 1969), 3) the abnormal helium abundance in some HII regions (Peimbert and Costero, 1969), etc.

Although observations of the scattered continuum suggest that the HII region cores may be dust-free, dust grains and gas must be well-mixed in view of the infrared observations. This difficulty may be solved by introducing globules with sizes $\sim$0.001 pc. These globules and the molecular clouds adjacent to HII regions are the main sources supplying dust to HII regions.

1. INTRODUCTION

Research into the nature of interstellar grains started with Stebbins, Huffer, and Whitford (1934) who found the wavelength dependence of interstellar extinction. Reliable extinction curves are now available for a large number of stars (for references see Wickramasinghe and Nandy, 1972). In order to provide good fits to these observed curves, many different kinds of grain and of grain mixtures have been proposed (Wickramasinghe and Nandy, 1971; Huffman and Staff, 1971; Gilra, 1971; Isobe, 1973), including: ice particles, iron particles, silicate particles, Platt particles, etc. Infrared and ultraviolet observations show band structures in the extinction curves: the 10-$\mu$ band due to silicate grains (Stein and Gillett, 1969), the 3-$\mu$ band of ice grains (Gillett and Forrest, 1973), and the 2200-A band of graphite grains (Stecher, 1969). Witt and Lillie (1973) have also derived the relative efficiencies for scattering and absorption (Figure 1) of interstellar grains from observations of galactic scattered light.

Although knowledge of the wavelength dependence of interstellar extinction and scattering makes the optical properties of interstellar grains clear, it does not give us enough information to decide about the

*Hugo van Woerden (ed.), Topics in Interstellar Matter, 61-79. All Rights Reserved.*
*Copyright © 1977 by D. Reidel Publishing Company, Dordrecht-Holland.*

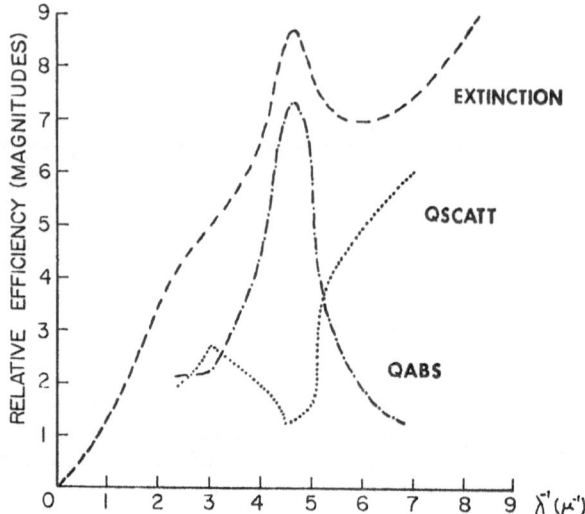

Figure 1. The relative efficiencies for extinction, absorption, and scattering of interstellar grains.

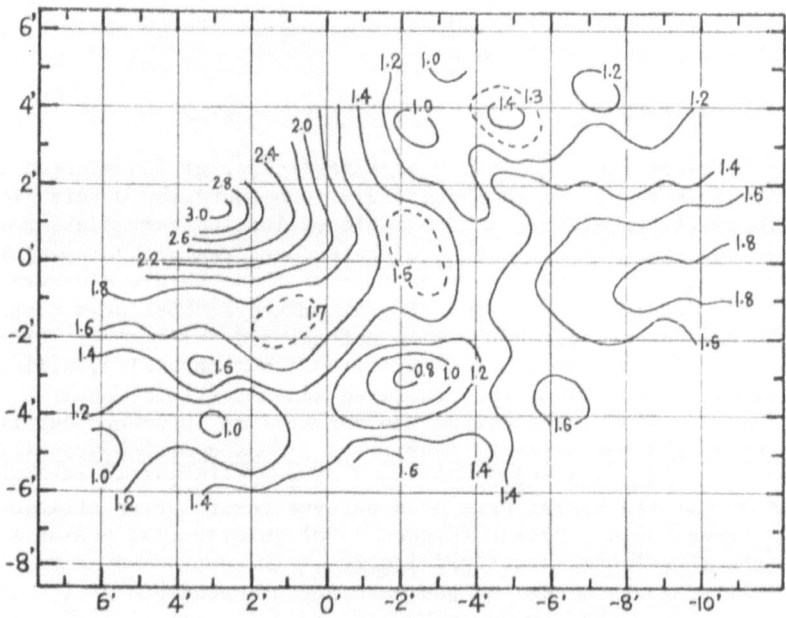

Figure 2. Interstellar extinction of the Orion Nebula at the wavelength 6563 A. Center is the Trapezium, west is right, and north is up. The contour unit is one magnitude.

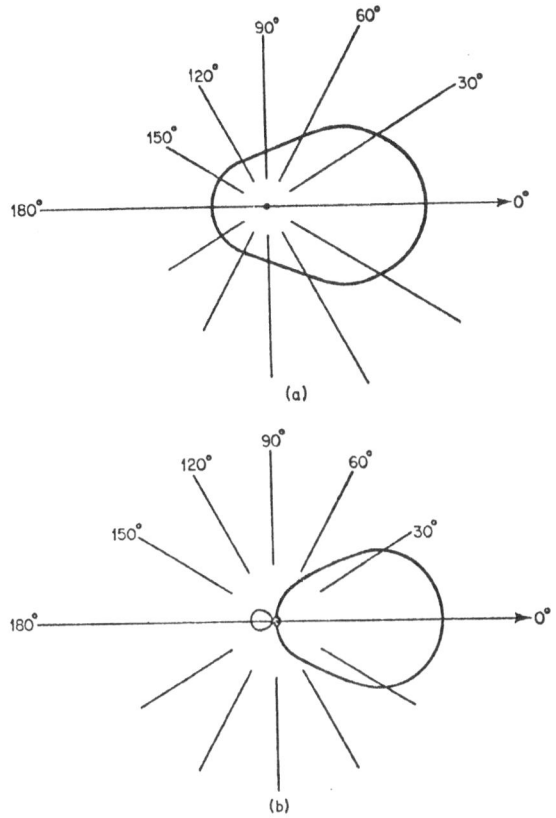

Figure 3. Scattering diagrams for dielectric
spheres with m = 1.33; upper: $2\pi a/\lambda - 1$; lower:
$2\pi a/\lambda = 1.4$. Radius vector represents intensity
of scattered light on an arbitrary scale.


Table 1. Gas/dust ratio in four HII regions.

The ratio of hydrogen gas column density, $N_H$, to dust column density,
$N_d$, multiplied by scattering cross section, $C_{sca}$, as observed by O'Dell
and his collaborators.

| Object | $N_H/N_d \; C_{sca}$ |
|---|---|
| M20 | $0.4 \times 10^{21} \; cm^{-2} \; mag^{-1}$ |
| M8 | 2.2 |
| M16 | 1.9 |
| M42 (inner parts) | 14.4 |
| M42 (outer parts) | 0.5 |
| General field | 2.0 |

main component of interstellar grains. For this purpose it is useful to consider the thermal properties which are different for various types of particle in HII regions.

In fact O'Dell and Hubbard (1965) and O'Dell, Hubbard, and Peimbert (1966) have found that the observed nebular continuum is stronger than expected from atomic processes and must be identified as light scattered by dust grains in HII regions. Since electron temperature and electron density are high enough, and the radiation from the exciting star is strong, dust grains in HII regions will rapidly evaporate by thermal evaporation and sputtering. The evaporation time scale depends strongly on the grain material; hence it provides a criterion for the composition of dust grains.

There are strong intensity fluctuations in HII regions. From the differences between densities found from optical and radio observations, Osterbrock and Flather (1959) suggested the existence of condensations in HII regions. If such condensations include dust grains, they can explain many observations (Dyson, 1968; Dopita, Isobe, and Meaburn, 1975) and supply dust grains to HII regions. Moreover, since many HII regions are connected with molecular clouds, dust grains will flow into HII regions from these clouds. Therefore, it is important to consider the role of dust, which is continuously supplied from the condensations in HII regions and from molecular clouds, in the evolution of HII regions.

In this review we will summarize the observational evidence for the existence of dust grains in HII regions, and discuss globules and molecular clouds as a supply source of dust grains to HII regions.

## 2. OPTICAL OBSERVATIONS OF DUST IN HII REGIONS

The most extensive observations of dust in the directions to HII regions have been done by Ishida and Kawajiri (1968), at the wavelength of Hα. Unfortunately, since the objects (W3, M16, M17, W4, M20, M8, W80) observed by them are far from the Sun, foreground interstellar extinction confuses the distribution of extinction in these HII regions. Moreover, the radio maps used for comparison with these Hα observations have rather poor resolution (∼11'). Isobe and Kurihara (1970) derived a contour map of interstellar extinction for the Orion Nebula, which has only a small amount of foreground extinction and for which radio observations with 1' resolution (Schraml and Mezger, 1969) are available.

In the last few years, many radio contour maps of resolution better than 1 arc minute have been published (Wink, Altenhoff, and Webster, 1973; Israel, Habing, and De Jong, 1973; Wink and Altenhoff, 1975; Gull and Martin, 1975; Israel, 1976). Further comparisons of Hα and radio observations are to be expected.

Direct information about dust in HII regions was given by the observations of continuum light scattered by dust grains (O'Dell and Hubbard, 1965; O'Dell, Hubbard, and Peimbert, 1966). Table 1 shows the

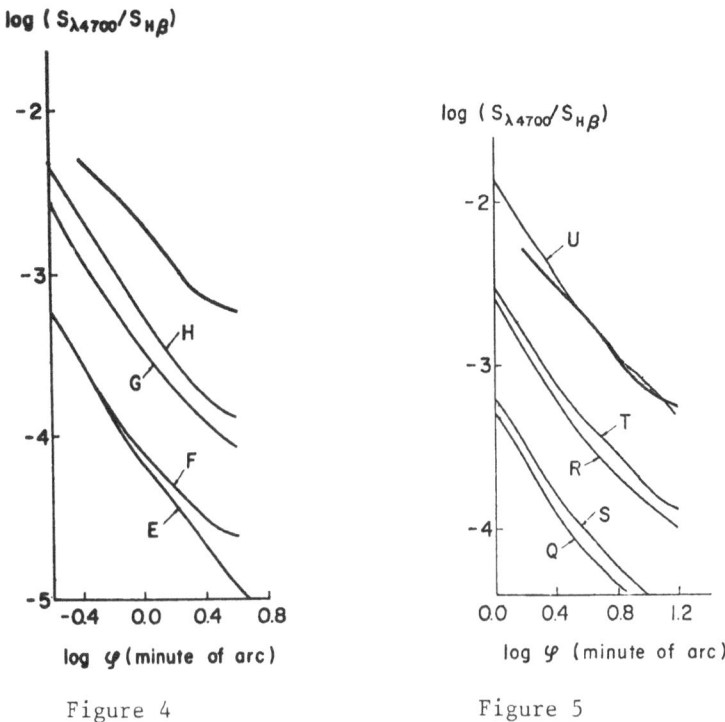

Figure 4                    Figure 5

Figure 4. Ratio of the surface brightness of scattered continuum light
to that of Hβ in NGC 6523 (M8). Thick line shows the observed ratio,
and thin lines show ratios calculated for several sets of various
parameters, when interstellar grains are composed of graphite grains.
The lines G and H show the cases that all carbon atoms obtained from
one and ten times cosmic abundance are condensed into graphite grains
with a radius 0.05 micron.

Figure 5. Same as Figure 4, but for silicate grains. The lines T and U
show the cases that all silicon atoms obtained from one and ten times
cosmic abundance are condensed into silicate grains with a radius 0.1
micron.

ratio of the column densities of hydrogen and dust, in terms of the
quantity $N_H/N_d$ $C_{sca}$, for 4 HII regions. Krishna Swamy and O'Dell (1967)
and Isobe (1970) have shown that the minimum value of $N_H/N_d$ $C_{sca}$ is
about $1.0 \times 10^{21}$ atom cm$^{-2}$ mag$^{-1}$, adopting uniform scattering efficiency
of dust grains and cosmic abundances of oxygen, carbon, and silicon
relative to hydrogen. Therefore, it was claimed that dust is over-
abundant in the outer parts of the Orion Nebula (M42) and in the Trifid
Nebula (M20). Spectroscopic observations by Simpson (1973) and photographic observations by Dopita, Isobe, and Meaburn (1975) support the
low value of $N_H/N_d$ $C_{sca}$ in the outer parts of the Orion Nebula.

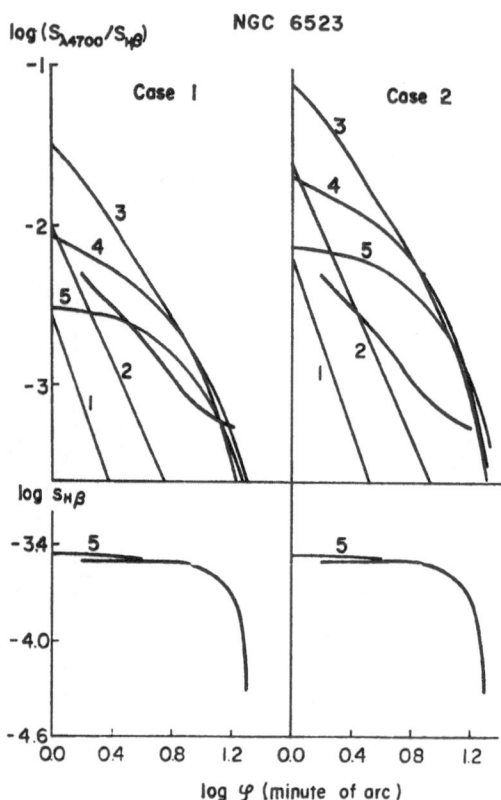

Figure 6. Same as Figure 4, but for silicate grains.
The difference of Case 1 and Case 2 originates in the
difference in the flux from the exciting stars. Line 3
shows the case that all oxygen atoms obtained from
cosmic abundance are condensed into ice grains. Lines
4, 5 and 1, 2 show the cases that ice grains evaporate
in the inner and outer regions, respectively.

However, the assumption of uniform scattering efficiency is correct
only for grains with radii  smaller than ∿0.05μ. The forward directivity
of scattering is evident for grains with radii of 0.1 to 1.0μ at visual
wavelengths (Figure 3). Adopting the scattering functions $S(\theta)$ for ice,
graphite, and silicate grains from the tables of Isobe (1975a) and
taking the distribution of electron density in an HII region indicated
by radio observations, one may calculate a predicted surface-brightness
ratio of continuum to Hβ, $S_{\lambda 4700A}/S_{H\beta}$, and compare this to observation.
Figures 4, 5, and 6 show the results for M8, taken from Isobe's (1975b)
paper. Since graphite particles are highly absorptive, an increase of
their particle density contributes little to scattering but mainly to
absorption. Therefore, the high values of $S_{\lambda 4700A}/S_{H\beta}$ observed in M8,
M16, M20, and M42 cannot be realized by graphite grains. Although the
situation for silicate particles is better than for graphite, because
of their high scattering efficiency, Figure 5 shows that the amount of

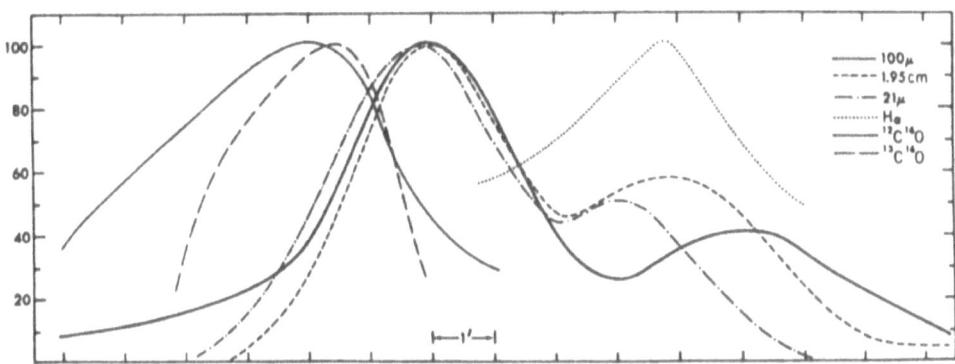

Figure 7. Scans through the 100-μ peak of M17 with an angular resolution of 2ʹ.5. The position angle is 126° (SW to NE from left to right) and all data have been normalized to a peak value of 100.

silicate grains needed to explain the observed $S_{\lambda 4700A}/S_{H\beta}$ ratio exceeds that calculated from cosmic abundances. Ice particles have the property of highly forward scattering; hence, most of the scattered continuum light may originate in a small amount of ice particles in an envelope surrounding the grain.

Johnson (1965, 1968) has found that the ratio of total to selective extinction ($R = A_V/E_{B-V}$) in several young HII regions is larger than the normal interstellar value, 3. These high values of R may be caused by infrared emission and by abnormally large grain sizes. The stars in the outer part of the Orion Nebula have normal values of R (Walker 1969), but stars within 3ʹ from the exciting stars (the Trapezium) have large values of R (Lee 1968). Our interim conclusion is that near the exciting star the grain character is changed by some effect(s).

3. INFRARED AND RADIO OBSERVATIONS

The development of infrared observational techniques brings much information about interstellar dust. Observations of the 3-μ absorption band of ice grains and the 10-μ absorption band of silicate grains give direct evidence for the existence of these grains in interstellar space. The objects which show these absorption bands are the Kleinmann-Low Nebula (Gillett and Forrest, 1973), IRS 5 in W3 (Aitken and Jones, 1973), CRL 2591 (Merrill and Soifer, 1974), NGC 2024 No. 2 (Grasdalen, 1974), four compact HII regions: K3-50, W51-IRS 1, W51-IRS 2, W49-OH(2) (Gillett et al., 1975), the compact radio-HII region, G29.9-0.0 (Soifer and Pipher, 1975), H₂O maser sources associated with the HII regions S255 and S257 (Pipher and Soifer, 1976), and carbon stars (Noguchi et al., 1977). All of these objects, except the carbon stars, are connected with HII regions.

High-resolution infrared observations show strong relationships between HII regions and IR sources (see references in Wynn-Williams and Becklin, 1974). Harper et al. (1976) show intensity distributions at

several wavelengths for M17 (Figure 7). There is a clear relation between
the infrared and 1.95-cm intensities in the central parts of the figure.
To the left there is a large molecular cloud containing much dust. The
contour map of space absorption by Ishida and Kawajiri (1968) shows a
steep increase towards the SW direction, which suggests that dust in the
molecular cloud extends towards the NE and absorbs visible light from
the HII region. From the same figure we find that the far-infrared
(100-μ) and radio (2-cm) pictures are significantly more extended, particularly to the north-east, than the near-infrared (21-μ). A thermal
spectrum of M17 was obtained by Lemke (1974), who found that at least
two dust components are mixed with the gas throughout the nebula. He has
derived dust temperatures, luminosities and masses, with the usual
assumptions about dust grains; these are shown in Table 2. The mass of
hot dust is quite low, while that of cold dust is about 1/100 of the
hydrogen mass. In some HII regions, infrared observations indicate less
dust than predicted from dust-to-gas ratios derived from interstellar
extinction measurements. The depletion of dust in the HII regions
G29.9-0.0 (Soifer and Pipher, 1975) and G30.8-0.0 (Pipher, Grasdalen,
and Soifer, 1974) is shown observationally, with some ambiguity of the
assumed grain parameters. Although Wright (1973) estimates a 1000-fold
depletion of dust in the ionized region, a 1 to 100-fold depletion may
be a more reasonable value in HII regions (Panagia, 1974).

In connection with these problems, the presence of an absorption
feature at 10μ is interpreted as due to the cold component which also
gives rise to the 100-μ intensity. Radio observations show molecular
clouds connected with HII regions. Dust in these molecular clouds radiates also in the far-infrared. Chaisson and Wilson (1975) have derived
a mass of 1300 $M_\odot$ for the molecular concentration surrounding M20.
Figure 8 shows the difference in extent of HII region and molecular
cloud; the hydrogen recombination line disappears at the edge of the
visual HII region. This type of molecular cloud is observed also in M8
(Lada et al., 1976), in IC 1396 (Loren et al., 1975), and in B35 (Lada
and Black, 1976), and is discussed in detail in Section 4.

Churchwell et al. (1974) have observed radio recombination lines of
helium and hydrogen for 39 HII regions, and found that 3 HII regions
near the Galactic Centre have low values of $N(He^+)/N(H^+) < 0.02$. Also,

Table 2. Two dust components in M17 after Lemke (1974).

|              | Hot dust        | Cold dust       |
|--------------|-----------------|-----------------|
| Temperature  | 240 $^\circ$K   | 75 $^\circ$K    |
| Luminosity   | $10^6$ $L_\odot$ | $6\times10^6$ $L_\odot$ |
| Mass         | $2\times10^{-3}$ $M_\odot$ | 4 $M_\odot$ |

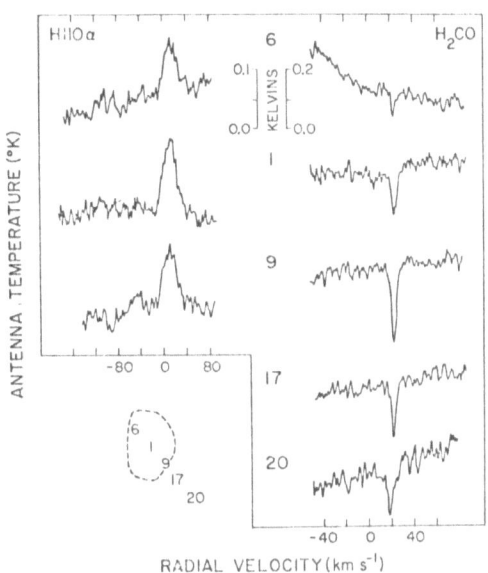

Figure 8. Representative spectra of 6-cm observations
of $H_2CO$ and H110$\alpha$ along a northeast-southwest diagonal
slice through the M20 complex. The insert at the lower
left side shows the positions of the observations
relative to the outer contour of 8 GHz nebular continuum radiation.

a relation is found between IR excess and the ratio, R, of helium and
hydrogen Strömgren-sphere radii (Mezger, Smith, and Churchwell, 1974),
that is, R is small if the IR excess is large. These results are interpreted as follows (Smith 1975); R is nearly 1 if the temperature of the
exciting star is high enough. However, if there are dust grains in HII
regions and the absorption cross-section of the dust at $\lambda < 504A$ (that
is, for helium-ionizing photons) is much larger than that at $912A > \lambda$
$> 504A$ (hydrogen-ionizing photons), the helium-ionizing photons are
selectively destructed. Since $N(He^+)/N(H^+)$ for most HII regions is about
0.1, the same value as the cosmic abundance, there must be few dust
grains in HII regions. These conclusions seem to conflict with the infrared results which indicate the presence of dust grains within HII regions.
We will discuss this problem further in Section 6.

## 4. CONNECTION BETWEEN HII REGIONS AND MOLECULAR CLOUDS

There are many HII regions associated with molecular clouds. The
estimated masses of molecular clouds are 10 - 100 times those for the
HII regions M8, M17, M20, and M42, which suggests that the HII region
is only a part of a whole cloud. In the molecular cloud star formation
is now going on, and we there find such IR sources as the KL and BN
objects in the Orion Nebula. The strong obscuration of the IR sources
(Becklin, Neugebauer, and Wynn-Williams, 1973) indicates that the mole-

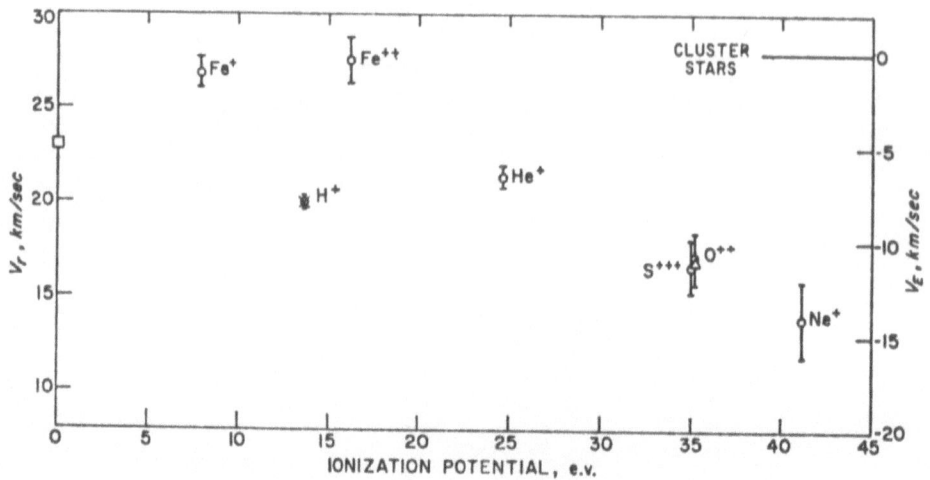

Figure 9. The velocity of various ions in the Orion Nebula. The abscissa gives the ionization potential of the ion in electron volts, the left-hand ordinate the radial velocity reduced to the Sun, and the right-hand ordinate the radial velocity reduced to the mean of the cluster stars.

cular cloud contains dust grains. Actually we know cases of close con-nection between molecular clouds and dark clouds which obscure optical HII regions. CO observations for IC 1396 (William and Vanden Bout, 1975), for NGC 6231 (Sherwood and Dachs, 1970), and for B35 (Lada and Black, 1976) show clear coincidences of molecular clouds with dark clouds.

Current calculations of the evolution of HII regions treat the case that density decreases outwards, or is constant. Assuming constant density as initial distribution, Mathews (1969) followed the evolution of an HII region and found that the density at the ionization front be-comes one or two orders higher than inside the HII region, but that such a high density disappears after $10^4$ years. Bohuski (1973) observed such a density increase near the dark cloud in M8 and M20. However, these objects are much older than $10^4$ years, and a high density at the ioni-zation front must be interpreted in some other way.

Kaler (1967) found that in the Orion Nebula ions of higher excita-tion are approaching the Sun faster than those of low excitation (Figure 9). Before the discovery of the molecular cloud behind the Orion Nebula, these results were interpreted as due to radiation of the exciting star, which dominates in the inner region where the ions are more highly exci-ted. Zuckerman (1973) suggests that radiation from ions of relatively low ionization potential originates primarily in regions close to the HI-HII boundary. Since the density in the molecular cloud is 10 to 100 times higher than in the HII region, the gas pressure at the HI-HII boundary must be very high, and ionized gas is accelerated and flows into the HII region. Similar phenomena are observed in other HI-HII boundaries. Observations of the H110α recombination line (Chaisson and Wilson, 1975) show anomalous velocities in the south-western part of M20, which originate in a gas flow from the HI to the HII region. Wellington, Sullivan, Goss, and Matthews (1976) find, from H109α aperture-synthesis observations of DR21, a velocity gradient of 20 km/sec from north-east to south-west. They interpreted this as rotation of the HII region.[*] However, Iguchi and Fukui (1976) suggested another possibility for the model of DR21, from the radial velocity observations of recombination lines and molecular lines shown in Table 3. Radial velocities of recombination lines are larger by 10 km/sec than those of molecular lines. Since the higher-frequency recombination lines are detected in regions with higher density, while the lower-frequency lines are preferentially detected in regions with lower density, the data given in Table 3 show that gas with lower density expands quickly in the radial direction whereas gas adjacent to the high-density molecular cloud has only started to leave the molecular cloud. This structure of DR21 is the same as that of the Orion Nebula. There are some other examples of gas outflow from molecular clouds to HII regions with less certainty.

Since molecular clouds include considerable dust emitting IR radiation, dust grains also flow into HII regions but some grains are blown back to the molecular cloud by the radiation pressure of the exciting star. Lortet-Zuckerman (1974) has shown that density and brightness peaks do not coincide at the brightnesss boundary and the peak density is located about 0.1 or 0.2 parsec in the obscured region, where the IR source is.

From the considerations discussed in this section we conclude that molecular clouds are sources of dust grains in HII regions.

Table 3.  Radial velocities and velocity dispersions of molecular and recombination lines in DR 21.

|            | V          | $\sigma_V$ |
| ---------- | ---------- | ---------- |
| $H_2CO$    | -3.2 km/s  | 3.9 km/s   |
| $NH_3$     | -2.3       | 2.8        |
| $C_2H$     | -2.4       | 2.5        |
| H158α      | +7.8       | 17.0       |
| H109α      | +2.6       | 20.0       |
| H 85α      | -1.0       | 24.6       |

[*] In a subsequent erratum Wellington et al. (1977) have indicated that the apparent H109α velocity gradient over DR 21 is not real, but due to an instrumental effect.

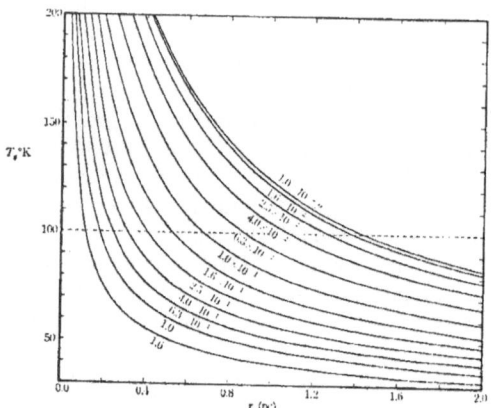

Figure 10. Equilibrium temperature Tg ($^{\circ}$K) of dust grains at
distance r(pc) from a central star of spectral type 05, with
temperature $5.53 \times 10^4$ K and radius $8.20 \times 10^{11}$ cm. The temperatures are shown for dust grains of radius 1.6, 1.0, 0.63,
0.40 ....., 0.010 μ.

## 5. PHYSICAL AND KINEMATICAL PROPERTIES OF DUST IN HII REGIONS

Following the discussion in Sections 2, 3, and 4, it is reasonable
to examine the properties of grains in HII regions.

The grain temperature is calculated by equating the energy absorbed
from stellar radiation with the energy emitted from the grain surface
if the wavelength dependence of the absorption coefficient is given
(Wickramasinghe 1967). In Figure 10, we show the dependence of the temperature of ice grains on grain radius and on distance from the exciting
star, for the case of the Orion Nebula (Isobe 1970). It is clear that
smaller grains have higher temperatures and evaporate more easily.
Graphite and silicate grains have low enough temperatures not to evaporate in most parts of HII regions (Isobe 1971). Sputtering of protons
is also an effective destruction mechanism. The time scale for destruction of an ice grain of typical radius $\sim 10^{-5}$ cm is of order $10^4$ years,
which is considerably less than the probable relaxation time for an HII
region of $10^{6-7}$ years. However, if we consider charged grains (Feuerbacher, Willis, and Fitton 1973), the time scale is 10 to 100 times
longer than that without charge.

The effect of grain charge is also important in the calculation of
expansion velocities of grains away from the exciting star. The terminal
velocity of expansion is determined by equating the radiation pressure
force to the drag force (Wickramasinghe and Nandy 1972) and is $\sim 100$ km/s.
Therefore, grains are expelled from the region near the exciting star to
1 pc distance in about $10^4$ years. If we consider grain charge, the time
scale is prolonged to $\sim 10^{5-6}$ years (Moorwood and Feuerbacher 1975). Since

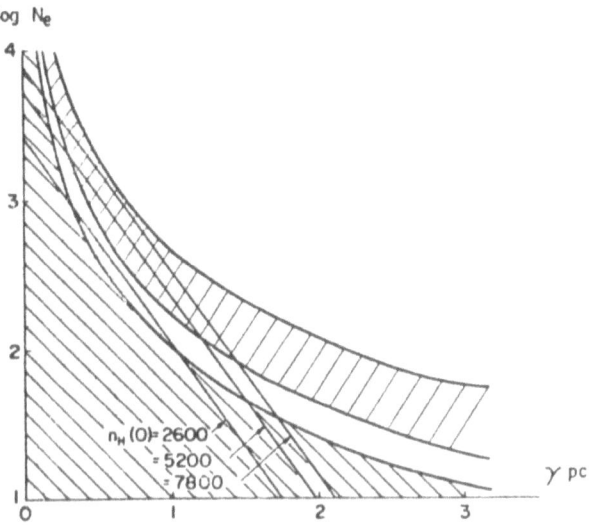

Figure 11. The region on the log $n_e$ - r plane where grains
have velocities higher than 1 km/s. The area shown by ///
is the region for graphite core-ice mantle grains, and \\\
is the region for graphite grains.

the amount of charge on the surface of grains depends on the electron
density and local radiation field in the HII region, and on the photoelectric properties of grain surfaces, the expansion velocities of
various types of grain vary over the range 1 - 100 km/sec in HII regions
(Figure 11, Isobe 1975c).

This kind of velocity difference of dust grains in HII regions is
expected to be the trigger of condensations. Theoretically, Nordsieck
(1971) and Capriotti (1971) have shown that the formation of condensations is possible in the ideal case.

6. THE ORION NEBULA

In Sections 2 - 5 we considered the general character of dust in
HII regions. Here, we concentrate on the dust in the Orion Nebula,
because of the abundant data available.

There are many observations of the Orion Nebula in radio molecular
lines. It is clear that there is a molecular cloud with a mass of $\sim 10^3 M_\odot$
behind the HII region. Harvey et al. (1974) obtained contour maps of
1-mm continuum radiation and of the $H_2CO$ emission line at 2 cm. The
similarity of the two maps suggests that most of the observed 1-mm radiation arises in the molecular cloud. They reasonably conclude that the
1-mm radiation is primarily thermal radiation from dust grains in the
molecular cloud. Similarities in the 1-mm and HCN emission are also ob-

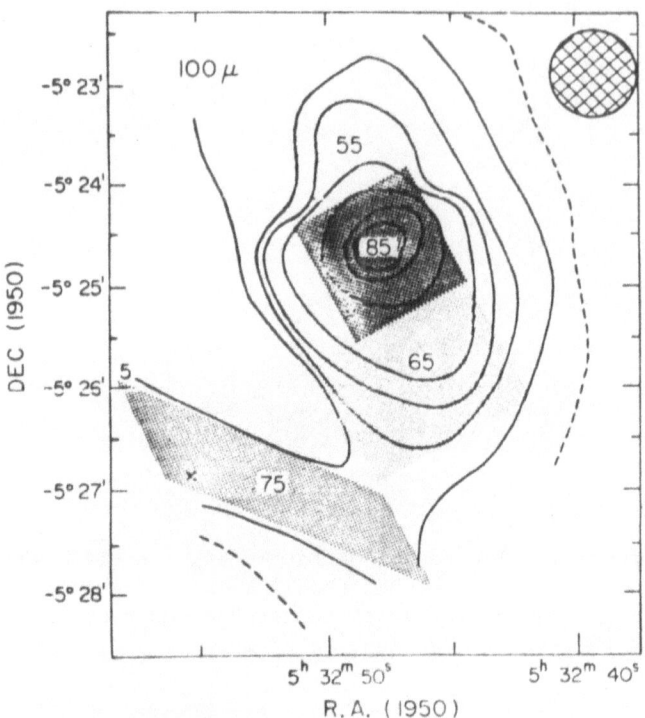

Figure 12. The average colour temperature between 50 and 100 μ
is shown for several regions of the Orion Nebula, superposed
on the contours of 100-μ emission. The temperatures have been
computed, assuming grey emission, from the observed ratio of
average 50-μ and 100-μ surface brightnesses.

served in DR 21 (Werner et al., 1975). However, Harper (1974) showed
that more than half of the 91-μ flux originates in a lower surface
brightness component, the distribution of which is similar to that of
the extended HII component, although the brightest feature of the 91-μ
map is a compact source centered on the KL nebula. This is clearly confirmed by comparing the 50-μ contour map with the 100-μ contour map
(Werner et al., 1976). Figure 12 shows colour temperatures between 50 μ
and 100 μ superposed on contours of 100-μ emission. The positions of
maximum 50- and 100-μ surface brightness and of maximum colour temperature coincide with the position of the KL nebula, but in the southern
portion of the ridge the temperature does not decrease as markedly as
to the north. Dust in the HI-HII boundary is heated by radiation from the
Trapezium in the southern ridge; this relatively high-temperature (∿75 K)
ridge coincides in position with a prominent optical filament (observations of [SII] by Gull (1974), Hα and [NII] by Dopita, Isobe, and Meaburn
(1975) ). This is consistent with the bright-rim structure given in
Figure 9 of Dopita et al. (1975), where the surface of the HI-HII boundary

is parallel to the line of sight and must extend at least 0.1 pc along
the line of sight in order to account for its observed luminosity of
$7 \times 10^3$ $L_\odot$. Zuckerman (1973) suggested that the distance between the
Trapezium and the molecular cloud is as small as 0.1 pc from consideration of the observed kinematics. The distance of $\sim$0.1 pc is also derived
from the Hβ surface distribution in the central portion (1' - 2' from
the Trapezium) of the Orion Nebula (Schiffer and Mathis 1974). Dust
grains at such a short distance from the Trapezium must have temperatures
much higher than 100 K (Isobe 1971b) and infrared radiation (10 - 20 μ)
should be observed. Ney and Allen (1969) observed the central portion of
the Orion Nebula at 11.6 μ and 20 μ and found an infrared component (the
NA Nebula) at the position of the Trapezium and with a higher 11.6-μ/
20-μ ratio than that of the KL Nebula.

From the fact that the molecular clouds adjacent to the HII regions
contain dust grains, there must have been dust grains in that part of the
molecular cloud which is now the HII region, before the exciting star
turned on. Dust grains near the exciting star are blown to the outer
part ($\sim$0.3 pc) of an HII region in about $10^5$ years. Mathis (1970) has
calculated the intensity ratio of the Balmer lines, taking into account
the effect of dust grains in HII regions; he found that in the Orion
Nebula there are dust grains causing 1.8 magnitudes of extinction, by
comparing the calculated line ratios with the observed ones. Münch and
Persson (1971) obtained more extensive results using their line and
continuum observations with high angular resolution. They found that
dust is well mixed with the gas and, statistically, fluctuations in gas
density are positively correlated with variations in dust density. They
also found that the position of the Hα ridge mentioned above is nearer
to the Trapezium than the ridge of the scattered continuum. This implies
some segregation of dust and gas at the ridge: the scattered continuum
(50-μ and 100-μ observations by Werner et al. (1976) ) mainly originates
in grains in the HI-HII boundary and the line radiation is emitted by
gas adjacent to this boundary.

Dopita et al. (1975) obtained a contour map of the surface-brightness
ratio of the scattered continuum to the Hβ emission, $S_{\lambda4700}/S_{H\beta}$, which
is consistent with the results of O'Dell and Hubbard (1965). It was
found that ice grains with a forward-scattering efficiency in the outer
shell (> 1 pc) could account for most of the scattered light, that is,
there is a dust-free core (< 1 pc) surrounding the Trapezium. They also
postulated the presence of some grains which scatter light from the
Trapezium perpendicular to the line of sight, in order to explain the
wavelength dependence of the scattered continuum light.

It is well known that there are condensations in the Orion Nebula;
these were first introduced by Osterbrock and Flather (1959) in order to
interpret the discrepancy between optical and radio data. Direct evidence
for such condensations has been clearly shown in the filter photographs
of the core region by Elliott and Meaburn (1974). Recently, Tamura (1976)
obtained the total number (about 400) and the size of condensations (of
the order of 0.001 pc) in the following way. Neglecting variations of

electron temperature, ionic helium abundance, and Hβ surface brightness
with angular distance in the region R < 3', he tried to account for the
intensity fluctuations of Hβ, [O III] λ 4363, [O III] λ 5007 + [O III]
λ 4959, and He λ 4471, in terms of condensations alone. The mean
intensity of each line was related to the parameters of the condensations
and the surrounding region: the total number N and the size $\ell$ of condensations, an electron density $n_c$ and electron temperature $T_c$ of condensations, and an electron density $n_s$ and electron temperature $T_s$ in the
surroundings. The statistical solution yielded the following results:

$$T_s/T_c \sim 0.8, \; n_s/n_c \sim 10^{-2 \; to \; -3}, \; N \sim 400, \; \ell \sim 0.001 \; pc.$$

These results agree well with those of Dyson (1968), which were obtained
in an attempt to explain the splitting of forbidden lines in the central
region.

Although the dust grains ejected into HII regions do not exist much
longer than $10^{3 \; to \; 5}$ years in the central regions, globules, some special
kinds of condensations, and molecular clouds are continuously supplying
dust grains into HII regions. Dopita et al. (1975) have calculated the
evolution of such globules. Since the velocity of grains towards the
centre of a globule caused by radiation pressure is only slightly higher
than that of ionization of the globule surface, most grains are concentrated in the core of the globule where star formation takes place, but
some grains flow into the HII region. The density in the region just
adjacent to the globule surface is $\sim 10^6$ cm$^{-3}$, which is greater than that
in the surroundings by 3 orders of magnitude. Dust in this partially-
ionized region probably scatters light from the Trapezium with a scattering angle $\sim 90°$; this scattered continuum light contributes to the observed scattered continuum light in the inner region and changes its wavelength dependence to that observed (O'Dell and Hubbard, 1965).

Although radio data give a normal helium abundance relative to
hydrogen, optical data indicate some depletion of helium in the central
region (Peimbert and Castero, 1969). These contradictions are solved by
considering such globules as proposed by Dopita et al. (1975). Optical
observations are weighted towards the high-density regions where helium-
ionizing photons are destroyed by grains recently ejected from globules,
whereas at radio wavelengths we look through the dust-free regions.

At present we cannot say much more, but the large value of R = $A_V$/
$E_{B-V}$ (Johnson, 1968) and the other optical structure (Gull, 1974) may
perhaps be explained by introducing globules.

To conclude this section we briefly comment on polarization observations. Dyck and Beichman (1974) observed infrared polarization in the KL
Nebula, and confirmed the existence of ice and silicate grains in it.
Hall (1974) has shown that the degree of polarization of the continuum
light increases from the Trapezium to the outer regions and that the
position angle of the electric vector is perpendicular to the radius
vector from the Trapezium (we call this positive polarization - the

normal scattering by dust grains). However, Isobe (1976) obtained negative polarization in the western portion (< 3' from the Trapezium). This
can be explained by back-scattering (Hanner, 1971): in this region the
continuum light scattered by grains at the HI - HII boundary situated
on the far side of the Trapezium dominates the observed continuum light.


7. CONCLUSION

We can now state with some confidence that dust exists in HII
regions. However, there do exist strong, small-scale density fluctuations in the dust. Such globules as are proposed in the Orion Nebula
may generally exist in other HII regions, and they may solve many observational discrepancies such as the anomalous helium abundance, temperature and density differences between optical and radio observations, and
so on. Since the Orion Nebula is nearest to the Sun and fortunately faces
us, we have considerable observational data with the best angular and
spectral resolution. However, in order to generalize the dust problems
discussed here, we should observe HII regions (for example M17 and DR21)
which do not wink at us but to the nice guy on the other stars.

The author would like to express his sincere thanks to Mr. Fukui
for discussions and information related to radio astronomy.


REFERENCES

Aitken, D.K. and Jones, B.: 1973, Astrophys. J. 184, 127
Becklin, E.E., Neugebauer, G. and Wynn-Williams, C.G.: 1973, Astrophys.
    J. 182, L7
Bohuski, T.J.: 1973, Astrophys.J. 184, 93
Capriotti, E.R.: 1971, Astrophys. J. 166, 563
Chaisson, E.J. and Wilson, R.F.: 1975, Astrophys. J. 199, 647
Churchwell, E., Mezger, P.G. and Huchtmeier, W.: 1974, Astron. Astrophys.
    32, 283
Dopita, M., Isobe, S. and Meaburn, J.: 1975, Astrophys. Space Sci. 34, 91
Dyck, H.M. and Beichman, C.A.: 1974, Astrophys. J. 194, 57
Dyson, J.E.: 1968, Astrophys. Space Sci., 1, 388
Elliott, K.H. and Meaburn, J.: 1974, Astrophys. Space Sci. 28, 351
Feuerbacher, B., Wills, R.F. and Fitton, B.: 1973, Astrophys. J. 181, 101
Gillett, F.C. and Forrest, W.J.: 1973, Astrophys. J. 179, 483
Gillett, F.C., Forrest, W.J., Merrill, K.M., Capps, R.W. and Soifer,
    B.T.: 1975, Astrophys. J. 200, 609
Gilra, D.P.: 1971, Nature 229, 237
Grasdalen, G.L.: 1974, Astrophys. J. 193, 373
Gull, S.F. and Martin, A.H.M.: 1975, in HII Regions and Related Topics,
    ed. by T.L. Wilson and D. Downes, p. 369
Gull, T.R.: 1974, in HII Regions and the Galactic Center, ed. by A.F.M.
    Moorwood, p. 1
Hall, R.: 1974, in Planets, Stars, and Nebulae, ed. by T. Gehrels, p. 881
Hanner, M.S.: 1971, Astrophys. J. 164, 425

Harper, D.A.: 1974, Astrophys. J. 192, 557.
Harper, D.A., Low, F.J., Rieke, G.H. and Thronson, Jr. H.A.: 1976,
    Astrophys. J. 205, 136.
Harvey, P.M., Gatley, I., Werner, M.W., Elias, J.H., Evans II, N.J.,
    Zuckerman, B., Morris, G., Sato, T., and Litvak, M.M.: 1974,
    Astrophys. J. 189, L 87.
Huffman, D.R. and Staff, J.L.: 1971, Nature Phys. Sci. 229, 45.
Iguchi, T. and Fukui, Y.: 1976, private communication.
Ireland, J.G.: 1970, Astrophys. Space Sci. 6, 107.
Ishida, K. and Kawajiri, K.: 1968, Publs. Astron. Soc. Japan 20, 95.
Isobe, S.: 1970, Publs. Astron. Soc. Japan 22, 429.
Isobe, S.: 1971a, Publs. Astron. Soc. Japan 25, 253.
Isobe, S.: 1971b, Annals Tokyo Astron. Obs., XII, 286.
Isobe, S.: 1975a, Annals Tokyo Astron. Obs., XIV, 141.
Isobe, S.: 1975b, Annals Tokyo Astron. Obs., XIV, 227.
Isobe, S.: 1975c, Annals Tokyo Astron. Obs., XIV, 258.
Isobe, S.: 1976, to be published.
Isobe, S.: Kurihara, H.: 1970, Tokyo Astron. Bull., Second Series,
    No. 203.
Israel, F.P., Habing, H.J. and DeJong, T.: 1973, Astron. Astrophys. 27,
    143.
Israel, F.P.: 1976, Astron. Astrophys. 48, 193.
Johnson, H.L.: 1965, Astrophys. J. 141, 923.
Johnson, H.L.: 1968, in Nebulae and Interstellar Matter, ed. by B.M.
    Middlehurst and L.H. Aller, p. 167.
Kaler, J.B.: 1967, Astrophys. J. 148, 925.
Krishna Swamy, K.S. and O'Dell, C.R.: 1967, Astrophys. J. 147, 529.
Lada, C.J. and Black, J.H.: 1976, Astrophys. J. 203, L 75.
Lada, C.J., Gull, T.R., Gottlieb, C.A. and Gottlieb, E.W.: 1976,
    Astrophys. J. 203, 159.
Lee, T.A.: 1968, Astrophys. J. 152, 913.
Lempke, D.: 1974, in HII Regions and the Galactic Center, ed. by A.F.M.
    Moorwood, p. 53.
Loren, R.B., Peters, W.L. and Vanden Bout, P.A.: 1975, Astrophys. J.
    195, 75.
Lortet-Zuckermann, M.C.: 1974, in HII Regions and the Galactic Center,
    ed. by A.F.M. Moorwood, p. 13.
Mathews, W.G.: 1969, Astrophys. J. 157, 583.
Mathis, J.S.: 1970, Astrophys. J. 159, 263.
Merrill, K.M. and Soifer, B.T.: 1974, Astrophys. J. 189, L 27.
Mezger, P.G., Smith, L.H. and Churchwell, E.: 1974, Astron. Astrophys.
    32, 269.
Moorwood, A.F.M. and Feuerbacher, B.: 1975, Astrophys. Space Sci. 34,137.
Münch, G. and Persson, S.E.: 1971, Astrophys. J. 165, 241.
Ney, E.P. and Allen, D.A.: 1969, Astrophys. J. 153, L 194.
Noguchi, K., Maihara, T. Okuda, H., Sato, S. and Mukai, T.: 1976,
    submitted to Publs. Astron. Soc. Japan.
Nordsieck, K.H.: 1971, Astrophys. J. 163, 287.
O'Dell, C.R. and Hubbard, W.B.: 1965, Astrophys. J. 142, 743.
O'Dell, C.R., Hubbard, H.B. and Peimbert, M.: 1966, Astrophys. J. 143,
    743.

Osterbrock, D. and Flather, E.: 1959, Astrophys. J. 129, 26.
Panagia, N.: 1974, in HII Regions and the Galactic Center, ed. by
    A.F.M. Moorwood, p. 163.
Peimbert, P. and Costero, R.: 1969, Bol. Obs. Ton. y Tac. 5, 3.
Pipher, J.L., Grasdalen, G.L., and Soifer, B.T.: 1974, Astrophys. J.
    193, 283.
Pipher, J.L. and Soifer, B.T.: 1976, Astron. Astrophys. 46, 153.
Sherwood, W.A. and Dachs, J.: 1976, Astron. Astrophys. 48, 187.
Schiffer, F.H. and Mathis, J.S.: 1974, Astrophys. J. 194, 597.
Schraml, J. and Mezger, P.G.: 1969, Astrophys. J. 156, 269.
Simpson, J.P.: 1973, Publs. Astron. Soc. Pacific 85, 479.
Smith, L.F.: 1975, in HII Regions and Related Topics, ed. by T.L. Wilson
    and D. Downes, p. 175.
Soifer, B.T. and Pipher, J.L.: 1975, Astrophys. J. 199, 663.
Stebbins, J., Huffer, C.H., and Whitford, A.E.: 1934, Publs. Washburn
    Obs. 15, Part V.
Stecher, T.P.: 1969, Astrophys. J. 157, L 125.
Stein, W.A. and Gillett, F.C.: 1969, Astrophys. J. 155, L 197.
Tamura, S.: 1976, Sci. Report Tohoku Uni., in press.
Walker, M.F.: 1969, Astrophys. J. 155, 447.
Wellington, K.J., Sullivan III, W.T., Goss, W.M., and Matthews, H.E.:
    1976, Astron. Astrophys. 47, 351.
Wellington, K.J., Sullivan III, W.T., Goss, W.M., Matthews, H.E.: 1977,
    Astron. Astrophys. 54, 319.
Werner, M.W., Elias, J.H., Gezari, D.Y., Hauser, M.G., and Westbrook,
    W.E.: 1975, Astrophys. J. 199, L 185.
Werner, M.W., Gatley, I., Harper, D.A., Becklin, E.E., Loewenstein, R.F.,
    Telesco, C.M., and Thronson, H.A.: 1976, Astrophys. J. 204, 420.
Wickramasinghe, N.C. and Nandy, K.: 1971, Monthly Notices Roy. Astron.
    Soc. 153, 205.
Wickramasinghe, N.C. and Nandy, K: 1972, Reports on Progress in Physics,
    35, 157.
Wink, J.E. and Altenhoff, W.J.: 1975, Astron. Astrophys. 38, 109.
Wink, J.E., Altenhoff, W.J., and Webster, W.J.: 1973, Astron. Astrophys.
    22, 251.
Witt, A.N. and Lillie, C.F.: 1973, Astron. Astrophys. 25, 397.
Wright, E.L.: 1973, Astrophys. J. 185, 569.
Wynn-Williams, C.G. and Becklin, E.E.: 1974, Publs. Astron. Soc. Pacific,
    86, 5.
Zuckerman, B.: 1973, Astrophys. J. 183, 863.

UNUSUAL, LARGE-SCALE, MOTIONS IN HII REGIONS

J. Meaburn,
Department of Astronomy,
The University, Manchester, England

ABSTRACT

    In many HII regions, huge volumes of ionized gas emit split lines.
Such splitting occurs exclusively over dark areas surrounded by bright
rims; these rims, which are produced by ionization fronts eating into
the adjacent neutral masses, emit single lines centred on the mean motion
of the nebular complex. Detailed observations of this situation are presented for the Orion (M42), Carina, Omega (M17), and 30 Doradus Nebulae.
In the latter, $Ca^+$ H and K absorption-line velocities suggest that a
HI/HII shell is collapsing toward the central ionizing stars.

1. INTRODUCTION

    Splitting of the optical and radio recombination lines has been
found over large areas of practically every HII region where a search
for it has been conducted (M42, M8, M16, M17, Carina, Rosette, 30 Doradus, NGC 2264, M20, etc.). It occurs exclusively over the darker areas,
apparent on optical photographs, which are associated with dusty neutral
intrusions; whereas the bright filaments, surrounding these, emit single
lines centred on the mean motion of each nebular complex. These filaments
are always found to be regions of low excitation (emitting predominantly
[NII], [OI], [SII] lines, etc.) and must be bright rims produced by
ionization fronts moving, away from the embedded ionizing stars, into
neutral material. The nature of this splitting, and its close relationship to the motions and presence of the neutral material will now be
illustrated for only the four most extensively observed complexes of HII,
HI, dust and molecules.

2. THE ORION NEBULA (M42)

    Wilson et al. (1959) discovered several small areas (0.02 to 0.07
pc across) in the dense ($10^4$ $cm^{-3}$) core of M42 emitting optical lines
split by 25 km/s. Larger-scale (0.5 pc across) splitting of the [NII]

*Hugo van Woerden (ed.), Topics in Interstellar Matter*, 81-88. *All Rights Reserved.*

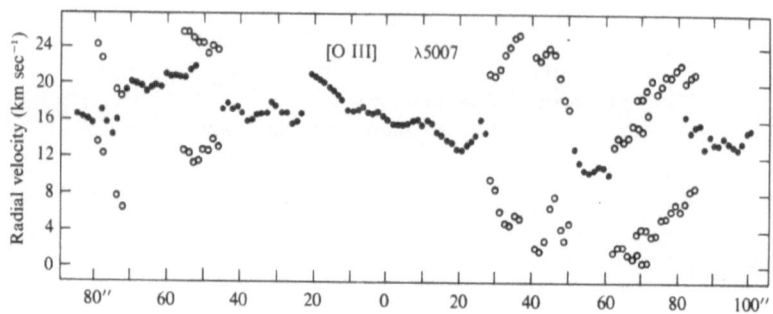

Figure 1.   Radial velocities of [OIII] in the Orion Nebula, along an East-West line 25" south of θ'Ori C. Scale: 100" = 0.2 pc.

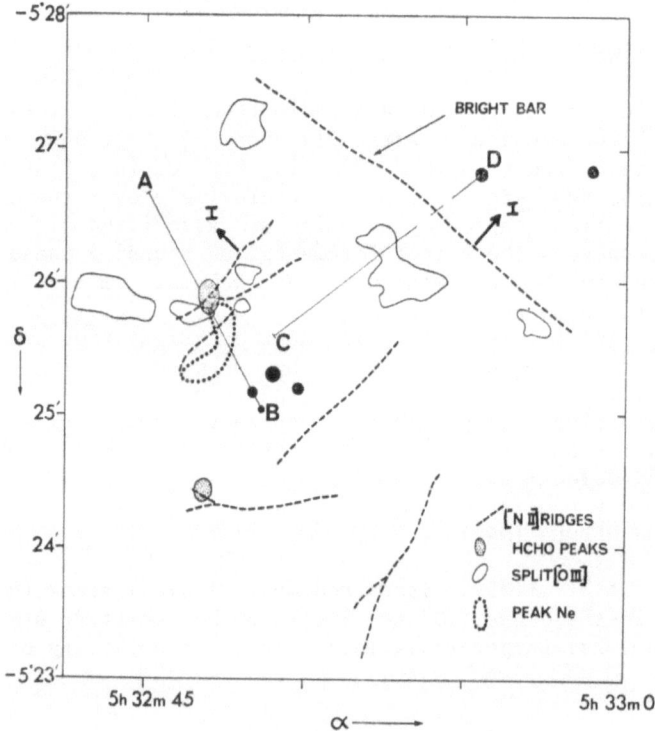

Figure 2.   Phenomena in the core of the Orion Nebula. The [NII] ridges are bright rims.

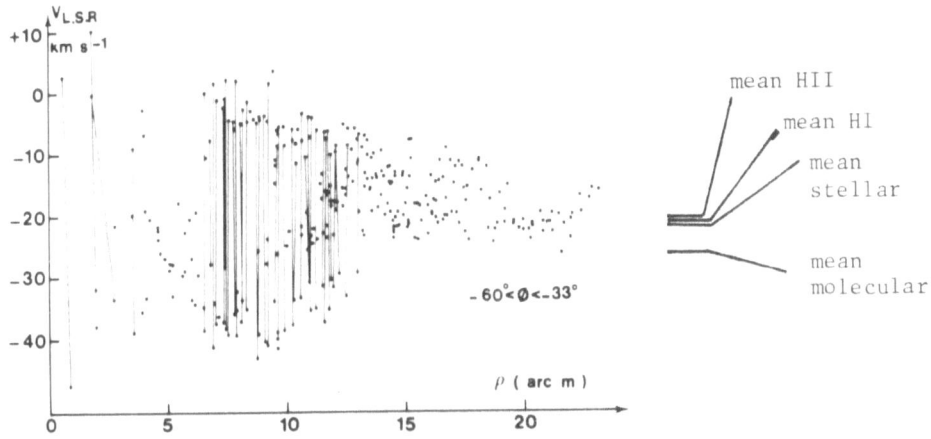

Figure 3.   Radial velocities of [NII] in the Carina Nebula, as a function of distance from η Carinae; 20 arcmin = 12 pc.

line was found by Deharveng (1973) in the lower-density ($2 \times 10^2$ cm$^{-3}$) outer areas. The radial velocities of [OIII] along an E-W line 25" south of θ' Ori C are shown in Figure 1. The areas where [OIII] is split in the core of M42 are compared in Figure 2 to the bright rims, producing [NII] ridges (Elliott and Meaburn, 1974), the peaks of the HCHO emission (Thaddeus et al., 1971) and the region of highest local electron density (Elliott and Meaburn, 1973). Likewise, the larger areas of [NII] splitting (Deharveng, 1973) occur near the more extensive [NII] ridges in the outer regions (Dopita et al., 1975).

3. THE CARINA NEBULA

Recently Deharveng and Maucherat (1975) have shown that large areas (8 pc across) of the core of the Carina Nebula are emitting [NII] lines split by up to 50 km/s. Such splitting has been found also in H109α and H90α profiles measured by Huchtmeier and Day (1975). The central, dark, area where the lines are split is again surrounded by bright rims, emitting single lines, whose radial velocities are near the mean motions of the neutral and ionized complexes also indicated in Figure 3.

Other, larger (100 pc across) areas of line splitting have been reported by Smith, Gull and Bohuski (see Walborn and Hesser, 1975), and a neutral/ionized shell expanding away from the stars was suggested by Dickel (1974).

However, most intriguing are the measurements by Walborn and Hesser (1975) of the H and K interstellar absorption lines, from stars deeply embedded in the core of this nebula. These reveal the presence of clouds of neutral gas, within this volume, with a remarkable range of radial

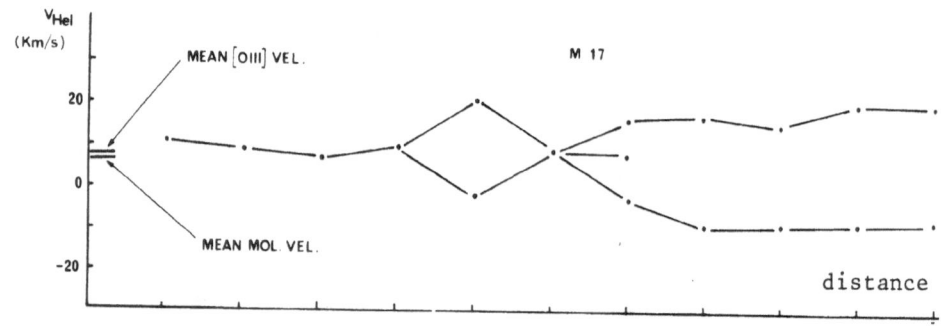

Figure 4.   Radial velocities of [OIII] in M17, over a range of 2.6 pc.

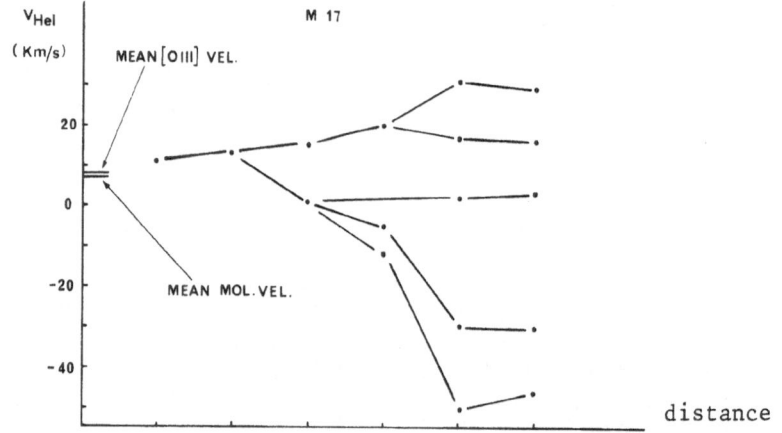

Figure 5.   Radial velocities of [OIII] in M17, over a range of 1.3 pc.

velocities. These occur at velocities of -199, -117, -63, -39, -6, +55 and +87 km/s (with respect to L.S.R.). The -39 and -6 km/s features are closely comparable to the individual velocities of the two components of the split optical lines.


4. THE OMEGA NEBULA (M17)

        In M17, the [OIII] line was found (Elliott and Meaburn, 1975 and Meaburn, 1975) to be split (up to five velocity components) over a large, dark area surrounded by a bright rim. The progression of this splitting, both along the length of M17 and from a bright rim to over an obvious neutral mass, is shown in Figures 4 and 5 respectively. This is compared to the mean radial velocities of the ionized and molecular complexes (Goudis and Meaburn, 1976). The [OIII] line again splits to both positive and negative radial velocities compared to these means.

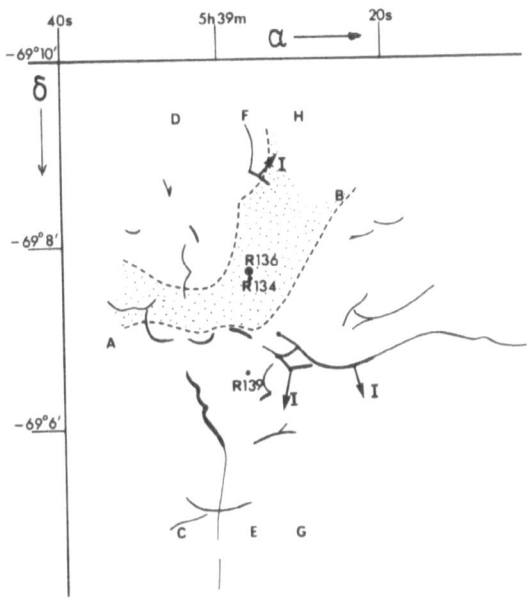

Figure 6.  The region of [OIII] line splitting is shown dotted, and compared to [NII] ridges (bright rims) in 30 Doradus.


At one position, the $^{12}C^{16}O$ emission line has components at +9.2 and +20.2 km/s (Wilson et al., 1974). These are closely correlated (Goudis and Meaburn, 1976) with two of the three [OIII] components from exactly the same area, which occur at -4, +8.5 and +18 km/s.


5. THE 30 DORADUS NEBULA

Similar, though more extensive, splitting of the [OIII] line was discovered by Smith and Weedman (1972) over several areas of this super-massive ($5 \times 10^5$ $M_\odot$) HII region, which appears to be the nucleus of the spiral structure of the extremely young objects in the Large Magellanic Cloud (Schmidt-Kaler and Isserstedt, 1975). Figure 6 compares the location of one region in the core of 30 Doradus, where [OIII] is split by about 50 km/s, to that of the bright rims (Elliott et al., 1976). The direction of motion of the ionization fronts, I, producing these rims, is also indicated, as are three of the ionizing stars. The heliocentric radial velocities from [OIII], measured at points between A and B on Figure 6, are shown in Figure 7 (Meaburn, 1977). Other, much larger, outer areas exhibit similar behaviour. The mean radial velocity of the whole HII region, derived from the single Hα and [OIII] lines, and that of the central cluster of ionizing stars (Feast, 1961) is also indicated.

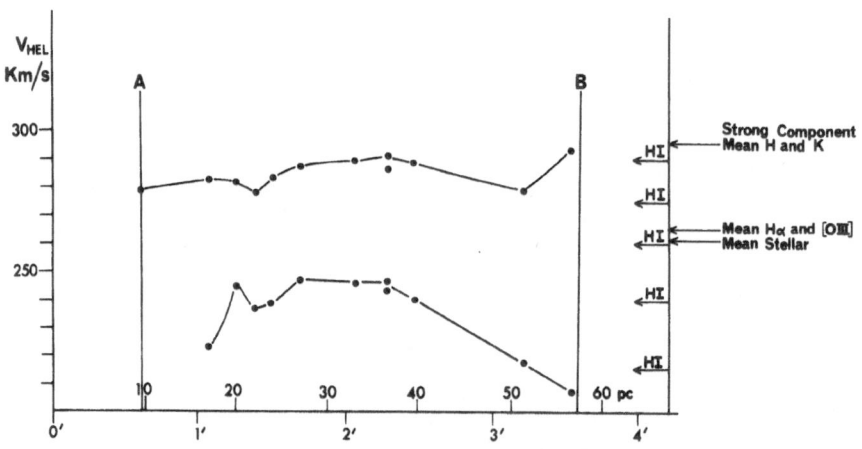

Figure 7.   Radial velocities of [OIII] between points A and B in Fig. 6.

Several ionized clouds, between 50 and 300 pc across, can be identified which have distinct positive and negative radial velocities with respect to i) the mean for the whole complex and ii) the radial velocity of single lines from the bright rims. These ionized clouds also have radial velocities which correspond to some of the components, indicated by inflexions in the 21-cm HI profile obtained by McGee and Milton (1966) with a 14.5' beamwidth, from the whole core of the nebula. These HI velocities are also marked on Figure 7.

Likewise it can be seen that the H and K interstellar absorption lines (Feast, 1961; Walborn and Hesser, 1975) have a strong component in the central stars (R136, 134 and 139) closely coincident with only the component at positive radial velocity in the [OIII] line. A collapsing HI/HII shell surrounding these stars is then suggested. A faint absorption component at negative velocity is only suspected (Walborn and Hesser, 1975).

6. DISCUSSION

Although different in scale, the phenomena occurring in these four nebulae are so similar in nature that a common explanation must be sought.

First, it is becoming increasingly clear that flows of ionized material, from relatively stationary neutral clouds, do not explain the large-scale splitting of the optical emission lines (see Meaburn, 1975 for instance). The radial velocities of the ionized and adjacent neutral material appear too closely correlated. Situation 2, depicted in Figure 8, is more plausible than situation 1.

The evidence is consistent with the view that these HII regions are composed of a series of irregular HI/HII shells, which become increasingly

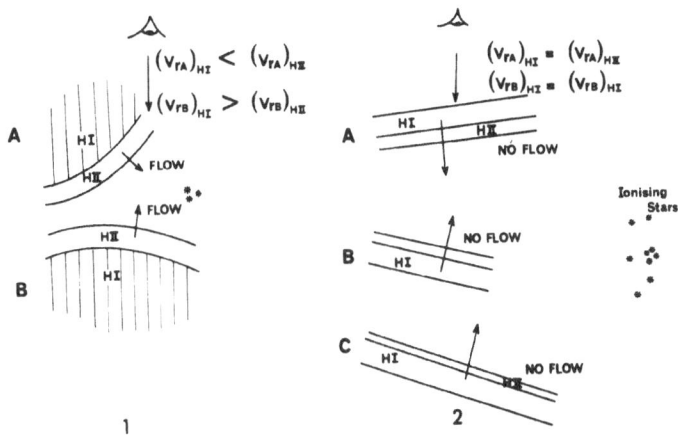

Figure 8.  Models for the motions of HII shells and adjacent
HI.  Model 2, an HI/HII shell with equal velocities in HI and
HII, is favoured over model 1, where the HII shell flows away
from the HI cloud.

large away from the dense nebular cores containing the ionizing stars.
The bright rims are then observed when the HI/HII interfaces, on the
inside edges of the shells, are viewed tangentially. The present evidence
favours a generally contracting situation, particularly for the core of
30 Doradus, though this is by no means yet conclusive.

More than two velocity components in the optical lines, as observed
in M17, could be produced by seeing different contracting (or expanding)
shells along the same line of sight, as illustrated in Figure 8.

It is possible that such contracting shells could be produced in the
periods of stellar formation. Alternatively, energetic stellar winds or
supernova explosions may be invoked to supply the energy for an expanding
situation.

Very large, faint, filaments surround 30 Doradus (2 kpc diameter),
M42, M17, etc.  These could also be bright rims of the outermost, low-
density, HI/HII shells within either a contracting or expanding configuration.

REFERENCES

Deharveng, L.: 1973, Astron. Astrophys. 29, 341.
Deharveng, L. and Maucherat, M.: 1975, Astron. Astrophys. 41, 27.
Dickel, H.R.: 1974, Astron. Astrophys. 31, 11.
Dopita, M.A., Isobe, S. and Meaburn, J.: 1975, Astrophys. Space Sci. 28,61
Elliott, K.H. and Meaburn, J.: 1973, Astron. Astrophys. 27, 367.
Elliott, K.H. and Meaburn, J.: 1974, Astrophys.Space Sci. 34, 473.

Elliott, K.H. and Meaburn, J.: 1975, Astrophys.Space Sci. 35, 81.
Elliott, K.H., Goudis, C., Meaburn, J. and Tebbutt, N.J.: 1976, Astron.
     Astrophys. (in press).
Feast, M.W.: 1961, M.N.R.A.S. 122, 1.
Goudis, C. and Meaburn, J.: 1976, Astron. Astrophys. 51, 401.
Huchtmeier, W.K., Day, G.A.: 1975, Astron. Astrophys. 41, 153.
McGee, R.X. and Milton, J.A.: 1966, Aust. J. Phys. 191, 343.
Meaburn, J.: 1975, "HII Regions and Related Topics" (Springer-Verlag)
     222.
Meaburn, J.: 1977, Astrophys.Space Sci. (in press).
Schmidt-Kaler, Th. and Isserstedt, J.: 1976, Astrophys.Space Sci. 41,139.
Smith, M.G. and Weedman, D.W.: 1972, Astrophys. J. 172, 307.
Thaddeus, P., Wilson, R.W., Kutner, M., Penzias, A.A. and Jefferts, K.B.:
     1971, Astrophys. J. Letts. 168, 59.
Walborn, N.R. and Hesser, J.E.: 1975, Astrophys. J. 199, 535.
Wilson, O.C., Münch, G., Flather, E.M. and Coffeen, M.F.: 1959, Astro-
     phys. J. Suppl. 4, 199.
Wilson, W.J., Schwarz, P.R., Epstein, E.E., Johnson, W.A., Etcheverry,
     R.D., Mori, T.T., Berry, G.G. and Dyson, H.B.: 1974, Astrophys. J.
     191, 357.

# PLANETARY NEBULAE

G.S. Khromov,
Astronomical Council, USSR Academy of Sciences, Moscow.

ABSTRACT

   While the theory of the radiation of planetary nebulae (PN) appears
satisfactory, the quality of observations often leaves much to be
desired. Even in close doublets, relative line intensities may be
seriously in error, probably owing to erroneous estimates of the continuum. These errors affect the determinations of electron density.
   The radio and optical images of PN are much alike. More surprises
may come from emission in the 21-cm and molecular lines. Infrared
observations suggest that dust is well mixed with the gas; in the younger planetaries the dust may be hotter and the infrared emission is
stronger.
   Determinations of the physical parameters of PN are hampered by
incomplete and unreliable spectrophotometric data, and by structural
peculiarities. Computer models have so far been too primitive. Both
density and ionization must vary with radius, and are interdependent;
hence, shell models are required. Studies of small-scale condensations
have centred on the comet-like structures in NGC 7293. The physical
conditions there are probably close to those in the main body of this
nebula, but spectra must resolve this matter.
   The evolution of PN shells is observed over 4 or 5 orders of magnitude in density. However, densities above $10^8$ cm$^{-3}$ are not observed and
the evolution of PN progenitors (out of old red giants?) remains speculative. Age estimates based on a constant rate of expansion are clearly
too primitive. The increase of the ratio of the HeII $\lambda$ 4686 to H$\beta$ with
time does not necessarily imply an evolutionary rise in temperature of
the PN nucleus.

## 1. INTRODUCTION

   Since the beginning of this century, planetary nebulae (PN) have
developed into a whole branch of astrophysical research. In a short
review one can not describe all the important results and ideas which
have appeared recently. Each year brings about 100 publications on

*Hugo van Woerden (ed.), Topics in Interstellar Matter, 89-96. All Rights Reserved.*
*Copyright © 1977 by D. Reidel Publishing Company, Dordrecht-Holland.*

planetary nebulae. All we can do is to outline briefly the general state
of the problem today, emphasizing difficulties and unanswered questions.*

## 2. THE QUALITY OF OBSERVATIONAL DATA

Traditionally a review like this starts with the theory of lumi-
nescence of planetary nebulae. There is no doubt that at present this
theory is quite perfect in its principal points. Actually the reliability
and completeness of the theory are greater than those of the observational data. That is why we cannot utilize all the opportunities offered by
theory for the determination of physical conditions in the nebulae, nor
for the future development of theory itself.

This pessimistic statement is based on an analysis of the precision
of the spectrophotometric data. We have investigated this precision by
comparing measurements of the relative intensities of lines published by
different authors. We used 38 publications over the last 30 years; they
contained the optical (mostly photographic) spectra of 77 bright planetary nebulae. The results are summarized in Table 1. Obviously the
precision of spectral data is astonishingly low. Of course some series
of measurements, especially photoelectric ones, actually can be very
good in the absolute sense. However, when dealing with an average spectrophotometric result one has every reason to use these pessimistic estimates.

The situation is not very much better for the intensity ratios in
close doublets, which are widely used for determination of electron
densities from the forbidden transitions in $p^3$ configurations. The
results presented in Table 2 were obtained for close doublets with common upper level, such as [OIII] $\lambda$ 4959+5007 A. There is a curious indication for a systematic error for the weaker lines. Perhaps this error is
due to mistakes in the level of the continuous spectrum.

Figure 1 demonstrates how these errors affect the determination of
electron density from a close doublet: the blue doublet of SII. The
scatter of points on this plot is due only to the low precision of line-
intensity measurements.

The precision is even poorer for the continuous nebular spectrum.
According to one of the most qualified estimates, the precision of data
on the continuum of the brightest planetary nebula, NGC 7027, is worse
than 20%.

---

*

For a recent, complete summary with many references, see G.S. Khromov
(1976), Trans. IAU XVI A, Part 3, p. 78 (in Reports on Astronomy, IAU
Commission 34).

TABLE 1.  Average precision of spectrophotometry of planetary nebulae.
Total number of the objects: 77.  Total number of publications used: 38.
Wavelength range:  3 700 – 10 830 A.  Figures in the Table are mean
errors (in %) of a single published intensity of a spectral line.

| Lines<br>Objects | Very bright | Bright | Medium | Weak |
|---|---|---|---|---|
| Bright | 32<br>(26 lines) | 43<br>(20 lines) | 57<br>(37 lines) | 81<br>(13 lines) |
| Medium | 40<br>(52 lines) | 55<br>(51 lines) | 70<br>(63 lines) | >50<br>( 3 lines) |
| Faint | >20<br>( 4 lines) | >54<br>( 7 lines) | -- | -- |

TABLE 2.  Average precision of close-doublet intensity ratios in planetary nebulae, for common spectrophotometric data.
Figures in the Table are mean errors (in %) of doublet intensity ratios
from a single published spectrum.

| Doublets<br>Objects | Very bright | Bright | Medium | Weak |
|---|---|---|---|---|
| Bright | 13<br>(49 doubl.) | 33<br>(10 doubl.) | -28<br>(8 doubl.) | -44<br>(1 doubl.) |
| Medium | 16<br>(97 doubl.) | -27<br>(11 doubl.) | -33<br>(2 doubl.) | -- |
| Faint | 21<br>(22 doubl.) | -- | -- | -- |

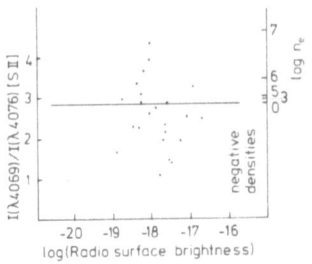

Figure 1.  Measured intensity ratio of the blue doublet of $[SII]$ , and
derived electron density $n_e$, versus radio surface brightness. The scatter is due to observational error.

## 3. RADIO AND INFRARED RADIATION

The time has not come yet to summarize observations of the far-ultraviolet emission from planetary nebulae. This new observational field is very promising indeed, and many interesting results will probably appear in future.

The mechanism of radio emission from planetary nebulae is well understood. It is clear that these objects are thermal emitters. The radio images of PN are quite similar to their pictures in optical hydrogen radiation. Also, the origin of the radio recombination lines will hardly pose serious theoretical questions. Thus the perspectives for radioastronomical studies of PN are rather obvious. It is worth-while to continue the search for exotic radio-spectral details, such as the 21-cm hydrogen line or molecular emissions. Also, it would be of interest to determine the frequencies of spectral maxima for many PN, as well as the shapes of spectral distributions around these maxima; and to measure flux densities for faint and stellar-like planetary nebulae.

The challenging problem of the infrared radiation of PN is now beyond the initial accumulation of data and ideas. Emission from hot dust can explain the main observed properties of the infrared radiation. We know now that in planetary nebulae dust must be well mixed with the gas, though there probably is a concentration of particles towards the outer borders of the gaseous shells. A certain amount of dust must be present in PN. In younger objects the dust is hotter, and statistically they are better infrared emittors. During the evolutionary expansion of the shell, the dust is cooled and the observed infrared emission in the 8 to 12 micron band fades rapidly. Figure 2 demonstrates this effect.

However some major problems remain. We do not yet know the shapes of the infrared spectra of PN. We have only a vague idea of the chemical composition and physical properties of the dust particles. Obviously, further progress of these studies depends primarily on infrared observations and laboratory experiments.

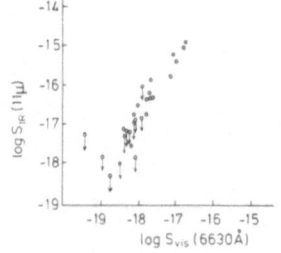

Figure 2.    Correlation of infrared (11-μ) and visual (6630-Å) fluxes of planetary nebulae. Both fluxes fade in the course of expansion of the nebula.

## 4. STRUCTURE AND PHYSICAL PROPERTIES

Studies of emission mechanisms are the basis for determination of physical characteristics of astronomical objects. In the case of PN we encounter many problems when attempting to apply the theory to real objects.

First, our spectrophotometric data are not complete. There are only a few cases, like NGC 7027 or NGC 7662, where we can use almost all possible combinations of spectral transitions with known theoretical atomic parameters for determinations of the basic physical characteristics of the nebulae. In more typical cases we are limited to a small number of combinations of the brighter lines.

Second, the reliability of typical spectral data is low as shown above. In fact, a further error is contributed by the reddening corrections.

Third, we do not know how to take into account the structural peculiarities of planetary nebulae. One of the basic defects of modern semi-empirical computer models of PN is their primitive structure. Simple morphological observations clearly show that the radiation of low-excited ions such as OII, SII, Cl III or NII comes from the outer layers of the nebulae, while the emission of ions with higher ionization potentials like OIII, AIV or NeIII originates in the inner regions. The latter have the toroidal shapes characteristic of the densest parts or "main structures" of planetary nebulae. Finally, the ions of highest excitation such as NeV or HeIII are concentrated in the innermost parts of the nebulae.

A freely expanding gaseous cloud or shell must have an outward density gradient. Physically it is also evident that the ionization of a cloud, illuminated from inside, should fall along its radius. Consequently we come to a "shell model" of PN, where density and ionization are correlated. A study of 77 PN has shown that this correlation does exist in typical objects:

$$n_e \ (K \ V) \simeq n_e(O \ III) \simeq 6 \ n_e(A \ IV) \simeq$$

$$\simeq 10 \ n_e(Cl \ III) \simeq 30 \ n_e(O \ II) \simeq 30 \ n_e(S \ II)$$

It is difficult to say how reliable these numbers are, but the general tendency is obvious.

Many authors have written about density fluctuations and small-scale condensations in PN. The majority of these papers were based upon one single observational example: the comet-like structures visible on photographs of the unique, large planetary nebula NGC 7293. Some theoreticians may not have had a chance to look carefully at a good picture of this object; if they had, some of the theories probably would not have

appeared. The filaments in NGC 7293 are really rather complex, they are
not always radial and sometimes have peculiar configurations. For a good
picture, see Aller's article in Sky and Telescope 38, 12, July 1969. My
own opinion is that the filaments are normally ionized, optically thin,
and their densities probably close to that in the main structure of NGC
7293. But, of course, such statements are speculative until we obtain
spectra of these intriguing details.


## 5. EVOLUTION

After the publication of the well-known Catalogue of Galactic Planetary Nebulae by Perek and Kohoutek, only small numbers of misidentifica-
tions and new discoveries have been reported. It looks as if we actually
know practically all the galactic planetary nebulae. The common method
of identification of stellar-like PN on low-dispersion objective-prism
spectra is crude; hence, the small percentage of false identifications
probably shows that the planetary nebulae are the brightest among the
compact galactic emission-line objects.

Our concept of the galactic distribution of PN is closely connected
with their distance scale. The method of relative determination of
distances, suggested by Shklovskij, is commonly used and no serious doubt
about its consistency has arisen. However, the absolute calibration of
such scales is an independent problem. The only reliable solution could
be provided by trigonometrical parallaxes for the central stars of the
nearest nebulae. Until now we have practically nothing of this kind.

The general features of the evolution of the gaseous shells are
quite obvious, though more sophisticated studies are limited by the
available observational information. Figure 3 illustrates the general
evolutionary picture. The scatter of points is due to the low quality of
observations; it would correspond to a dispersion of nebular masses of
6 orders of magnitude! One may conclude that we observe the evolution of
PN through a very large (4-5 orders of magnitude) range of densities.

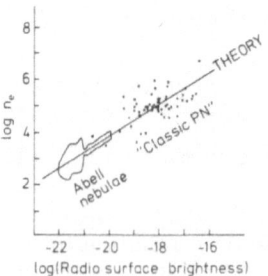

Figure 3.   Evolution of planetary nebulae. Both electron density and
surface brightness decrease in the course of time.

The number of objects denser than $10^6$ cm$^{-3}$ is small, and no superdense
nebulae with $n_e > 10^8$ cm$^{-3}$ are observed. The probable reason is that
the spectra of such nebulae are very different from those of the "classic"
planetary nebulae. Actually, almost all the typical forbidden lines will
be degenerated there. Moreover, at these very high densities the nebular
shells may not be completely ionized.

As long as we have not studied the possible observational properties
of superdense PN, our attempts to identify the progenitors of these
objects are entirely speculative. Recent attempts to approach the origin
and formation of PN by extrapolating the evolution of old red giants with
energy sources in the shell encounter major difficulties.

The age estimates of PN, based upon an assumed expansion at a
constant rate, are too primitive. While predominantly neutral in its
first evolutionary stages, a nebula probably ionizes at a density of
order $10^8$ to $10^9$ cm$^{-3}$. After that, the interior gas pressure rises 100
times and the characteristic expansion velocity, determined by the velocity of sound in the gas, jumps up by an order of magnitude. The process
as a whole should be quite similar to the explosion of a gas cloud.

What we do observe are the later evolutionary stages, when the
dynamics of the nebula is determined by its isothermal expansion.
Analysis shows that the principal observational data on the expansion of
PN can be naturally described by an appropriate gas-dynamical theory. No
primeval, "inherited" velocity field of expansion need be postulated.

Of course, the gas-dynamical expansion begins at birth of the planetary nebula; however, as long as the greater fraction of its mass remains
neutral, this expansion is very slow. Qualitatively the growth of radius
of a PN with time is illustrated by Figure 4. Clearly, an evolutionary
time scale derived from the velocity of expansion in the ionized stage
would be seriously in error.

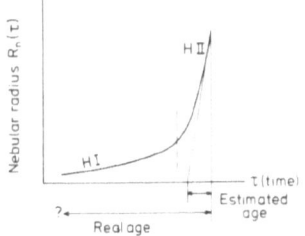

Figure 4.   Growth of nebular radius with time. The growth rate in the
ionized stage defines a time scale (estimated age) considerably smaller
than the real age.

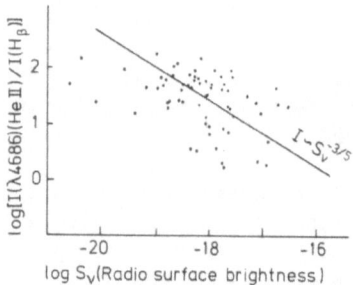

Figure 5.   Intensity of HeII λ 4686 (relative to Hβ) versus radio surface
brightness.


     A second problem is the well-known hypothesis that the temperatures
of nuclei of PN increase with time. Figure 5 shows the intensity of the
HeII λ 4686 line, relative to Hβ, versus the radio surface brightness
for numerous planetary nebulae. The excitation of the nebulae, characterized by the HeII emission, does grow from the younger to the older objects
and at first glance it can be naturally explained by an evolutionary
warming-up of the exciting star. However, there is another, even more
natural, explanation. The line on the graph shows a theoretical dependence, evaluated from a simple theory of nebular evolution. The radiation
flux in the Hβ line is proportional to the emission measure, consequently
it decreases with increasing age. On the other hand the (HeII) radiation
comes from the HeIII Strömgren zone; hence its flux does not depend on
the evolution of the density of this zone. The agreement between the
simple theory and the observations seems to be quite satisfactory. If so,
there is no need for an additional assumption about evolutionary variations of the ionizing radiation field of the central star.


6. CONCLUSION

     Many problems in the study of planetary nebulae remain to be solved.
Further observational work is required first of all.

# CIRCUMSTELLAR MASERS

Lewis E. Snyder
Department of Astronomy
University of Illinois

ABSTRACT

The newest circumstellar maser, silicon monoxide, is discussed, and its relationship to the other known circumstellar masers, hydroxyl and water, is briefly explored. Most silicon monoxide masers are associated with post-main sequence objects which are also water and hydroxyl masers and from this association new interpretations of maser velocities, geometries and distances have been found.

1. INTRODUCTION

Hydroxyl (OH), water ($H_2O$) and silicon monoxide (SiO) are the molecules found in maser emission from circumstellar shells associated with late-type stars. Silicon monoxide is the newest and, in several ways, the simplest circumstellar maser; hence our discussion will concentrate on the SiO maser. This maser was not known to exist until 1973 December when a group of molecular emission lines near 3.4 mm wavelength was detected in the direction of the Orion molecular cloud and tentatively identified as maser emission from the J=2-1 transition of $^{28}$SiO in its first vibrationally excited state (v=1) at 86,243 MHz rest frequency (Snyder and Buhl 1974). Subsequently, a series of significant observations yielded detections which confirmed the identification of vibrationally excited SiO beyond doubt: v=1, J=3-2 at 129,363 MHz from Ori A (Davis et al. 1974); v=1, J=1-0 at 43,122 MHz from W Hya and Ori A (Thaddeus et al. 1974); and v=2, J=1-0 at 42,821 MHz from o Cet, Ori A, VY CMa, R Leo and W Hya (Buhl et al. 1974). While the molecular identification was still in the process of being confirmed, a survey of the SiO v=1, J=2-1 transition demonstrated that most of the emission sources are stars, either Mira or semiregular variables, which also display OH/$H_2O$ maser emission, and a rough correlation was demonstrated between the existence of SiO maser emission and the measured infrared flux (Kaifu, Buhl and Snyder 1975). No SiO maser emission was detected from OH/$H_2O$/IR sources in HII regions.

The laboratory work on SiO has been extended as a direct result of

*Hugo van Woerden (ed.), Topics in Interstellar Matter, 97-104. All Rights Reserved.*

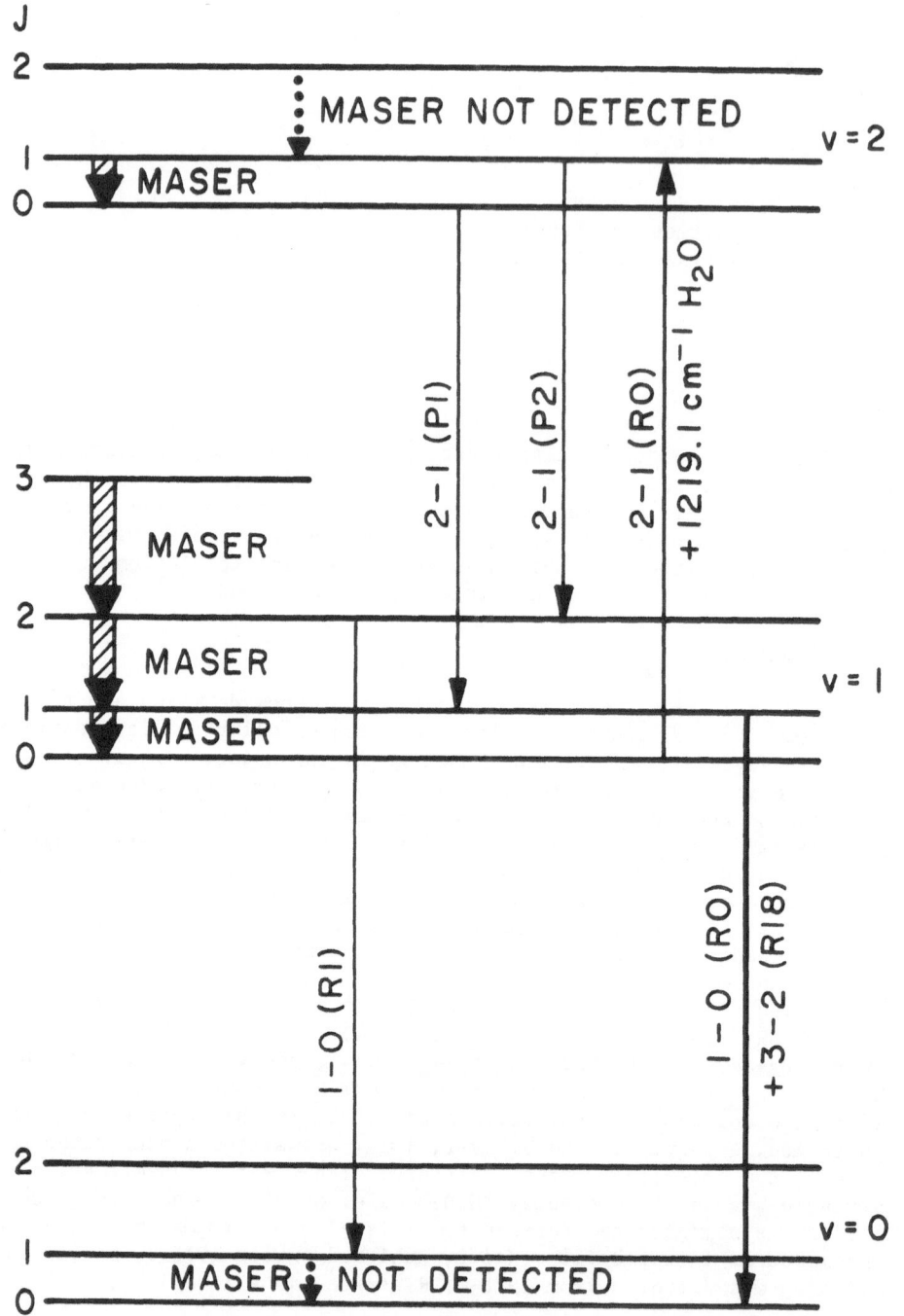

FIG. 1. - Schematic diagram showing SiO energy levels (not drawn to scale). Drawing reproduced with permission of the Astrophysical Journal.

the astronomical detection. Prior to the maser discovery, Törring (1968) had measured the J=1-0 transitions of $^{28}$SiO for the vibrational states v=0, 1, 2, and 3. The original maser identification at 86,243 MHz by Snyder and Buhl (1974) was based on a best fit of the astronomical spectra to a theoretically predicted spectrum by Lovas and Johnson (1974) which utilized Törring's (1968) measurements. Very recently, direct laboratory measurements of all SiO transitions of astrophysical importance were made by Manson et al. (1977). Using the best values of the rest frequencies determined by Manson et al. (1977), the astronomically observed SiO transition frequencies are:

| v | 0 | 1 | 2 |
|---|---|---|---|
| J=1-0 | 43,423.798 MHz | 43,122.027 MHz | 42,820.539 MHz |
| J=2-1 | 86,846.891 MHz | 86,243.350 MHz | |
| J=3-2 | | 129,363.262 MHz | |

Note that the SiO maser is observed only in the v=1 and 2 transitions. At present most v=0 transitions are not believed to be in maser emission (Buhl et al. 1975).

## 2. PROPERTIES OF SIO MASERS

SiO masers morphologically resemble $H_2O$ and OH masers in that they (a) have small source sizes (Moran et al. 1977); (b) show some evidence for circular polarization (Johnson and Clark 1975); (c) show time-varying spectra (Spencer and Schwartz 1975) and have non-thermal population distributions (Buhl et al. 1975). There is at least an 80 percent chance that a circumstellar SiO maser is also an $H_2O$ maser and a 75 percent chance that it is an OH maser. (Snyder and Buhl 1975).

Several possible pumping mechanisms for the SiO maser have been proposed. Figure 1 is a schematic diagram taken from Snyder and Buhl (1975) which shows SiO energy levels (not drawn to scale) and transitions likely to be important for explaining the maser mechanism. The broad, shaded arrows in Figure 1 show the detected SiO maser transitions, dotted arrows show those not detected, and the thin, solid arrows show allowed infrared transitions.

Geballe and Townes (1974) have proposed several possible resonance mechanisms for SiO maser emission and it appears that one variation of their resonance infrared line pumping scheme would give a qualitative explanation of the observational results. If an infrared transition of $H_2O$ at 1219.1 $cm^{-1}$ produces excess radiation, its close coincidence with the SiO v=2-1 (R0) line would depopulate the v=1, J=0 level and over-populate the SiO v=2, J=1 level. Excess photons would leave the J=1 level, cascading down and branching as follows: (1) v=2, J=1-0 and then v=2-1 (P1) to overpopulate the v=1, J=1 level; (2) v=2-1 (P2), over-populating the v=1, J=2 level, and through v=1, J=2-1, giving the 86,243

MHz SiO maser, and overpopulating the v=1, J=1 level. Thus both photon routes (1) and (2) overpopulate the v=1, J=1 level, finally giving rise to the v=1, J=1-0 SiO maser. This variation of the Geballe and Townes (1974) pumping scheme might be preferred over their proposed pumping mechanism which utilizes strong coincidences within a single molecular spectrum because (a) it does not necessarily predict a CO maser; (b) it is supported by the high correlation between H2O and SiO maser sources; and (c) it would not require every SiO infrared source to be also a maser source. This scheme also predicts a weak v=1, J=3-2 SiO maser because the v=2, J=3 level is not directly pumped but rather is inverted due to underpopulation of the J=2 level. Thus underpopulation would be caused by a saturated v=2, J=2-1 transition coupled with fast relaxation to v=0 through the resonance pair v=1-0 (R0) + v=3-2 (R18), as suggested by Geballe and Townes (1974).

Kwan and Scoville (1974) have taken a different approach by proposing a continuum pumping mechanism which requires that radiative de-excitation of v=1 to v=0 must be more rapid than collisional de-excitation and that the v=0-1 vibrational transitions be optically thick. When these conditions are met, radiative trapping in the v=0-1 vibrational transitions decreases the rate of de-excitation of a v=1 rotational level as its J value increases. This leads to population inversions in the v=1 rotational levels if population is transferred from v=0 into v=1 by collisions or through absorption from v=0 into v=2 and back to v=1 by spontaneous cascade.

The diverse possibilities proposed to explain the SiO maser mechanism highlight the importance of making spectral line interferometric measurements of SiO maser sources. These measurements will be critically important for determining the true brightness temperatures of the SiO emission lines and for making source models which will permit the next stage of detail in modeling of the maser mechanism.

## 3. KNOWN SOURCES

At present there are 47 known SiO maser sources. They are, in order of increasing right ascension:

| | | | |
|---|---|---|---|
| IRC+10011 | IRC+60169 | WX Ser | RT Aql |
| W And | $L_2$ Pup | RU Her | GY Aql |
| Mira | VY CMa | U Her | $\chi$ Cyg |
| S Per | R Cnc | RR Sco | RR Aql |
| R Hor | RL Mi | AH Sco | IRC-10529 |
| NML Tau | R Leo | IRC+10322 | IRC+40442 |
| W Eri | RT Vir | IRC+20328 | NML Cyg |

| R Dor | R Hya | IRC-20424 | T Cep |
|---|---|---|---|
| TX Cam | S Vir | VX Sgr | IRC+10523 |
| IRC+50137 | W Hya | IRC+10365 | R Peg |
| Ori A | RX Boo | R Aql | R Cas |
| U Ori | S CrB | IRC-20540 | |

The above sources were gathered from published references and unpublished data kindly furnished by radio observers (Blair, Thaddeus and
Davis 1975; Churchwell and Winnewisser 1976; Dickinson and Blair 1976;
Robinson 1976; Schwartz 1977; Snyder and Buhl 1974, 1975; Kaifu, Buhl
and Snyder 1975; Spencer et al. 1976). It should be noted that Ori A
is the only SiO maser source which is also associated with a molecular
cloud region. Most of the other sources are associated with known late-
type stars.

4. VELOCITIES

Reid and Dickinson (1976) have used ground-vibrational state SiO
data to argue quite convincingly that the true stellar velocity of long-
period variable stars lies between the optical absorption and emission
lines near the midpoint of the OH radial velocity pattern. This shat-
ters the traditional view, held for over 35 years, that the stellar
velocity is associated with the velocity of the optical absorption lines.
The stellar radial velocities determined by Reid and Dickinson (1976)
argue strongly for the shell model rather than for the shock model of
circumstellar envelopes. That is, it appears more likely that OH maser
emission originates in the front and back halves of an expanding shell
of circumstellar material, perhaps ejected during shell oscillation,
rather than from shock waves which give rise to blueshifted OH and from
more distant quiescent gas which supports OH emission near the absorption
line and stellar radial velocities.

5. MASER GEOMETRIES

Van Blerkom and Auer (1976) noted that the high excitation energies
required for SiO force the population inversion to occur close to the
stellar photosphere (perhaps within a few stellar radii) but the OH
maser, in contrast, is formed in a region more than 100 stellar radii
distant. Thus occultation of the far side of a disk or sphere and
direct amplification of stellar radiation at the maser frequency can be
large effects for SiO but not for OH. Their solution of the SiO
radiative transfer problem for a spherical shell gives an asymmetric
emission doublet; for a rotating disk, it gives an emission triplet.
Thus Van Blerkom and Auer (1976) find that the distinctive triplet
structure in the velocity pattern of the VY CMa SiO maser lines observed by Snyder and Buhl (1975) could be caused by emission from a
rotating equatorial disk seen nearly edge-on and suggested that both
VY CMa and NML Cyg, which has a somewhat similar velocity pattern,

may be young objects still surrounded by solar nebulae. Their conclusion supports Herbig's (1972) model for VY CMa. Hence while it
appears that most SiO maser emission sources are associated with post-
main sequence stars, at least two sources with triplet velocity patterns may be young stars still in the process of forming.

## 6. DISTANCES

Cahn (1977) has found an interesting correlation for Mira variable stars between the absolute SiO maser luminosity at maximum light
and mean spectral type at normal maximum using data from observed v=1,
J=1-0 maser sources. His method suggests that knowing the spectral
type of a Mira variable at zero phase and the distance will allow an
upper bound on the integrated SiO flux to be predicted; thus Cahn (1977)
has found a criterion for selecting Mira variable stars to search for
SiO maser emission. On the other hand, the correlation of spectral
type with SiO maser flux makes it possible to assign distances to SiO
maser sources if the spectral class and type are known. Using the
latter correlation, Cahn (1977) finds that it is very difficult to detect SiO maser emission from Miras more distant than approximately
2 kpc (assuming the spectral type at maximum is M8).

## 7. ORION

The Orion $H_2O$ and OH maser emission pattern has rather amorphous
structure, but the SiO emission has the doublet structure typical of
the spherical shell surrounding a late-type star. The $H_2O$ maser
emission pattern contains features at $\sim 16$ km s$^{-1}$ and $\sim -6$ km s$^{-1}$
which generally match the velocity pattern of the SiO doublet. The
doublet structure of the SiO profile is very similar to that of 1612
MHz OH masers observed in the circumstellar shells of late-type stars.
This similarity suggested to Snyder and Buhl (1974) that the Orion
SiO maser is associated with a late-type star. This view also was
advocated by Allen and Penston (1974) and is strengthened by the fact
that almost all of the other SiO masers which have been found so far
are associated with late-type stars and no SiO masers, other than Orion,
have been found in the direction of known OH/$H_2O$/IR sources associated
with HII regions and molecular cloud regions. The picture of a late-
type object in an otherwise young cluster of Orion objects is controversial, however, because the SiO observations might possibly be explained in another manner. For example, Habing et al. (1975) support
the SiO maser in a cocoon around the Becklin Neugebauer object in Orion.
Except for the fact that they assume that the Becklin Neugebauer object
is a star which has just reached the main sequence and is still accreting
matter, the conditions are similar to those found in expanding envelopes
of late-type stars. This argues for an expanding envelope to explain
the doublet structure of the Orion SiO maser and also makes the Orion
object somewhat unique in that with the possible exception of VY CMa
and NML Cyg, discussed earlier, almost all other SiO masers are thought
to be associated with post-main sequence objects.

In order to attack the questions raised by the Orion SiO maser, Moran et al. (1977) made a series of SiO and $H_2O$ observations and performed a three-station VLBI experiment on the Orion region. They found evidence that the maser features in SiO and $H_2O$ near 16 and -6 km s$^{-1}$ are associated with the same object. Their size limit for the Orion SiO features is > 1.5 a.u. and their position measurements place the SiO maser emission closest to a compact source IRc2 (Rieke et al. 1973) in the Kleinmann-Low infrared nebula but they found that the Becklin Neugebauer object is also within their error bars. Thus while their data suggest that the heart of the Orion molecular cloud contains an object with a shell structure which is similar to the circumstellar shell surrounding a late-type star, they can not exclude the possiblity of a pre-main sequence object. They conclude that the final answer to the Orion problem probably awaits the use of millimeter wave interferometers utilizing baselines between 1 and 50 km to determine SiO maser sizes and absolute positions.

## 8. CONCLUSION

There is a high degree of correlation, perhaps 80% or more, between OH, $H_2O$ and SiO circumstellar maser stars. The majority of SiO maser sources are associated with late-type stars. The high SiO excitation energies introduce a degree of uniqueness in the radiative transfer problem that is not available either in OH or $H_2O$ and which, for the first time, allows interesting solutions of maser geometry and distance problems. The failure to detect SiO masers in known OH/$H_2O$/IR sources associated with HII regions and molecular cloud regions (other than Orion) could be due to beam dilution and/or time-varying intensities. If not, the absence of SiO maser emission would serve as an observational indication that particular OH/$H_2O$/IR sources are not associated with late-type stars and hence may be much younger, protostellar regions.

The author's research is supported in part by NSF Grant AST 76-04541 to the University of Illinois.

## REFERENCES

Allen, D.A. and Penston, M.V.: 1974, Nature 251, 110.

Blair, G.N., Thaddeus, P. and Davis, J.H.: 1975, preprint.

Buhl, D., Snyder, L.E., Lovas, F.J. and Johnson, D.R.: 1974, Astrophys. J. Letters 192, L97.

Buhl, D., Snyder, L.E., Lovas, F.J. and Johnson, D.R.: 1975, Astrophys. J. Letters 201, L29.

Cahn, J.H.: 1977, Astrophys. J. Letters, in press (15 March).

Churchwell, E. and Winnewisser, G.: 1976, personal communication.

Davis, J.H., Blair, G.N., Van Till, H. and Thaddeus, P.: 1974, Astrophys. J. Letters 190, L117.

Dickinson, D.F. and Blair, G.N.: 1976, Bull. A. A. S. $\underline{8}$, 347.

Geballe, T.R. and Townes, C.H.: 1974, Astrophys. J. Letters $\underline{191}$, L37.

Habing, H.J., Olnon, F.M., Bedijn, P.J. and de Jong, T.: 1975, Lecture Notes in Physics $\underline{42}$, (ed. T.L. Wilson and D. Downes, New York: Springer-Verlag), 205.

Herbig, G.H.: 1972, Astrophys. J. $\underline{172}$, 375.

Johnson, D.R. and Clark, F.O.: 1975, Astrophys. J. $\underline{197}$, L69.

Kaifu, N., Buhl, D. and Snyder, L.E.: 1975, Astrophys. J. $\underline{195}$, 359.

Kwan, J. and Scoville, N.: 1974, Astrophys. J. Letters $\underline{194}$, L97.

Lovas, F.J. and Johnson, D.R. 1974, personal communication.

Manson, E.L., Jr., Clark, W.W., DeLucia, F.C. and Gordy, W.: 1977, Phys. Rev., in press.

Moran, J., Johnston, K.J., Spencer, J.H. and Schwartz, P.R.: 1977, Astrophys. J., submitted.

Reid, M.J. and Dickinson, D.F.: 1976, Astrophys. J. $\underline{209}$, 505.

Rieke, G.H., Low, F.J., and Kleinmann, D.E.: 1973, Astrophys. J. Letters $\underline{186}$, L7.

Robinson, B.J.: 1976, personal communication.

Schwartz, P.R.: 1977, personal communication.

Snyder, L.E. and Buhl, D.: 1974, Astrophys. J. Letters $\underline{189}$, L31.

Snyder, L.E. and Buhl, D.: 1975, Astrophys. J. Letters $\underline{197}$, 329.

Spencer, J.H. and Schwartz, P.R.: 1975, Astrophys. J. $\underline{199}$, L111.

Spencer, J.H., Schwartz, P.R., Bologna, J.M. and Waak, J.A.: 1976, Bull. A. A. S. $\underline{8}$, 372.

Thaddeus, P., Mather, J., Davis, J.H. and Blair, G.N.: 1974, Astrophys. J. Letters $\underline{192}$, L33.

Törring, T.: 1968, Zs. f. Naturforschung $\underline{23A}$, 777.

Van Blerkom, D. and Auer, L.: 1976, Astrophys. J. $\underline{204}$, 775.

Chapter 3


INTERSTELLAR MOLECULES AND DUST




Papers presented in a joint session of IAU Commissions 34
(Interstellar Matter), 40 (Radio Astronomy) and 44 (Space
Research), Grenoble, 25 August 1976.

Session Chairmen: Brian J. Robinson, Arthur D. Code

# OBSERVATIONS OF MOLECULAR CLOUDS

B. Zuckerman
University of Maryland

ABSTRACT

   Observations of interstellar molecules in regions of
star formation are summarized.  It is concluded that kinematics in molecular clouds are still poorly understood.  The
field is in need of some clever new ideas and/or observations.

   At the time of the 1973 General Assembly of the I.A.U. approximately
25 molecules were known to exist in the interstellar medium.  These molecules contained as many as seven atoms and as many as four "heavy"
atoms (i.e. non-hydrogens).  Various unidentified lines had been acci-
dentally discovered, the most interesting of which were X-ogen and U90.7.
Tentative identifications for these two lines included $HCO^+$ and HNC
respectively.

   At present, January 1977, a total of over 40 molecules have been
discovered and identified in interstellar clouds (see Table 1).  These
molecules contain as many as nine atoms and as many as six heavy atoms.
The latter record is held by $HC_5N$, discovered through the beautiful
efforts of a joint Anglo-Canadian group (Avery et al. 1976).  X-ogen
has been positively identified as $HCO^+$ both astronomically and in the
laboratory and $DCO^+$ has been seen in large abundance in some molecular
clouds.  U90.7 has been identified as HNC in different laboratories in
at least three countries around the world (Germany, Australia, U.S.A.).
The interesting molecule $N_2H^+$ has been discovered, by accident, and
identified both astronomically and in the laboratory.  Many other
unidentified lines have taken the place of those mentioned above.  For
example, Dr. B. E. Turner informs me that he has 26 unidentified lines
filed away in his desk drawer!

   Considering that the interstellar D/H ratio is only $\sim 10^{-5}$ it is
remarkable that already seven deuterated species have been positively
or possibly identified in molecular clouds (HD, DCN, $DCO^+$, DNC, and

*Hugo van Woerden (ed.), Topics in Interstellar Matter,* 107-112. *All Rights Reserved.*

TABLE 1:
Known Interstellar Molecules as of January 1977

| Number of Atoms | Molecule | Optical/UV | Radio Spectrum | Detected in Astronomical Objects |
|---|---|---|---|---|
| 2 | $H_2$ – hydrogen | ✓ | – | 1, 2(?) |
| | $OH$ – hydroxyl radical | ✓ | $\Lambda$ – doubling | 1, 2, 3, 4 |
| | $SiO$ – silicon monoxide | | Rotational | 2, 3 |
| | $SO$ – sulfur monoxide | | Rotational | 3 |
| | $SiS$ – silicon monosulfide | | Rotational | 2 |
| | $NS$ – nitrogen sulfide | | Rotational | 3 |
| | $CH^+$ – methylidyne ion | ✓ | – | 1 |
| | $CH$ – methylidyne | ✓ | $\Lambda$ – doubling | 1, 3 |
| | $CN$ – cyanogen radical | ✓ | Rotational | 1, 2, 3 |
| | $CO$ – carbon monoxide | ✓ | Rotational | 1, 2, 3, 4 |
| | $CS$ – carbon monosulfide | | Rotational | 1, 2, 3 |
| 3 | $H_2O$ – water | | Rotational | 2, 3 |
| | $H_2S$ – hydrogen sulfide | | Rotational | 3 |
| | $SO_2$ – sulfur dioxide | | Rotational | 3 |
| | $HCN$ – hydrogen cyanide | | Rotational | 1(?), 2, 3 |
| | $HNC$ – hydrogen isocyanide | | Rotational | 1(?), 3 |
| | $OCS$ – carbonyl sulfide | | Rotational | 3 |
| | $HCO^+$ – formyl ion | | Rotational | 1, 3 |
| | $HCO$ – formyl radical | | Rotational | 3 |
| | $CCH$ – ethynyl radical | | Rotational | 1(?), 2, 3 |
| | $N_2H^+$ – | | Rotational | 1(?), 3 |
| 4 | $NH_3$ – ammonia | | Inversion | 1, 3 |
| | $H_2CO$ – formaldehyde | | K-doubling | 1, 3, 4 |
| | | | Rotational | 1, 3 |
| | $HNCO$ – isocyanic acid | | Rotational | 3 |
| | $H_2CS$ – thioformaldehyde | | K-doubling | 3 |
| | | | Rotational | 3 |
| | $C_3N$ – cyanoethynyl radical | | Rotational | 2 |
| 5 | $HC_3N$ – cyanoacetylene | | Rotational | 1(?), 2, 3 |
| | $HCOOH$ – formic acid | | K-doubling | 3 |
| | $CH_2NH$ – methanimine | | K-doubling | 3 |
| | $H_2CCO$ – ketene | | Rotational | 3 |
| | $NH_2CN$ – cyanamide | | Rotational | 3 |
| 6 | $CH_3OH$ – methyl alcohol | | K-doubling | 3 |
| | | | Rotational | 3 |
| | $CH_3CN$ – methyl cyanide | | Rotational | 3 |
| | $NH_2CHO$ – formamide | | K-doubling | 3 |
| | | | Rotational | 3 |

TABLE 1 (contd.)

| Number of Atoms | Molecule | Optical/UV | Radio Spectrum | Detected in Astronomical Objects |
|---|---|---|---|---|
| 7 | $CH_3C_2H$ - methyl acetylene | | Rotational | 3 |
|   | $CH_3CHO$ - acetaldehyde | | K-doubling | 3 |
|   |  | | Rotational | 3 |
|   | $NH_2CH_3$ - methylamine | | Rotational | 3 |
|   | $CH_2CHCN$ - vinyl cyanide | | K-doubling | 3 |
|   | $HC_5N$ - cyanodiacetylene | | Rotational | 3 |
| 8 | $HCOOCH_3$ - methyl formate | | K-doubling | 3 |
| 9 | $(CH_3)_2O$ - dimethyl ether | | Rotational | 3 |
|   | $CH_3CH_2OH$ - ethyl alcohol | | Rotational | 3 |
|   | $CH_3CH_2CN$ - ethyl cyanide | | Rotational | 3 |

1 - Dark Dust Cloud,  2 - Circumstellar Envelopes,  3 - H II Region and/or Galactic Center, 4 - External Galaxies

deuterated water, ammonia and methylamine). This indicates both the power of the observational techniques as well as fortunate circumstances that produce very large isotopic fractionations. At the upper end of the abundance scale, a variety of independent arguments suggest that the ubiquitous CO molecule contains between 10 and 50% of the interstellar carbon.

Molecular clouds appear to be fragmented on at least three scales: $10^5$, $10^3$, and 10 $M_{\odot}$. Many OB associations are composed of subgroups separated by 10-20 pc and strung out along the galactic plane (e.g., Blaauw 1964). Elmergreen and Lada (1977a, b) have identified strings of giant $10^5$ $M_{\odot}$ molecular clouds, for example near M17, as the precursors of these OB sub-groups. Detailed studies of many giant molecular clouds near, for example, Orion (Kutner et al. 1976), M17 (Lada 1977), and W49 (Mufson and Liszt 1977) have been carried out. These giant clouds appear to fragment into $10^3$ $M_{\odot}$ pieces separated by distances ~1 pc, which may be precursors of galactic clusters such as the Trapezium. These fragments have been studied in many regions, in many molecules, and by many astronomers (see, for example, the 2-mm $H_2CO$ map of Kutner et al. 1976). Some $10^3$ $M_{\odot}$ fragments are located near bright H II regions but others are classical dark clouds conspicuous on the Palomar survey prints and not associated with obvious H II regions. In addition to the molecular studies these dark clouds have been observed in the infrared by Strom and co-workers and in the radio continuum by Gilmore and Brown.

Up until the time of the 1973 I.A.U. meeting, observations of broad lines in OH, $H_2CO$ and CO were generally ignored or sloughed over as due to "turbulence" (by broad we mean observed linewidths are much greater than thermal widths for T ≤ 100 °K). No systematic thought was given to understanding the causes of the large linewidths. This situation was altered by the systematic (radial) motions models of Liszt et al. (1974) and Goldreich and Kwan (1974). They noted that although $^{12}CO$ lines are very optically deep, essentially none showed a self-reversed profile. This is not easy to understand in a micro-turbulent cloud with hot CO in the interior and cold CO near the edges. Therefore, they suggested a radial motions (free-fall collapse) model where, for a uniform density cloud, $v \propto r$. In such a cloud, cold foreground gas is blue-shifted with respect to hot background gas and it is possible to see through entire clouds even in the centers of very optically deep CO lines.

There are problems with such models. Zuckerman and Evans (1974) worried about excessive star formation rates and agreement of $H_2CO$ absorption and CO emission velocities. Milman (1975) noted that the front parts of molecular clouds near reflection nebulae have the same velocities as the center of the CO lines. These authors tended to favor models with macroscopic turbulence rather than free-fall collapse. But such turbulence will die out unless it is regenerated or the clouds are very clumpy so that the supersonically moving clumps do not collide very often. Some recent observations of ammonia by Barrett et al. (1977) and by Matsakis (1977) and of cyanoacetylene by Morris et al. (1977) have been interpreted to require extreme clumpiness in a few clouds. However, alternative explanations still appear possible and the clumpiness question cannot yet be regarded as settled.

A few clouds with self-reversed CO profiles have not been observed; Snell and Loren (1977) summarize data for four such clouds. These observations suggest that, at least in some clouds, collapse is not very fast or v is not proportional to r or both. At first glance, these profiles might be considered to support a mostly turbulent model. However, Snell and Loren argue that certain asymmetries in the $^{12}CO$ and $^{13}CO$ profiles imply that the four clouds are collapsing. Both they and Lucas (1976) have attempted to model such clouds assuming comparable systematic (collapse) and turbulent motions. Lucas assumes a constant density cloud and therefore a velocity field $v \propto r$, whereas Snell and Loren consider collapse on to a point mass, so $v \propto r^{-1/2}$. However, there may be observational and/or computational problems with these models. For example, if systematic and turbulent motions are comparable they cannot be easily decoupled as in the large velocity gradient approximation (e.g., Goldreich and Kwan 1974). For the specific clouds and collapse model considered by Snell and Loren there appear to be ambiguities in interpretation of data for NGC 1333, Mon R2 and W3. NGC 1333 may represent a complicated situation of clouds in collision according to Loren (1976). $H_2CO$ absorption profiles show that in both Mon R2 and W3 there are at least two clouds with different radial velocities along the line of sight to the H II regions. For the ρ Oph cloud there is no increase of CO linewidth near positions where self-reversal is observed as predicted in a model with $v \propto r^{-1/2}$.

Other examples of self-reversed profiles exist in the literature. In a few cases such as NGC 2024 (Penzias, discussion remark at this General Assembly) and NGC 2071 (Lada, discussion remark at I.A.U. Symposium 75) the $^{12}$CO and $^{13}$CO asymmetries are in the opposite direction from the clouds discussed by Snell and Loren. Are NGC 2024 and NGC 2071 expanding?

In summary, it is still not possible to make a clearcut choice between collapse and "turbulence". There are some recent observations of self-reversed lines suggestive of modest collapse velocities but even here the turbulent velocities are comparable and the observational evidence for collapse is ambiguous. We still do not have a clear understanding of motions in molecular clouds.

Recent results on the Orion molecular cloud concern both the very large and very small scale. Kutner et al. (1976) have made maps of $^{12}$CO, $^{13}$CO, and $H_2$CO (2-mm emission) which extend from the Orion Nebula in the south to NGC 1977 in the north. The molecular contours cut-off very sharply at NGC 1977, suggesting that it is ionization bounded by the molecular cloud. Optical, infrared and radio measurements could be made near the H II region-molecular cloud interface. Useful measurements might include: optical determinations of ionization stratification and radial velocity; infrared searches for a bright bar like the one seen in the Orion Nebula (Becklin et al. 1976); radio continuum interferometry of NGC 1977; radio recombination line measurements of hydrogen and carbon. Determination of the velocity of the ionized gas could, for example, resolve the Trumpler effect problem of the Orion Nebula (e.g., Zuckerman 1973). The presumed exciting star for NGC 1977 has the same radial velocity as the Trapezium. If NGC 1977 is found to have almost the same radial velocity as the molecular cloud, which should be the case if the H II region is flowing off the cloud transverse to the line of sight, the implication would be that the Trumpler effect is small for the Trapezium stars.

At the Kleinmann-Low nebula (KL) in Orion $^{12}$CO emission is detectable over a total velocity range of at least 150 km/s (Zuckerman et al. 1976). The high velocity gas appears to be localized to a region <1' in diameter (Kuiper et al. 1977a), which might be comparable in size to KL. The high velocity CO is probably associated with pre- rather than post-main sequence objects. Free-fall on to compact objects appears unlikely because of the very large masses implied in a small region. Outflow models are probably to be preferred but the implied mass ($\sim 10^{-3}$ $M_\odot$/yr) and momentum outflow is quite large if it is to be supplied by a cluster of young stars. Recent observations by Kuiper et al. (1977b) indicate that HCN and SO lines from KL also have extended wings. However, they and Gilmore (1977) were unable to find any other H II regions or infrared clusters with CO line wings as extended as those in Orion.

Future prospects for observations of molecular clouds look very bright. New large telescopes and interferometers sensitive to millimeter wavelength radiation will permit substantially more detailed study of molecules in regions of stellar birth and death in the Milky Way and in external galaxies. New instrumental techniques have already pushed the short wavelength frontier to submillimeter wavelengths (e.g. observation of the $J = 3 \to 2$ transition of CO on the 200-inch Hale telescope by Phillips et al. 1976). Portions of the mm and sub-mm spectrum not accessible from the ground will be observable from aircraft, balloons, and satillites. Initial steps in this direction have been taken with a 183 GHz radiometer flown on the G.P. Kuiper Airborne Observatory (Waters et al. 1976).

REFERENCES

Avery, L.W., Broten, N.W., MacLeod, J.M., Oka, T., and Kroto, H.W. 1976, Astrophys. J., 205, L173
Barrett, A.H., Ho, P.T.P., and Myers, P.C. 1977, Astrophys. J., in press
Becklin, E.E., Beckwith, S., Gatley, I., Matthews, K., Neugebauer, G., Sarazin, C., and Werner, M.W. 1976, Astrophys. J. 207, 770
Blaauw, A. 1964, Ann. Rev. Astron. Astrophys. 2, 213
Elmergreen, B.G. and Lada, C.J. 1977a, Astrophys. J., in press
Elmergreen, B.G. and Lada, C.J. 1977b, Astrophys. J., in press
Gilmore, W. 1977, in preparation
Goldreich, P. and Kwan, J. 1974, Astrophys. J. 189, 441
Kuiper, T.B.H., Kuiper, E.N.R., Zuckerman, B. 1977a, Astrophys. J., in press
Kuiper, T.B.H., Kuiper, E.N.R., Zuckerman, B. 1977b, in preparation
Kutner, M.L., Evans, N.J. and Tucker, K.D. 1976, Astrophys. J. 209, 452
Lada, C.J. 1977, Astrophys. J. Suppl., in press
Liszt, H.S., Wilson, R.W., Penzias, A.A., Jefferts, K.B., Wannier, P.G., and Solomon, P.M. 1974, Astrophys. J. 190, 557
Loren, R.B. 1976, Astrophys. J. 209, 466
Lucas, 1976, Astron. Astrophys. 46, 473
Matsakis, D. 1977, private communication
Milman, A.S. 1975, Astrophys. J. 202, 673
Morris, M., Snell, R.L., Van den Bout, P. 1977, Astrophys. J., in press
Mufson, S.L. and Liszt, H.S. 1977, Astrophys. J., in press
Phillips, T.G., Huggins, P.J., Neugebauer, G., and Werner, M.W. 1976, Bull. Am. Astron. Soc. 8, 565
Snell, R.L. and Loren, R.B. 1977, Astrophys. J., in press
Waters, J.W., Gustincic, J.J., Kakar, R.K., Kuiper, T.B.H., Swanson, P.N., Kerr, A.R., and Thaddeus, P. 1976, Bull. Am. Astron. Soc. 8, 564
Zuckerman, B. 1973, Astrophys. J. 183, 863
Zuckerman, B. and Evans, N.J. 1974, Astrophys. J. 192, L149
Zuckerman, B., Kuiper, T.B.H., and Kuiper, E.N.R. 1976, Astrophys. J. 209, L137

# ISOTOPIC ABUNDANCES IN INTERSTELLAR CLOUDS[*]

C.H. Townes
University of California

ABSTRACT

   The observation of microwave spectra of molecules in interstellar clouds allows separation and detection of the lines of isotopes of many of the more common elements. Comparison of intensities of isotopic lines shows that the relative isotopic abundances for C, O, S, N, and Si are generally rather similar to those found on Earth. However, there are interesting and provocative differences.

   Special conditions of opacity, of cloud structure, of excitation, or of chemical fractionation of isotopes can impair the determination of isotopic abundances from the intensities and shapes of molecular lines. High opacity is commonly encountered, and so is chemical fractionation at least in the case of the hydrogen isotopes. However, careful selection and interpretation of measurements and comparison of different spectra seem to allow the determination of relative isotopic abundances to a useful precision and degree of certainty. The $^{12}C/^{13}C$ ratio is generally about 45, one-half that found on Earth, but is as low as about 20 in the Sgr A and Sgr B clouds, and as high as about 80 in some other clouds. Such apparent variations are probably real, and do not depend simply on distance of the cloud from the Galactic Center, as might be expected if interstellar clouds at a given distance are intermingled. The $^{17}O/^{18}O$ abundance ratio is slightly greater in interstellar clouds than on Earth. Both deuterium and $^{15}N$ are substantially depleted in the Sagittarius clouds as compared with most other parts of the Galaxy. This provides some evidence that deuterium in the Galaxy is a relict of events other than stellar activity.

   Most of these results fit rather well current views of the nucleosynthesis and evolving stellar history of the Galaxy. However, the variation in isotopic ratios seems to show that the large molecular clouds,

---

[*] Work supported in part by NASA Grant NGL 05-003-272 and NSF Grant AST 75-13501.

*Hugo van Woerden (ed.), Topics in Interstellar Matter, 113-123. All Rights Reserved.*
*Copyright © 1977 by D. Reidel Publishing Company, Dordrecht-Holland.*

of mass $10^5 - 10^6$ $M_\odot$, retain their integrity for approximately $10^9$ years or longer. The Earth may have been formed in a cloud which was somewhat poorer than average in $^{13}C$ relative to $^{12}C$. The long lifetime of massive clouds is not surprising, except in view of possible gravitational collapse, for which important details of the dynamics involved are obscure.

---

Detection of a number of molecular lines in interstellar clouds almost immediately raised the hope of substantial new information on relative isotopic abundances, and by now there is a large amount of observational results on relative intensities of isotopic lines in the microwave region. While interpretation of these observations is still subject to much debate, they now provide a considerably clearer picture of isotopic abundances throughout our Galaxy and of its nucleosynthetic history than what has been previously available. Molecular rotational spectra are particularly suited for observation of different isotopes, since the molecular moment of inertia changes substantially with any change in isotopic weight, producing a calculable frequency change for isotopic rotational lines which allows them to be easily isolated and identified. The separation between isotopic lines is in fact sometimes large enough to make awkward an accurate intensity comparison. A number of isotopes of the more abundant atomic species such as H, C, N, O, and S, have now been detected and techniques available raise the hope of good determination of isotopic abundances in external galaxies as well as in clouds throughout our own Galaxy. We discuss here primarily the results of microwave measurements in dense interstellar clouds. Optical spectroscopic measurements in the hotter, less dense interstellar material differ both in technique and possibly also in subject matter, since it will be seen that isotopic abundances may vary between different regions.

In spite of the present abundance of good measurements, there are serious pitfalls in the derivation of relative abundances from observation of the relative intensities of isotopic lines. Of these, the most obvious and troubling is the problem of optical depth or saturation of line intensity. Frequently, the optical depth of a spectral line of an interstellar cloud is quite uncertain, partly because its effective excitation temperature is not known and unlikely to be close to any other temperature measured, and partly because the cloud cannot be expected to fill the antenna beam uniformly. For example, even though the total intensity of an emission line may correspond to an antenna temperature far less than what is expected to be the excitation temperature of the levels involved, it can still be optically quite thick since it may occur in small patches scattered over the solid angle observed by the antenna rather than in a continuous cloud completely filling the antenna's field of view. Even in the absence of such an extreme case, the optical depth and temperature can hardly be expected to be uniform over the region subtended by the antenna beam, nor can the Doppler-broadened line be expected to have any simple shape. Another uncertainty which is perhaps more subtle, and can be very troublesome in particular cases, is the question whether a given isotope may be much more concentrated in a particular molecular species than its average abundance. There is also an uncertainty whether the different isotopic species may be excited to quite different degrees in a given rotational transition. Chemical fractionation--the concentration of an isotopic species in a particular molecule--does not usually seem large enough to degrade the present rather coarse precision of measurement of isotopic abundance ratios except for the case of deuterium. However, some chemical fractionation for other isotopes must certainly be present, and raises serious uncertainties, perhaps particularly for the C isotopes. This will be considered in more detail in connection with the experimental data.

Excitation by collisions should not usually make any troublesome difference in excitation of the various isotopic species. Excitation by resonant absorption or by resonant trapping of radiation, on the other hand, can produce substantial differences in excitation. This problem, radiation trapping, is closely associated with that op optical depth. Resonant absorption or stimulated emission produce particularly large uncertainties when a continuum background is present. This is because stimulated emission or absorption intensities are proportional to the difference in populations between upper and lower states, which is less than the population of each by a factor approximately $h\nu/kT$. Spontaneous emission, on the other hand, which produces the emission from a thin cloud in the presence of no continuum, is simply proportional to the population of the upper state. Thus it is less sensitive to disturbance of population of a single state than is stimulated emission (or absorption) by a factor $kT/h\nu$, which is often as large as 10 or 100. Maser amplification represents stimulated emission in the presence of such a disturbance. It occurs in many molecular resonances, and can make impractical any interpretation of relative line intensities for isotopic abundance determination even when the amplification is not large. As a rough guide, the optical depth must be much less than $h\nu/kT$ before stimulated radiative effects can be expected to be free of distortion by trapping or emission of radiation. Absorption in $H_2CO$ is an exceptional case which is more reliable than most even though the relative populations are anomalous, since its excitation temperature is very low ($\sim 1^{\circ}K$) and the factor $kT/h\nu$ is not large.

Fortunately, there are some measures available for coping with the above complications and thereby obtaining reasonably trustworthy measurements of relative isotopic abundances. These measures nevertheless rely on the assumption that no very extreme unknown peculiarities exist in the nature of clouds, or in excitation, or in differential isotopic concentration. To obviate problems with optical depth, it is frequently possible to use relatively weak lines. The optical depth may be judged to some extent by the intensity of the line, but perhaps more reliably by comparison of the line shapes for two different isotopic species of widely different abundance or of two related lines with widely different intensities. If the lines are essentially identical in shape, the total line intensities may give a reasonably accurate view of the relative abundances. However, it is clear that some cases are quite deceiving. It appears that rather commonly, clouds are composed of a number of clumps, each considerably smaller than the antenna beamwidth (Mayer et al. 1973, Townes 1976). If each clump is optically deep, line shapes are then determined primarily by the relative velocities of these clumps and all lines which are optically rather deep appear to have approximately the same shape, regardless of their actual depth.

A velocity gradient through the cloud can give similar effects. Lines which are optically deep in the center of the line should still allow determination of isotopic abundances if the relative intensities are measured rather far into the wings of the lines, where they are unlikely to be optically deep. Extension of such measurements as far into the wings as possible and extrapolation as a function of distance from the line center should allow reasonable determinations of isotopic ratios (Bertojo et al. 1974, Wannier et al. 1976). An alternative, and almost equivalent, system of obviating problems with optical depth is to use very high angular resolution so that a cloud can be explored on a scale over which it is more likely to be uniform and so that relatively thin parts of the cloud can be used for line intensity measurements (Fomalont and Weliachew 1973). This latter technique is broadly similar to use of line wings, since some of the optically thin parts might be presumed to be involved in the weakest part of a line from a larger region.

A technique which is frequently useful in helping to eliminate problems of optical depth is the comparison of line intensities only for relatively rare isotopes. For example, the $^{12}C/^{13}C$ ratio can be measured by examining the very weak lines of CO in the rare species $^{13}C^{18}O$ and $^{12}C^{18}O$. Usually, however, the combination of such rare isotopes as $^{13}C$ and $^{18}O$ in a single molecule makes its lines so weak that detection is either impossible or the intensities measured are quite uncertain. Both this technique and use of the far wings of lines make the ultimate available sensitivity very important; the needed resolution of complex clouds makes high angular resolution also important.

A very useful technique, and one which may give rather definite but more complex information, involves ratios between two rare isotopes. Thus, the two rare isotopes $^{17}O$ and $^{18}O$ may be compared in the spectrum of a molecular species, or a hybrid isotopic ratio may be obtained by comparing a $^{13}C^{16}O$ line intensity with that of a $^{12}C^{18}O$ line. If these lines are optically thin, which is not always the case even for such rare isotopes, their ratio of intensities immediately gives the ratio of the fractional abundance of $^{13}C$ to the fractional abundance of $^{18}O$, since in both cases $^{12}C$ and $^{16}O$ are very close to 100% abundance. Similarly, the ratio of fractional abundances of $^{13}C$ and of $^{15}N$ may be obtained from the hybrid ratio of $H^{13}C^{14}N$ to $H^{12}C^{15}N$.

"Chemical" fractionation of isotopes will in principle always occur, since the binding energies, reaction rates, and spectra associated with different isotopes are never identical. The largest differences occur between H and D. For example, the difference in zero-point vibrational energy between HCN and DCN is about 600 $cm^{-1}$; that between

$H_2$ and HD is about 280 $cm^{-1}$. Thus, if equilibrium occurs between these two pairs of molecules at a temperature of 100 K (kT = 70 $cm^{-1}$), the deuterium will be more concentrated in DCN than in HD by a factor of about 36; in fact concentration of D of this magnitude or more is frequently observed in interstellar clouds. The difference in $^{12}C$ and $^{13}C$ energies in various molecules, however, is only about 10 $cm^{-1}$, so that chemical fractionation due to energy differences should not be so striking for such heavier isotopes. An alternate mode of concentration for deuterium has been suggested by Watson (1973) based on preferential UV dissociation of HD over $H_2$. Watson et al. (1976) have also suggested an exchange between ionized carbon and CO of the form

$$^{13}C^+ + {}^{12}CO \rightarrow {}^{12}C^+ + {}^{13}CO + \Delta E,$$

where $\Delta E \approx 23$ $cm^{-1}$. Such an exchange producing an equilibrium distribution of $^{13}C$ in a cloud of low temperature could concentrate $^{13}C$ in CO by a factor of two or more, and raises troublesome questions about how representative the $^{13}C$ abundance in CO may be. Fortunately, isotopes of other common chemical species such as O, N, or S, do not seem so likely to have a fractionation mechanism which is energetically so favoured.

It should be noted that most of the molecules in interstellar space are unlikely to be formed under equilibrium conditions at low temperature, and that most efforts to verify specific reaction paths by experimental observation have been unsuccessful. It seems likely that a more varied complex of processes than the simple types outlined above are responsible for isotopic distributions and hence that fractionation, not being dominated by a single simple mechanism, may not be very large in many cases. Nevertheless, such fractionation will occur on occasion, and needs careful consideration.

As can be seen from the above, it is not hard to construct conditions in idealized clouds which would give very misleading information about isotopic abundances as measured by any specific criterion. Since our information on specific conditions in clouds is in fact quite limited, whether determinations of isotopic abundances are misleading must for the present be determined primarily from detailed examination of consistency of the results coming from a variety of methods and a variety of clouds. In the process of such examination, some methods and some clouds will certainly be identified as examples where interesting phenomena occur which do in fact give misleading isotopic information when simply interpreted. Enough consistency and cross-checks, however, already seem to be available to give assurance of some reliability of the more careful isotopic determinations.

The various studies of isotopic ratios have yielded considerable interesting information on the relative abundance of the rare isotopes $^{13}C$, $^{17}O$, $^{18}O$, $^{15}N$, $^{30}Si$, $^{33}S$, $^{34}S$, and D of the more common chemical elements, though the precision of this information must be treated with caution as indicated above. A first approximation to the results is that the abundances of these isotopes in interstellar space are very similar to those found in terrestrial sources. A somewhat more detailed description is that these ratios are not far from terrestrial abundances but there are noticeable deviations, and also variations between clouds in the Galaxy. $^{13}C$, one of the most easily detected rare isotopes, is usually about twice as abundant as on Earth, but appears to vary from this figure by about a factor of two in each direction, as is illustrated in Table 1.

The $^{12}C/^{13}C$ ratios in various clouds have been derived for Table 1 by two different and somewhat complementary techniques--from the hybrid emission intensity ratio $^{12}C^{18}O/^{13}C^{16}O$ and from the direct absorption intensity ratio $H_2^{12}C^{16}O/H_2^{13}C^{16}O$. For the former case, with one exception the third column lists the ratio $^{12}C^{18}O/^{13}C^{16}O$ times 500, the terrestrial $^{16}O/^{18}O$ abundance ratio. This would give the correct $^{12}C/^{13}C$ ratio if the $^{16}O/^{18}O$ isotopic ratio were everywhere terrestrial, as has been suggested by some to be an adequate approximation. The one exception is the case of Sgr B2, where there seems to be fairly clear evidence (Whiteoak and Gardner, 1975) that the $^{16}O/^{18}O$ ratio is nearer to 250, and this number has hence been used rather than 500. For both columns, dependent respectively on CO and $H_2CO$ spectra, sources and data have been selected and treated in a way to minimize errors due to optical depth. The data come from a wide variety of sources with the particular values and errors based on my own judgement of all presently available information. Most of the CO results listed are derived from data of Wannier et al. (1976). Origins of the $H_2CO$ data are more varied (Bertojo et al. 1974, Fomalont and Weliachew 1973, Zuckerman et al. 1974, Wilson et al. 1976, Matsakis et al. 1976, Evans et al. 1975, Gardner et al. 1971).

Reliability of the data and its subjectivity to optical depth, isotopic fractionation, or other disturbing factors is perhaps best judged from the consistency of results from the two molecules, which at this stage may be viewed as reasonably satisfactory. Unsuspected high optical depths would make $^{12}C/^{13}C$ ratios obtained from CO in column 3 of Table 1 too high and those from $H_2CO$ in column 4 too low, but there seems to be no substantial systematic difference of this type. The most potent mechanism for chemical fractionation of $^{13}C$ so far suggested (Watson et al. 1976) is expected to concentrate $^{13}C$ in CO. However, unless contrary to most expectations $H_2CO$ is formed from CO, this mechanism would not concentrate $^{13}C$ in $H_2CO$. The fact that no large systematic difference of this type appears is some reassurance that chemical fractionation is not unusually large for $^{13}C$. A few particular cases, such as NGC2024 and DR21, may represent such a fractionation. Sgr B2 and Sgr A, where the largest deviations from average $^{12}C/^{13}C$ ratios occur, also show deviations in other isotopes (cf. Table 3) for which no very potent mechanism of chemical fractionation at present seems likely.

That chemical fractionation of isotopes can occur is well illustrated by the case of deuterium, where the critical vibrational energy

Table 1 - Abundance ratio $^{12}C/^{13}C$ in a variety of interstellar clouds.

| Source | Distance from Galactic Center (kpc) | $\dfrac{^{12}C^{18}O}{^{13}C^{16}O}$ x 500 | $H_2^{12}CO/H_2^{13}CO$ |
|---|---|---|---|
| | | R' | R |
| Sgr B2 (62 km/sec) | ∿0.1 | 15 + 6* | 14 + 5 |
| Sgr A   (42 km/sec) | ∿0.1 | | 18 + 8 |
| Sgr A   (-4 km/sec) | ∿8 | | 50 + 4 |
| W33 | 6 | | 30 + 8 |
| W43 | ∿7 | | 50 + 10 |
| W51 | 8 | 64 + 4 | 63 + 12 |
| M17 | 8 | 57 + 9 | >70 |
| W31 | 8 | 54 + 10 | 39 + 3 |
| M8 | 9 | 34 + 5 | |
| NGC 6334 | 9 | 36 + 3 | |
| W49 | 9 | 32 + 11 | 41 + 4 |
| DR21 | 10 | 36 + 5 | 47 + 4 |
| L134N | 10 | | 37 + 7 |
| Solar System | 10 | 89 | 89 |
| Orion A | 11 | 38 + 2 | |
| NGC 2024 | 11 | 28 + 5 | 48 + 6 |
| NGC 2264 | 11 | 56 + 6 | |
| Cas A | ∿11 | | >90 |
| W3 | ∿12 | 61 + 10 | 68 + 12 |
| NGC 7538 | 13 | 43 + 8 | |

* This number uses an abundance ratio $^{16}O/^{18}O$ of 1/250 rather than 1/500 (see text).

The ratio R' is obtained by comparing the $^{13}C^{16}O$ and $^{12}C^{18}O$ intensities and assuming the $^{16}O/^{18}O$ abundance ratio is equal to that on Earth, 1/500. The ratio R is obtained from the relative intensities of $H_2^{12}CO$ and $H_2^{13}CO$ lines, using what corrections can be made for optical-depth effects. For high $^{12}C/^{13}C$ ratios, R is probably more accurate; for low ratios R' may be more accurate. Errors listed are statistical. CO measurements come largely from Wannier et al. (1976), $H_2CO$ data from a variety of sources (Bertojo et al. 1974, Fomalont and Weliachew 1973, Zuckerman et al. 1974, Wilson et al. 1976, Matsakis et al. 1976, Evans et al. 1975, Gardner et al. 1971).

---

differences for light and heavy isotopes of hydrogen are about 320 cm$^{-1}$ rather than the approximately 23 cm$^{-1}$ found in carbon. The relative atomic abundance D/H in interstellar material is about 1.4 x 10$^{-5}$ (York and Rogerson 1976), not far from the ratio in the Jovian atmosphere (Beer and Taylor 1973), whereas the relative abundances of deuterated molecules can be orders of magnitude larger, as shown in Table 2. This table also indicates that, unless the Sgr B2 and Sgr A clouds have a

Table 2 - Relative abundances of molecules containing the rare isotopes D and $^{13}$C.

| Source | $\dfrac{DCN}{H^{13}CN}$ | $\dfrac{D^{12}CO^{+}}{H^{13}CO^{+}}$ |
|---|---|---|
| Sgr B2 | 0.018 | |
| Sgr A | | |
| M17 | 0.126 | |
| W31 | 0.126 | |
| DR21 | 0.084 | |
| DR21 (OH) | 0.21 | 0.9 |
| Solar System | 0.013 | |
| Orion A | 0.16 | |
| W51 | 0.084 | |

Data listed come from Wannier et al. and Hollis et al. (1976).

Table 3 - Relative abundances of rare isotopes of O, S, and the hybrid abundance ratios $^{13}C^{32}S/^{12}C^{34}S$ and $^{12}C^{15}N/^{13}C^{14}N$.

| | $\dfrac{C^{17}O}{C^{18}O}$ | $\dfrac{C^{33}S}{C^{34}S}$ | $\dfrac{^{13}CS}{C^{34}S}$ | $\dfrac{H^{12}C^{15}N}{H^{13}C^{14}N}$ |
|---|---|---|---|---|
| Sgr B2 | 0.31 $\pm$ .02 | 0.19 $\pm$ .13 | 0.61 $\pm$ .18 | 0.02 $\pm$ .02 |
| Sgr A | | | | 0.01 $\pm$ .02 |
| K39 | 0.27 $\pm$ .05 | | | |
| M17 | | | 0.62 $\pm$ .11 | |
| DR21 | 0.21 $\pm$ .02 | | | |
| Solar System | 0.186 | 0.18 | 0.26 | 0.33 |
| Orion A | 0.27 $\pm$ .06 | 0.21 $\pm$ .06 | 0.38 $\pm$ .07 | 0.25 $\pm$ .02 |
| NGC2264 | 0.17 $\pm$ .03 | | 0.37 $\pm$ .24 | |
| W51 | | | 0.38 $\pm$ .21 | 0.21 $\pm$ .02 |

Data listed here come primarily from Wilson et al. (1976), Linke et al. (to be published), and Wannier et al. (1976).

---

very different amount of chemical fractionation than other regions, deuterium abundance in them is exceptionally low. It has been suggested (Wannier et al., to be published) that this indicates that stellar activity, which must have been greater in these clouds than in most other regions, destroys rather than produces deuterium. This is the normal expectation, and provides some experimental confirmation that the deuterium detected in our Galaxy was produced during very early states of the universe.

Measured isotopic abundance ratios for oxygen, sulphur, and nitrogen isotopes are shown in Table 3. The $^{15}N/^{14}N$ ratio is seen to be substantially less in Sgr B2 and Sgr A than in other regions, again

indicating a very different nuclear history for these clouds. Since
they are very dense, large, and near the galactic center where much
stellar formation must have occurred, a relative decrease in $^{15}$N abundance and a relative increase in $^{13}$C and $^{17}$O abundances are in fact what
is expected theoretically (Wollmann 1973, Audouze et al. 1975).

Circumstellar material around the strong infrared star IRC + 10216
provides an interesting comparison with that found in interstellar
clouds. Around this carbon star, which must represent some of the stellar
activity that gradually changes interstellar isotopic abundances, the
relative abundance of $^{17}$O also seems to be substantially enhanced (Rank
et al. 1974) as in Sgr B2. Unpublished information of Hall et al. (private communication) that $^{12}$C/$^{13}$C > 33 in this object and the H$^{13}$C$^{14}$N/
H$^{12}$C$^{15}$N ratio of Morris et al. (1971) shows that $^{14}$N/$^{15}$N > 1000. This
represents depletion of $^{15}$N and increase in $^{17}$O in IRC + 10216, as is
found in the Sagittarius clouds.

The isotopic abundance measurements of interstellar clouds available so far seem to establish the following:
1. The $^{13}$C and $^{17}$O relative abundances are generally higher than
those in the solar system, with $^{13}$C higher by about a factor of 2.
In at least a few specific cases, the $^{15}$N and $^{2}$H abundances are lower
and the $^{18}$O abundance possibly higher than those in the solar system.
2. There are substantial variations in isotopic abundances between
different large interstellar clouds, with some of these variations not
dependent on distance from the galactic center alone.

It is not easy to understand how $^{13}$C can have a relative abundance
at present as much as twice that which occurred when the solar system
was formed (Wollmann 1973, Audouze et al. 1975). A qualitative description of expectations is that many rare isotopes, such as $^{13}$C, are secondary products formed from other primary heavy nuclei such as $^{12}$C, and
that if stellar formation has proceeded at a uniform or slightly decreasing rate their relative abundances should change approximately
linearly with time. Since the solar system is thought to have been
formed towards the latter part of the history of our Galaxy, a doubling
of the $^{13}$C relative abundance since then is surprising. The explanation
may simply be that the cloud from which our solar system formed was not
an average one, but relatively poor in $^{13}$C. This conclusion is consonant
with observation number two above, that interstellar clouds are not very
thoroughly mixed and differ in isotopic composition even at a given
radius from the galactic center.

Because the Galaxy rotates and there is considerable relative
motion of clouds in a time short compared to the several billion years
required to change average chemical or isotopic composition, it has
generally been expected that chemical abundances and isotopic composition would be a function of distance from the galactic center but not
of galactic longitude. Observed isotopic abundances appear to challenge
that expectation, and in fact there seems to be no good quantitative
argument why the massive and dense clouds, where many molecules are

found, should be very thoroughly mixed with other galactic matter. For example, clouds of about $10^5$ solar masses, density $\sim 10^5$ molecules/cm$^3$, and a few parsecs in linear dimension, should have a very long ($>> 10^9$ years) lifetime against collisions with each other since there are only a few hundred of them in the Galaxy. Collisions or mixing with normal interstellar material of density $\sim 0.2$ molecules/cm$^3$ would occur, but would dilute the large dense clouds only on a time scale which is also somewhat longer than $10^9$ years. Hence, these large cloud units may indeed have a long-term integrity and a lifetime comparable to the age of the Galaxy provided they do not collapse permanently into stars. Their gravitational collapse time, based on a simple, uniform and initially static model, would be of the order $10^5 - 10^6$ years and quite short. It is evident that nonuniformities, turbulence, magnetic fields, stellar formation itself, and supernovae all tend to extend the lifetime against collapse. Furthermore the relative abundance of large clouds and paucity of groups of recently formed stars without surrounding clouds gives some empirical evidence against a lifetime as short as $10^6$ years for the massive clouds. Unfortunately, their dynamics is too complex for any detailed theory of gravitational collapse and relatively complete condensation into stars to be available. The accumulation and clarification of isotopic abundance measurements may provide the clearest evidence on the long-term history of these clouds, as well as providing insight into the rates and types of nucleosynthesis which have occurred in our Galaxy or others since they were formed.

REFERENCES

Audouze, J., Lequeux, J., and Vigroux, L.: 1975, Astron. Astrophys. 43, 71.

Beer, R. and Taylor, F.W.: 1973, Astrophys.J. 182, L131.

Bertojo, M., Chui, M.F., and Townes, C.H.: 1974, Science 184, 619.

Evans, N.G., Zuckerman, B., Morris, G., and Sato, T.: 1975, Astrophys.J. 916, 433.

Fomalont, E.B. and Weliachew, L.: 1973, Astrophys.J. 181, 781.

Gardner, F.F., Ribes, J.C., and Cooper, B.F.C.: 1971, Astrophys. Letters, 9, 181.

Hall, D., Ridgway, S., Kleinman, S., and Weinberger (private communication).

Hollis, J.M., Snyder, L.E., Lovas, F.J., and Buhl, D.: 1976, Astrophys. J. 204, L83.

Linke, A.A., Goldsmith, P.F., Wannier, P.G., Wilson, R.W., and Penzias, A.A. (to be published).

Matsakis, D.N., Chui, M.F., Goldsmith, P.F., and Townes, C.H.: 1976, Astrophys.J. 206, L63.

Mayer, C.H., Waak, J.A., Cheung, A.C., and Chui, M.F.: 1973, Astrophys. J. 182, L65.

Morris, M., Zuckerman, B., Palmer, P., and Turner, B.E.: 1971, Astrophys.J. 170, L109.

Rank, D.M., Geballe, T.R., and Wollman, E.R.: 1974, Astrophys.J. 187, L111.

Townes, C.H.: 1976, Mem. Soc. Roy. de Science de Liège, X, 453.

Wannier, P.G., Penzias, A.A., Linke, R.A., and Wilson, R.W.: 1976, Astrophys.J. 204, 26.

Wannier, P.G., Lucas, R., Linke, R.A., Encrenaz, P.J., Penzias, A.A., and Wilson, R.W.: 1976, Astrophys.J. 205, L169.

Wannier, P.G., Penzias, A.A., Linke, R.A., and Wilson, R.W. (to be published).

Watson, W.D.: 1973, Astrophys.J. 181, L129.

Watson, W.D., Anicich, V.G., and Huntress, W.T. Jr.: 1976, Astrophys. J. 205, L165.

Whiteoak, J.B. and Gardner, F.F.: 1975, Proc. ASA 2, 360.

Wilson, R.W., Penzias, A.A., Wannier, P.G., and Linke, R.A.: 1976, Astrophys.J. 204, L135.

Wilson, T.L., Bieging, J.H., Downes, D., and Gardner, F.F.: 1976, Astron. Astrophys. 51, 303.

Wollman, E.R.: 1973, Astrophys.J. 184, 773.

York, D.G. and Rogerson, J.B. Jr.: 1976, Astrophys.J. 203, 378.

Zuckerman, B., Buhl, D., Palmer, P., and Snyder, L.: 1974, Astrophys. J. 189, 217.

# FORMATION AND EXCITATION OF MOLECULAR HYDROGEN

A. Dalgarno
Center for Astrophysics, Harvard College Observatory,
Smithsonian Astrophysical Observatory, Cambridge,
Massachusetts, U. S. A.

## Abstract

The observational data from the Copernicus satellite on the relative abundances of atomic and molecular hydrogen are generally consistent with a theory that postulates an equilibrium between formation of $H_2$ on grain surfaces and destruction by fluorescent dissociation induced by the interstellar radiation field.

$H_2$ is detected in excited rotational levels. The rotational populations can be explained by a combination of ultraviolet pumping and excitation during the formation process. The derived densities range from 10 $cm^{-3}$ to 1000 $cm^{-3}$ and the gas pressures from $10^3$ $cm^{-3}$ K to well over $10^4$ $cm^{-3}$K and there is little evidence for a uniform cloud pressure supported by an intercloud medium. In some of the clouds the derived radiation field is unusually large suggesting that the cloud is close to the parent star and presumably physically associated with it. There is also observational evidence for clouds that are sheets 0.01 pc thick with densities between 100 $cm^{-3}$ and 1000 $cm^{-3}$, produced presumably by shock waves associated with expanding HII regions or old supernova remnants.

The Copernicus data also reveal the presence of HD in amounts which show that there must be a source of HD in addition to grain formation, which is probably the reaction sequence
$H^+ + D \rightarrow H + D^+$, $D^+ + H_2 \rightarrow H^+ + HD$. From the measured abundance of HD, the proton density can be derived and from it the ionizing flux within the cloud. Ionizing fluxes can also be derived from the observed abundances of OH. For $\zeta$ Oph the value is $1.2 \times 10^{-17}$ $sec^{-1}$ which if correct excludes the possibility of low energy cosmic ray ionization in the cloud.

Emission lines of the 1-0 band of $H_2$ have been detected recently in Orion and in NGC 7027. Emission from higher vibrational levels was not detected and the origin of the excitation is uncertain.

*Hugo van Woerden (ed.), Topics in Interstellar Matter, 125-133. All Rights Reserved.*
*Copyright © 1977 by D. Reidel Publishing Company, Dordrecht-Holland.*

Whether it is ultraviolet pumping or collision excitation, densities
of order at least $10^6$ cm$^{-3}$ appear to be required in the case of
Orion, suggesting the occurrence of a shock.  The H$_2$ in NGC 7027
may be formed by negative ion reactions and not by grain catalysis.
The effects of H$_2$ formation in collapsing clouds are mentioned
briefly.

---

The observations with the Princeton ultraviolet spectrometer
on the Copernicus satellite of interstellar absorption lines
arising from molecular hydrogen (Spitzer et al. 1973, Spitzer et al.
1974, Spitzer and Jenkins 1975, York 1976) have advanced significantly
our knowledge of interstellar clouds.  The observational data on
the relative abundances of atomic and molecular hydrogen are
generally consistent with the theory of Hollenbach et al. (1971),
a theory that postulates an equilibrium between formation of H$_2$
on grain surfaces and destruction by fluorescent dissociation
induced by the interstellar radiation field between the threshold
of the process at 1108 Å and the Lyman limit at 912 Å.

The detection of H$_2$ in excited rotational levels as high for
some clouds as the level with rotational quantum number J=7
(Spitzer and Morton 1976) is of particular interest.  The distribution
of rotational populations provides an important diagnostic probe
of the physical environment in which the molecules exist.  The
enhanced rotational populations of high J can be achieved by a
combination of ultraviolet pumping (Black and Dalgarno 1973)
and excitation during the formation process (Spitzer and Zweibel
1974).

The rotational levels with high values of J exceeding J=4 have
relatively short radiative lifetimes (Dalgarno and Wright 1972).
Their populations are unaffected by collisions for ambient densities
less than $10^4$ cm$^{-3}$ and they provide accordingly a direct measure of
the intensity of the ultraviolet radiation field to which the
molecules are subjected.  The rotational levels at low values of
J have radiative lifetimes that are larger than the mean times
for collisions except at very low densities and they provide a
direct measure of the kinetic temperature (Black and Dalgarno 1973).
Because the proton interchange reaction

$$H^+ + H_2 \text{ (o) } \overset{\rightarrow}{\leftarrow} H_2(p) + H^+$$

between ortho and para-hydrogen is rapid, the ratio of the populations
of the J=0 and J=1 levels is also a measure of the kinetic
temperature (Dalgarno et al. 1973) at the gas densities of most
interstellar clouds.  At very low gas densities, the low J levels

are destroyed by photoabsorption of the ultraviolet radiation
field and in such clouds their populations can be analyzed to
infer the initial rotational distribution produced by the formation
mechanism (Jura 1975a).

Application of these theoretical considerations requires the
solution of the vibrational-rotational radiative cascading
problem and cascade tables have been constructed (Black and Dalgarno
1976). They were used by Black and Dalgarno (1973) to infer
the density, temperature and ultraviolet radiation field intensity
for the cloud lying towards ζ Ophiuchi and more extensively by
Jura (1975a, b) for many other clouds. The formation rates
derived by Jura are consistent with a rate coefficient of
$3x10^{-17}$ $cm^3$ $sec^{-1}$. The kinetic temperatures lie between 60 K and
100 K and the densities range from 10 $cm^{-3}$ to $10^3$ $cm^{-3}$. For the
nearby unreddened stars α Vir, λ Sco and β Cen, York (1976)
derived densities between 0.3 $cm^{-3}$ and 0.8 $cm^{-3}$.

The gas pressures range from $10^3$ $cm^{-3}$ K to $5x10^4$ $cm^{-3}$ K and
there is little evidence for a uniform cloud pressure supported by
an intercloud medium. For the ζ Oph cloud studied by Black and
Dalgarno (1973) and for several of the clouds studied by Jura
(1975 a, b) the derived radiation field is unusually large as is
the gas pressure, suggesting that the cloud is close to the parent
star and presumably physically associated with it. The lowest
pressures, which occur for clouds with normal radiation fields,
are about $2x10^3$ $cm^{-3}$ K, comparable to that expected from the warm
or the hot intercloud medium.

There are several clouds in which the rotational populations
cannot be explained by a single component cloud and multicomponent
or multiple clouds must be invoked. Spitzer and Morton (1976) have
studied seven low velocity stars towards which the absorption lines
appear to broaden at high rotational quantum numbers. They
attribute the phenomenon to the existence of a second cloud moving
towards the Earth that is close to the star so that the high J
levels are strongly populated. The ultraviolet pumping theory
leads to the conclusion that the clouds are sheets of order
0.01 pc thick with densities between 100 $cm^{-3}$ and 1000 $cm^{-3}$.
These thin dense sheets are presumably shock waves produced by
expanding HII regions or old supernova remnants (Spitzer and Morton
1976). Steigman et al. (1975) and Castor et al. (1975) have
suggested that the stellar winds of early-type stars produce
thin dense circumstellar shells, and Hollenbach et al (1976) have
shown that the sheets will contain rotationally excited $H_2$.

There is no indication of any broadening of the high J lines
for the cloud towards ζ Ophiuchi but it also appears to be a dense
sheet (Black and Dalgarno 1973) as Herbig (1968) had originally
argued. The rotational structure cannot be reproduced by a single
component cloud. The table compares the observed populations with

Table 1.   ζ Oph Cloud:   Rotational Populations:  _log $N_J$ $cm^{-2*}$

| Quantity | | | Observation | Total | Hot Zone | Cold Zone |
|---|---|---|---|---|---|---|
| N(H) | | | 20.72 | 20.73 | 20.70 | 19.54 |
| N($H_2$) total | | | 20.62 | 20.62 | 20.25 | 20.38 |
| | J=0 | | 20.46 | 20.47 | 19.80 | 20.37 |
| | J=1 | | 20.10 | 20.07 | 20.04 | 18.90 |
| | J=2 | | 18.56 | 18.47 | 18.47 | 15.64 |
| | J=3 | | 17.07 | 16.94 | 16.94 | 14.90 |
| | J=4 | | 15.68 | 15.19 | 15.12 | 14.37 |
| | J=5 | | 14.63 | 14.75 | 14.70 | 13.75 |
| | J=6 | | 13.69 | 13.83 | 13.77 | 12.98 |
| | J=7 | | 13.55 | 13.58 | 13.53 | 12.64 |
| v=1 | J=0 | < | 12.84 | 12.47 | | |
| v=1 | J=1 | < | 13.08 | 12.90 | | |
| v=1 | J=2 | < | 13.00 | 12.92 | | |
| v=2 | J=0 | < | 12.64 | 12.13 | | |

* from Black and Dalgarno (1977)

---

those predicted by a two-component cloud consisting of an interior cold zone of density 2500 $cm^{-3}$ at a temperature of 22 K and an exterior hot zone of density 500 $cm^{-3}$ at a temperature of 110 K (Black and Dalgarno 1977). The derived radiation energy density at 1000 Å is $2 \times 10^{-16}$ ergs $cm^{-3}$ $Å^{-1}$, substantially smaller than the values derived from single component models (Black and Dalgarno 1973, Jura 1975 b) but probably consistent with the results of a more recent direct analysis of the J=6 population (Wright and Morton 1977). The radiation field and the gas pressure are large and the cloud may be responding to the influence of ζ Oph which is a runaway star undergoing mass loss.

The Copernicus data also reveal the presence of deuterated
hydrogen HD with a relative abundance to HD in the range of $10^{-5}$ to
$10^{-7}$ (Spitzer et al. 1973). There is sufficient $H_2$ in these clouds
that the $H_2$ shields itself from the ultraviolet radiation field
but because the photodestruction of $H_2$ is initiated by discrete
line absorption the $H_2$ does not shield the HD. Thus the measured
abundance rates of HD to $H_2$ imply a formation rate for HD somewhat
greater than $10^{-3}$ that of $H_2$. The measured ratio of D/H in
directions in which there is little molecular hydrogen is
$(1.8\pm0.4) \times 10^{-5}$ locally (York and Rogerson 1976) so that the formation
of HD must proceed some hundred times more rapidly than the formation
of $H_2$, and there must be a source of HD in addition to grain
catalysis. There seems little doubt that the source is the reaction
sequence

$$H^+ + D \rightarrow H + D^+$$

$$D^+ + H_2 \rightarrow HD + H^+$$

(Dalgarno et al. 1973, Black and Dalgarno 1973, Watson 1973). It
is interesting that it is the presence of $H_2$ that confirms the
importance of grain surface chemistry in interstellar clouds and the
presence of HD the importance of gas phase chemistry.

From the measured abundance of HD, the proton density can be
derived and if the chemical processes that lead to the destruction
of $H^+$ are understood the ionizing flux within the cloud can be
calculated (Black and Dalgarno 1973, Watson 1973). In analysis of
the HD observations, O'Donnell and Watson (1974) obtained upper
limits to the ionizing fluxes as a function of the gas density and
Barsuhn and Walmsley (1977) adopted the temperatures and densities
of the single component models of Jura (1975 b) to obtain ionizing
fluxes that range from $10^{-17}$ to $10^{-15}$ $s^{-1}$ for different clouds.
The estimates are sensitive to the adopted models and to the
chemistry. For $\zeta$ Oph, Black and Dalgarno (1977) have derived a
flux that is an order of magnitude less than that found by Barsuhn
and Walmsley (1977).

The variations found by O'Donnell and Watson (1974) and by
Barsuhn and Walmsley (1977) could be interpretated alternatively
as variations in the [D]/[H] ratio. However the chemical sequence
(see figure 1) that removes the $H^+$ ions leads to the formation of
OH (Black and Dalgarno 1973, Watson 1973) and measurements of OH
abundances provide additional estimates of the ionizing flux. OH
abundances have been measured towards o Per (Snow 1976) and towards
$\zeta$ Oph (Crutcher and Watson 1976, Chaffee and Lutz 1976). The HD
and OH abundances are consistent with a [D]/[H] ratio of $1.8 \times 10^{-5}$
and ionizing fluxes of $1.2 \times 10^{-17}$ $s^{-1}$ for $\zeta$ Oph (Black and Dalgarno

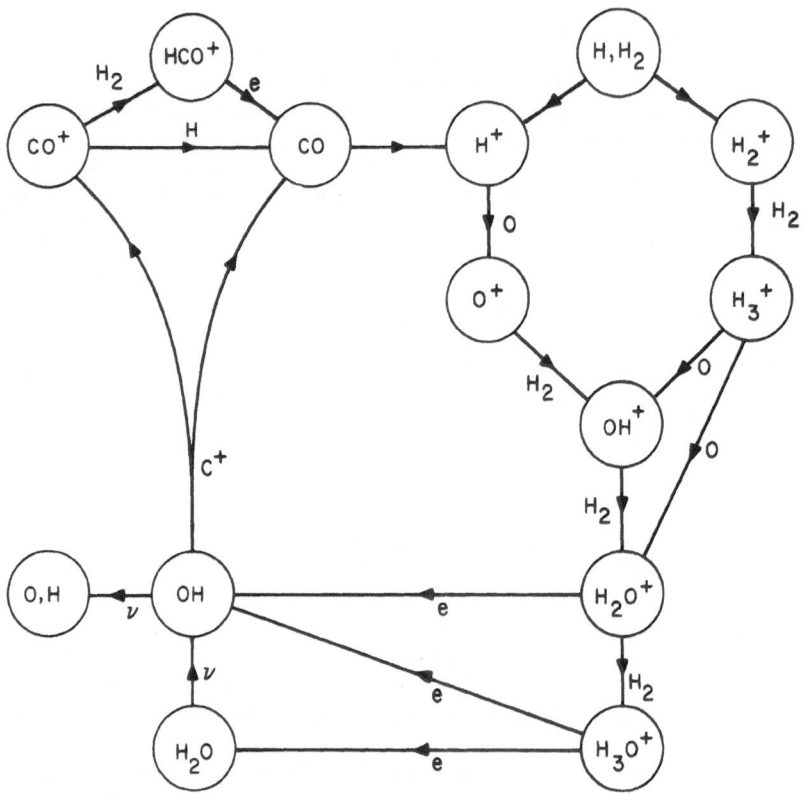

Figure 1:  Chemical reaction scheme for the removal of $H^+$ ions.

1977) and $2.0 \times 10^{-16}$ $s^{-1}$ for o Per (Black et al. 1977).

    Although molecular hydrogen is formed predominantly on grain surfaces under the physical conditions prevailing in most inter-stellar clouds, a gas phase reaction sequence may be more rapid at higher temperatures and at higher electron densities.  The sequence consists of radiative attachment

$$e + H \rightarrow H^- + h\nu$$

followed by associative detachment

$$H + H^- \rightarrow H_2 + e$$

(McDowell 1961, Dalgarno and McCray 1973). Hill and Silk (1975) have calculated that column abundances of $H_2$ of order $10^{13}$-$10^{15}$ cm$^{-2}$ are produced by the $H^-$ reactions in the transition zones around HII regions. The molecular hydrogen detected towards $\tau$ Sco may be an expample (Spitzer and Jenkins 1975).

The radiative attachment and associative detachment mechanisms are also effective in the warm component of the interstellar medium, if such a phase exists. The resulting column abundances of $H_2$ approach $10^{13}$ cm$^{-2}$. Barlow and Silk (1976) have argued that hydrogen atoms may be absorbed on to graphite surfaces with a binding energy of about 1 eV in which case, contrary to earlier expectations, formation of $H_2$ on grains may be efficient at the high temperatures of the warm intercloud medium. If so, the predicted column abundances are of the order of $10^{14}$ cm$^{-2}$ and intercloud $H_2$ should be detectable. According to Hill and Hollenbach (1976), the warm intercloud $H_2$ will be reflected by a large value for the ratio of the J=1 and J=0 populations. The argument is based upon some very uncertain rate coefficients for rotational energy transfer.

The vibrational-rotational cascade that follows the production of molecular hydrogen, either on grain surfaces or by negative ion reactions, and that follows the ultraviolet fluorescence leads to the emission of an infrared spectrum (Black and Dalgarno 1976). Quadrupole emission lines of the 1-0 band of $H_2$ have been detected recently in Orion (Gautier et al. 1976) and in NGC 7027 (Treffers et al. 1976). Gautier et al. argue however that the radiative intensities are inconsistent with ultraviolet pumping and suggest that the levels are excited by thermal collisions at a temperature of order 2000 K. If thermal equilibrium prevails, the ambient density must be at least $10^6$ cm$^3$. Shock models have been proposed by Kwan and Scoville (1976) and by Hollenbach and Shull (1977). At high densities, emission from higher vibrational levels is suppressed by collisional deactivation and ultraviolet pumping becomes a source of heating.

In the planetary nebula NGC 7027 the emission is weak and the observations are not inconsistent with the low density ultraviolet pumping model. Black (1977) has argued that the $H_2$ can be produced in NGC 7027 by the negative ion reaction sequence involving $H^-$.

Photodissociation of $H_2$ is a source of heating in an interstellar cloud (Milgrom et al. 1973, Stephens and Dalgarno 1973), as may be $H_2$ formation (Spitzer and Cochran 1973). In one model of $H_2$ formation (Barlow and Silk 1976), the newly formed molecule is released with substantial kinetic energy and heats the ambient gas whereas in another (Allen and Robinson 1976) the newly formed molecule is in thermal equilibrium with the grain and departs from it carrying but a small fraction of the bond energy. Depending upon the mechanism, either the gas or the grain is heated. In a

dense cloud, the thermal coupling between the gas and the grains
is close (cf. Scoville and Kwan 1976) and Oppenheimer (1977) has
argued that hydrogen atom recombination during cloud collapse
affects significantly the thermal structure of the cloud.  The
hydrogen atoms are assumed to exist in excess of their equilibrium
abundances because in the evolution of a dense cloud the time scale
for molecular hydrogen formation may be long compared to the cloud
lifetime (cf. Shu 1973, Kiguchi et al. 1974, Allen and Robinson 1976).
Knaap (1974) has observed substantial column densities of atomic
hydrogen in many dark clouds which are otherwise difficult to
explain. The formation of molecular hydrogen may also cause pressure
instabilities leading to fragmentation of the interstellar clouds
and to the formation of protostars (Reddish 1975).

## References

Allen, M. and Robinson, G. W.: 1976, Astrophys. J. 207, 745.
Barlow, M. J. and Silk, J.: 1976, Astrophys. J. 207, 131.
Barsuhn, J. and Walmsley, C. M.: 1977, Astron. Astrophys. in press.
Black, J. H.: 1977, private communication.
Black, J. H. and Dalgarno, A.: 1973, Astrophys. J. Letters 184, L101.
Black, J. H. and Dalgarno, A.: 1976, Astrophys. J. 203, 132.
Black, J. H. and Dalgarno, A.: 1977, Astrophys. J. Suppl., in press.
Black, J. H., Dalgarno, A. and Hartquist, T.: 1977, in preparation.
Castor, J., McCray R. and Weaver, R.:1975, Astrophys. J. Letters
       200, L107.
Chaffee, F. and Lutz, B. L.: 1977, Astrophys. J. in press.
Crutcher, R. M. and Watson, W. D.: 1975, Astrophys. J. Letters 203,
       L123.
Dalgarno, A. and McCray, R.:1972, Ann. Rev. Astron. Astrophys. 10, 375.
Dalgarno, A. and McCray, R.: 1973 Astrophys. J. 181, 95.
Dalgarno, A., Black, J. H. and Weisheit, J. C.: 1973, Astrophys.
       Letters 14, 77.
Dalgarno, A. and Wright, E. L.: 1972, Astrophys. J. Letters 174, L49.
Gautier, N., Fink, U., Treffers, R. R. and Larson H. P.: 1976,
       Astrophys. J. Letters 207, L129.
Herbig, G.: 1968, Zs. f. Astrophys. 68, 243.
Hill, J. K. and Silk, J.: 1975, Astrophys. J. Letters 202, L97.
Hill, J. K. and Hollenbach, D. J.: 1976, Astrophys. J. 209, 445.
Hollenbach, D., Chu, S.-I. and McCray, R.: 1976, Astrophys. J.
       208, 458.
Hollenbach, D. J., and Salpeter, E.E.: 1971, Astrophys. J. 163, 155.
Hollenbach, D. J., and Shull, M. J.: 1977, preprint.
Hollenbach, D. J., Werner, M. W. and Salpeter, E. E.: 1971 Astrophys.
       J. 163, 165.
Jura, M.: 1975a, Astrophys. J. 197, 575.
Jura, M.: 1975b, Astrophys. J. 197, 581.
Kiguchi, M., Suzuki, H., Sato, K., Miki, M., Tomamitsu, A. and
       Nakagawa, Y.: 1974, Publ. Astron. Soc. Japan 26, 499.

Knaap, G. R.: 1974, Astron. J. 79, 527 and 541.

Kwan, J. and Scoville, N.: 1976, Astrophys. J. Letters 210, L97.

McDowell, M. R. C.: 1961, Observatory 81, 240.

Milgrom, M., Panagia, N. and Salpeter, E. E.: 1973, Astrophys. Letters 14, 73.

O'Donnell, E.J. and Watson, W. D.: 1974, Astrophys. J. 191, 89.

Oppenheimer, M.: 1977, preprint.

Oppenheimer, M. and Dalgarno, A.: 1975, Astrophys. J. 200, 419.

Reddish, V. C.: 1975, Monthly Notices Roy. Astron. Soc. 170, 261.

Scoville, N. Z. and Kwan, J.: 1976, Astrophys. J. 206, 718.

Shu, F. H.: 1973, "Interstellar Dust and Related Topics," J. M. Greenberg and H. D. van de Hulst, eds. (Dordecht: Reidel).

Snow, T. P.: 1976, Astrophys. J. Letters 204 L127.

Spitzer, L., and Cochran, W. D.: 1973, Astrophys. J.Letters 186, L23.

Spitzer, L. Cochran, W. D. and Hirshfeld, A.: 1974, Astrophys. J. Suppl. 28, 373.

Spitzer, L., Drake, J. F., Jenkins, E. B., Morton, D. C., Rogerson, J. B. and York, D. G.: 1973, Astrophys. J. Letters 181, L116.

Spitzer, L. and Jenkins, E. B.: 1957, Ann. Rev. Astron. and Astrophys. 13, 133.

Spitzer, L. and Morton, W.: 1976, Astrophys. J. 204, 731.

Spitzer, L. and Zweibel, E. G.: 1974, Astrophys. J. Letters 191, L127.

Stephens, T. L. and Dalgarno, A.: 1973, Astrophys. J. 186, 165.

Treffers, R., Fink, U., Larson, H. P. and Gautier, T. N.: 1976, Astrophys. J. 209, 793.

Watson, W. D.: 1973, Astrophys. J. Letters 182, L73.

York, D. G. and Rogerson, J. B.: 1976, Astrophys. J. 203, 378.

# PROGRESS IN INTERSTELLAR MOLECULE FORMATION

William D. Watson
Departments of Physics and Astronomy
University of Illinois

ABSTRACT

Selected observational, laboratory and theoretical results from the past few years which are significant for delineating interstellar molecule reactions are presented. Molecular species for which reaction mechanisms are considered include OH, NH, CH, and $CH^+$ in diffuse interstellar clouds, and $HCO^+$, $N_2H^+$, HCN, HNC, and $H_2CO$ in dense clouds.

## I.  INTRODUCTION

In the brief report here, I will not attempt to give a comprehensive review of interstellar molecular formation.  Instead I will describe selected new observational, laboratory and theoretical results from the past few years that seem to be especially significant for delineating reaction mechanisms for molecules in the interstellar medium (ISM).  Most research that has quantitative applicability has been oriented toward gas phase reactions and this will be the emphasis here.  More extensive, recent reviews by Dalgarno and Black (1976), Herbst and Klemperer (1976), and Watson (1976) are available.  Formation and destruction of $H_2$ and HD will not be considered explicitly as this topic is reviewed by Professor Dalgarno in the foregoing discussion.

In order to place the recent progress in context, I will first recount some of the history of the topic.  The first comprehensive study of interstellar molecule formation seems to be that of Bates and Spitzer (1951) directed toward understanding the abundances of CH and $CH^+$. These molecules, along with CN, had been detected in the ISM about 1940.  In the early 1950's, the ISM was thought to be mainly in atomic form so that the Bates and Spitzer investigation was oriented toward processes that form these molecules directly from elements in atomic form.  These authors recognized the difficulties in reproducing the observed

*Hugo van Woerden (ed.), Topics in Interstellar Matter, 135-147. All Rights Reserved.*
*Copyright © 1977 by D. Reidel Publishing Company, Dordrecht-Holland.*

abundances of CH and CH$^+$ directly from the atoms in the gas if the uncertain cross sections have their most likely values.  In particular,
they pointed out the key role of CH$^+$ + e $\rightarrow$ C + H and the severe difficulties if the rate coefficient for this reaction is typical for such
processes ($\sim$10$^{-7}$cm$^3$s$^{-1}$).  This remains the major problem in understanding
the abundance of CH$^+$, and is one of the outstanding questions in molecular astrophysics.

Beginning in about 1960 and culminating with the studies of
Hollenbach and Salpeter (1971) [also, Hollenbach et al. 1971], the investigation of the formation of H$_2$ represents another phase in the
study of molecule formation in the ISM.  Under normal conditions in the
ISM, H$_2$ cannot be produced in appreciable quantities except by reactions
on grain surfaces.  Due to the weak binding of an H-atom to an inert
surface such as ice, which seems most likely to cover grains in the
ISM, it was unclear whether H-atoms from the gas will stick to the
grains and remain on the surface until enough other atoms stick to make
reaction probable.  The detailed analysis of Hollenbach and Salpeter
(1971) answered the question in the affirmative concluding that essentially every collision of an H-atom with a grain leads to an H$_2$
molecule.  Analysis of Copernicus satellite data verifies this prediction (O'Donnell and Watson 1974; Jura 1975).

Finally, in response to the rapid increase in the detection of
molecular species that began in 1968, reinvestigations of gas phase
and surface reactions of molecules were undertaken.  As a result reactions initiated by positive ions have been found to play a major, and
probably dominant, role in the chemistry of small, interstellar molecules other than H$_2$ (Herbst and Klemperer 1973; Watson 1973, 1974;
Dalgarno et al. 1973).  Unlike reactions between neutral particles,
reactions of positive ions with molecules normally have no activation
energy barrier that will retard their rates at interstellar temperatures.  For small ions and molecules, these rate coefficients almost
always are temperature independent so that measurements at laboratory
temperatures can be utilized directly.  All the "measured" rate coefficients for ion-molecule reactions that will be employed here are
obtained at 300$^{\circ}$ K.  It seems that certain molecules observed in the
ISM (e.g., HD, HCO$^+$, DCO$^+$, N$_2$H$^+$, HNC, and CCH) can only be produced in
adequate quantities by gas phase, as opposed to surface, reactions.
The key to achieving sufficient rates of ionization is the efficient
transfer of the cosmic ray ionization from hydrogen and helium to less
abundant elements.  By this mechanism, the effective rate for the
ionization of N, O, etc. atoms can be increased by a factor of 100 to
1000 over that for direct ionization of the atom by cosmic rays.  The
cosmic ray intensity normally employed as a basis for studies of interstellar reactions is that of the observed, high energy ($\gtrsim$ 100 MeV/nucleon)
cosmic ray nuclei in the neighborhood of the sun.  A minor uncertainty
is the variation of cosmic ray intensity with location in the galaxy,
and a more serious uncertainty is the degree to which the interiors of
dense gas clouds are screened from cosmic rays by magnetic fields.  In
a dense interstellar cloud, an appreciable fraction (perhaps $\simeq$ 1) of the

ionizations of the gas by cosmic rays lead directly to the molecular ion $HCO^+$. Recent observational and laboratory investigations that have conclusively identified the $J = 1-0$ microwave transition with one of the stronger interstellar emission lines have provided the strongest support for the importance of positive ion-molecule reactions for the chemistry of small molecules in dense clouds of the ISM.

In Section II, I give the assumed physical conditions for the two extreme type interstellar clouds, designated as the diffuse and dense clouds. Intermediate parameters are more complicated so that consideration of these extreme cases, for which several simplifications can be made, allow one to focus on the basic molecular reactions. The selected results of primary interest for the diffuse and dense cloud conditions are presented separately in Sections III and IV. A brief summary is provided in Section V.

## II. PHYSICAL CONDITIONS IN "DIFFUSE" AND "DENSE" CLOUDS

From the viewpoint here, the essential difference between the "diffuse" and "dense" cloud conditions is the presence of relatively unattenuated galactic starlight at ultraviolet wavelengths (912 to 2000 A) in the former. In the dense clouds, ultraviolet radiation is assumed to be negligible as suggested by the large optical depths at visual wavelengths and the high particle densities. In diffuse clouds, the ultraviolet radiation causes the gas, except for $H_2$, to be mainly in atomic form and for atoms with ionization potential below 13.6 eV to be fully ionized. This ionization provides the chief contribution to the electron density. Abundances of the heavy elements of interest here (C, N, O) are not drastically (less than a factor of ten) different from cosmic values. Number densities $n_0$ are roughly $10-10^3 cm^{-3}$ and gas kinetic temperatures are $50 \lesssim T(K) \lesssim 100$. The classic example of a diffuse cloud is that toward Zeta Ophiuchi. In dense clouds, the gas is assumed to be mainly molecular with the low electron density ($\lesssim 10^{-7}$) presumably provided by the cosmic rays. Element abundances are quite uncertain due to the possibility for depletion onto the cold ($\lesssim 20 K$) dust grains. In fact, a serious question is what prevents elements heavier than helium from being completely frozen out of the gas. The time to stick to a grain is always shorter than the cloud lifetime at high gas densities ($n_0 \gtrsim 10^3 cm^{-3}$). A reasonable possibility is that an appreciable fraction of the C, N, and O are converted into CO and $N_2$ early in the life of the cloud. These have high vapor pressures and probably do not freeze at the temperatures of the grains. The other, less abundant molecules are then produced from the free atoms that are released by the slow dissociation of CO and $H_2$ (see Section IV). Number densities and gas kinetic temperatures in dense clouds are roughly $10^4 \lesssim n_0 (cm^{-3}) \lesssim 10^6$ and $10 \lesssim T(K) \lesssim 70$. Examples are the dense, neutral gas clouds in Orion and in Sgr B2.

I will also consider molecular abundances to be locally steady-state. This is certainly valid in the diffuse clouds where the time to reach steady-state is short and the time scale for disturbing events (e.g., gravitational collapse, cloud-cloud collisions) is long ($\gtrsim 10^7$ yrs.). For dense clouds, the accuracy of the steady-state assumption is less clear because the lifetime of some molecular species against rearrangement by reactions are comparable with possible lifetimes for the clouds. Langer and Glassgold (1976) have quantitatively examined the time-scales for chemical reactions to reach steady-state. Unless dense clouds are collapsing at near their free-fall rate, which seems unlikely, the steady-state approximation is probably adequate for the abundances of the dominant molecular species.

III.   SELECTED REACTIONS IN DIFFUSE CLOUDS

(A)   CH, $CH^+$

As noted in Section I, understanding the abundances of these molecules has been a problem since the earliest investigations of chemical reactions in the ISM. Of all interstellar molecules, their formation processes have been the most extensively studied. At present there seems to be no serious problem in finding formation processes that can produce enough CH. The difficulty is with $CH^+$ because of its probable rapid destruction either by,

$$CH^+ + e \rightarrow C + H \tag{1}$$

or more certainly in regions with appreciable $H_2$ by,

$$CH^+ + H_2 \rightarrow CH_2^+ + H \tag{2}$$

The rate coefficient k (=⟨cross section times velocity⟩) for reaction (2) has been measured in the laboratory to be rapid ($k_2 = 1 \times 10^{-9} cm^3 s^{-1}$; Kim et al. 1974). Reaction (2) thus excludes $CH^+$ from being present in appreciable abundance in the interiors of clouds with large abundances of $H_2$. Attention in the $CH^+$ problem has focused primarily on whether the dissociative electron recombination of equation (1) has a rate coefficient similar to that for almost all molecules for which data exist ($k \gtrsim 10^{-7} cm^3 s^{-1}$), If its value were smaller by a factor $\sim 10^2$, there would be no serious problem in understanding the $CH^+$ abundance. Some calculations indicate a value $k_1 \simeq 10^{-7} cm^3 s^{-1}$ (Bardsley and Junker 1973; Krauss and Julienne 1973), though a more recent investigation of the excited states of CH suggests a much smaller value for $k_1$ (Giusti-Suzor and Lefebvre-Brion 1977).

The formation in the gas phase of $CH^+$ is a result in part of,

$$C^+ + H \rightarrow CH^+ + h\nu \tag{3}$$

and of reactions initiated by,

$$C^+ + H_2 \rightarrow CH_2^+ + h\nu \ . \tag{4}$$

Photodissociation and photoionization of CH, $CH_2$, $CH_3$ and $CH_4$ produced on grain surfaces will also contribute to the $CH^+$ abundance. None of these seem to be adequate, however, if $k_1 \simeq 10^{-7} cm^3 s^{-1}$. The most recent calculations for the rate of reaction (3) find $k_3 \simeq 1.3 \times 10^{-17} cm^3 s^{-1}$ at 100 K (Giusti-Suzor et al. 1976). A rate coefficient $k_4 \simeq 10^{-14} cm^3 s^{-1}$ may lead to enough $CH^+$ if the astrophysical conditions and uncertain reaction rates are favorable (Black et al. 1975) through reaction (4) followed by,

$$CH_2^+ + H_2 \rightarrow CH_3^+ + H \tag{5}$$

$$\begin{aligned} CH_3^+ + h\nu &\rightarrow CH_2^+ + H \\ &\rightarrow CH^+ + H_2 \end{aligned} \tag{6}$$

$$\begin{aligned} CH_3^+ + e &\rightarrow CH_2 + H \\ &\rightarrow CH + H_2 \end{aligned} \tag{7}$$

$$\begin{aligned} CH_2 + h\nu &\rightarrow CH_2^+ + e \\ &\rightarrow CH + H \end{aligned} \tag{8}$$

$$CH_2^+ + h\nu \rightarrow CH^+ + H \tag{9A}$$

$$CH_2^+ + e \rightarrow CH + H \tag{9B}$$

$$CH + h\nu \rightarrow CH^+ + e \tag{10}$$

A theoretical investigation (Herbst, Schubert and Certain 1977; see also Pearson and Roueff 1976) does find that $k_4$ should be significantly larger than $k_3$, but is not sufficient to determine whether $k_4$ is as large as the required value. Molecular structure calculations (Blint, Marshall and Watson 1976) indicate that the photodissociation rate for $CH_3^+$ is negligible in diffuse clouds in comparison with reaction (6). Even for the appropriate value of $k_4$, the reactions beginning with equation (5) seem unable to produce enough $CH^+$ without producing excessive CH in the diffuse clouds.

Because of the impasse that seems to have been reached in the $CH^+$ problem, at least before the very recent calculations relevant to $k_1$, the idea of production through dissociation and ionization of methane evaporated from grains in the intense radiation field of a nearby, approaching star (Bates and Spitzer 1951) has begun to be reconsidered.

An extensive review of the CH, $CH^+$ problem has recently become available (Dalgarno 1976).

(B)  OH, $H_2O$

For a number of years radio observations of OH indicated that it
should be present in diffuse clouds with a fractional abundance
$n(OH)/n_0 = 10^{-7} - 10^{-8}$. However, upper limits for the ultraviolet lines
near 1220 A obtained with the Copernicus satellite, along with an estimated upper limit, yielded an upper limit $\sim 10^{-9}$. Radio and optical
line studies are now in harmony as a result of a calculation for the
oscillator strength relevant for the ultraviolet line (Ray and Kelly
1975) and a detection of the OH line at 3058 A (Crutcher and Watson
1976a) for which the oscillator strength is known from laboratory
measurements. More sensitive observations have also detected the OH
lines near 1220 A (Snow 1976).

Detection of OH in diffuse clouds at near the observed abundance is
especially significant because it is expected to be produced both in
reactions on grains and as a result of the charge exchange,

$$H^+ + O \rightarrow O^+ + H \tag{11}$$

for which the rate coefficient at interstellar temperatures is uncertain.
Reaction (11) is a first step toward incorporating oxygen into molecules
after which,

$$O^+ + H_2 \rightarrow OH^+ + H \tag{12}$$

$$OH^+ + H_2 \rightarrow OH_2^+ + H \tag{13}$$

$$OH_2^+ + H_2 \rightarrow OH_3^+ + H \tag{14}$$

$$OH_3^+ + e \rightarrow OH + H_2 \tag{15}$$

$$\rightarrow H_2O + H \tag{16}$$

$$H_2O + h\nu \rightarrow OH + H \tag{17}$$

produce OH and $H_2O$. The abundance of HD in a cloud with OH provides
data on the $H^+$ abundance and the intensity of ultraviolet starlight
which are needed to predict the OH abundance for a specified value $k_{11}$.
Observed abundances and predictions are in agreement for likely values
of $k_{11}$, though the comparison is not sufficiently close to determine
whether formation of OH is primarily a result of gas phase or surface
reactions (see Crutcher and Watson 1976a).

Although the branching ratios for neither the surface reactions nor
equations (15) and (16) are known, comparable production rates for OH
and $H_2O$ seem likely for either. Destruction of $H_2O$ might reasonably
be a factor of ten more rapid than for OH when photodissociation dominates. Nevertheless, it is somewhat surprising that $H_2O/OH \lesssim 1/50$ as
deduced by Smith and Zweibel (1976) from the observation of Snow (1975).
There is some uncertainty about the oscillator strength for the $H_2O$
transition under examination.

(C)   NH

The absence of NH at a detectable level is significant for the rate
of surface reactions in the ISM.  Within the uncertainties, both CH and
OH could equally well be produced by gas phase or surface reactions.
In contrast, it seems to be possible to produce NH in comparable quantities only through surface reactions.  Of course, the catalytic properties
of grain surfaces are uncertain.  It seems highly probable, however,
that the cold grains in diffuse clouds are covered with at least a thin
mantle ($\gtrsim$ 5 monolayers) of ices and that atoms are only weakly bound to
this inert surface (Watson and Salpeter 1972).  Most likely the grains
then serve simply as a location where H-atoms are attached to the heavy
atoms (C, N, O, etc.) to produce CH, NH, OH, $CH_4$, $NH_3$, $H_2O$, etc. in
proportion to the abundances of the heavy atoms in the gas.  The chief
problem for understanding whether surface reactions play a major role
other than for the formation of $H_2$ has been determining if a mechanism
exists to return efficiently the products to the gas.  If so, it is
still uncertain whether the heavy atoms tend to be returned after a
single H-attachment as OH, NH, etc. or in the chemically saturated
forms $H_2O$, $NH_3$, etc.  In any case, we expect,

$$[NH]/[OH] \simeq [N]/[O]\big|_{gas} \stackrel{\sim}{>} [N]/[O]\big|_{cosmic} \simeq 1/7 \tag{18}$$

where the inequality is suggested by the higher vapor pressure of
ammonia in comparison with water.  Any significant alteration of the
[N]/[O] ratio in the gas from the "cosmic" abundance value is thought
to be due to freezing onto grains.

In contrast the observed limit is (Crutcher and Watson 1976b),

$$[NH]/[OH] \leq 1/100 \ . \tag{19}$$

Equation (18) is only a rough approximation and must be refined by considering corrections due to the possible differences in photodissociation
rates for OH and NH, as well as in the processes that convert $H_2O$ and
$NH_3$ into OH and NH if the chemically saturated form is mainly ejected
from grains.  When this analysis is performed, it still seems necessary
that (Crutcher and Watson 1976b),

$$[NH]/[OH] \geq 1/25 \tag{20}$$

if surface reactions dominate and have the characteristics suggested
above.  The conclusion is that OH production by gas phase reactions is
indicated for diffuse clouds.


IV.   SELECTED REACTIONS IN DENSE CLOUDS

(A)   $HCO^+$, $N_2H^+$

As noted in Section I, the initial investigations (Herbst and
Klemperer 1973; Watson 1973; 1974) of ion-molecule reactions in the ISM

noted the position of HCO$^+$ as a cornerstone of the proposed reaction
mechanisms.  Due to the abundance of hydrogen, most cosmic ray ionizations in a dense cloud are,

$$\text{cosmic ray} + H_2 \rightarrow \text{cosmic ray} + H_2^+ + e \; . \tag{21}$$

Reaction (21) is followed immediately by,

$$H_2^+ + H_2 \rightarrow H_3^+ + H \; , \tag{22}$$

after which the H$_3^+$ is destroyed mainly by,

$$H_3^+ + e \rightarrow H_2 + H, \; 3H, \tag{23}$$

$$H_3^+ + CO \rightarrow HCO^+ + H_2, \tag{24}$$

and,

$$H_3^+ + N_2 \rightarrow N_2H^+ + H_2. \tag{25}$$

Although the computed electron density and rate coefficient $k_{23}$ are
somewhat uncertain, the most likely values of these and the CO abundance suggested from observation are such that reaction (24) is expected to be competitive with reaction (23), as well as with reaction
(25).  Destruction of HCO$^+$ is mainly a result of,

$$HCO^+ + e \rightarrow H + CO \; . \tag{26}$$

From the above reactions, the abundance of HCO$^+$ was predicted with
moderate confidence to be sufficient for detection if the high energy
cosmic rays penetrate dense clouds (ionization rate $\simeq 10^{-17}$ per $H_2$ molecule sec$^{-1}$) and ion-molecule chemistry proceeds as expected.  At the
time of this prediction (1973), one of the strongest molecular lines
from dense clouds was the unidentified "X-ogen" line occurring within
the frequency neighborhood where the HCO$^+$ ($J = 1-0$) emission would be
expected.  For the past few years, the key test for the proposed ion-
molecule scheme has been the presence or absence of the HCO$^+$ line from
dense clouds.  This has now been resolved with the unambiguous identification of the "X-ogen" line with HCO$^+$ as a result of (1) observation
of another line with the appropriate strength at the frequency predicted for H$^{13}$CO$^+$ (Snyder et al. 1976) and subsequently (2) an accurate
laboratory measurement for the frequency of the H$^{12}$CO$^+$ transition
(Woods et al. 1975).  The high abundance of DCO$^+$ provides further
evidence that the HCO$^+$ actually occurs in the low temperature gas as
predicted by ion-molecule chemistry.  Low temperatures ($\lesssim 20$ K) are required to produce the necessary fractionation.  Although tests of the
proposed reactions did not initially focus upon N$_2$H$^+$ due to lack of
knowledge about the abundance of N$_2$, its identification at a comparable
intensity (Turner 1974; Green et al. 1974) is additional support because
of the parallel rates of N$_2$H$^+$ and HCO$^+$ if $[N_2]/[CO] \simeq 1$.

At higher gas densities, the rates of $HCO^+$ and $N_2H^+$ are not quite the same. Destruction of $N_2H^+$, but not $HCO^+$, can then occur through additional processes,

$$N_2H^+ + CO \rightarrow HCO^+ + N_2 \tag{27A}$$

and,

$$N_2H^+ + Y \rightarrow YH^+ + N_2 \tag{27B}$$

where Y is a molecule with a proton affinity greater than that of $N_2$ but less than CO. In addition, both will be destroyed by reactions with molecules X which have proton affinities greater than CO,

$$HCO^+ + X \rightarrow XH^+ + CO \tag{28A}$$

$$N_2H^+ + X \rightarrow XH^+ + N_2 \ . \tag{28B}$$

Observationally, the $[N_2H^+]/[HCO^+]$ is found to decrease in going from the outer to the inner, more dense part of the best studied cloud (the Orion molecular cloud). This is just what is expected if the abundance of CO is greater than that of molecules of type X as is likely. These observations are considered to be further support for the importance of ion-molecule reactions (Snyder et al. 1977). The requirement that the abundance of [CO] plus [Y] be greater than that of type X can be utilized to find a probable upper limit for the abundance ratio,

$$1/6 \geq [X]/[CO] \geq [H_2O]/[CO] \ . \tag{29}$$

A limit on the abundance of water, which is otherwise unobservable, is thereby obtained.


(B) HCN, HNC, $NH_3$

Another of the stronger emission lines due to molecules at micro-wave frequencies is that of HCN. At first glance, the formation of HCN in the ion-molecule scheme would appear to be clear, as recognized in the initial investigations. A basic aspect of the ion-molecule reactions is that the helium which is ionized by the cosmic rays reacts mainly with CO, $N_2$ and other "heavy" molecules, e.g.,

$$He^+ + CO \rightarrow C^+ + O + He \tag{30}$$

to liberate reactive atoms and ions. Ammonia is observed to be present in appreciable quantities ($[NH_3]/n_o \simeq 10^{-6}$) so that the reactions,

$$C^+ + NH_3 \rightarrow H_2CN^+ + H \tag{31}$$

$$H_2CN^+ + e \rightarrow HCN + H \tag{32}$$

should efficiently produce HCN. Fortunately, there exists a clean test of whether this mechanism dominates that is based on the fact that ionized carbon also destroys HCN,

$$C^+ + HCN \rightarrow C_2N^+ + H \ . \tag{33}$$

Hence,

$$[HCN]/[NH_3] < k_{31}/k_{33} \simeq 0.6 \tag{34}$$

where $k_{31}$ and $k_{33}$ have been measured in the laboratory (Huntress and Anicich 1976). Recent observational studies (Cheung, 1976) find that $[HCN]/[NH_3] \gtrsim 0.5$ so that the requirement (34) is only barely satisfied. There are other, less direct formation mechanisms for HCN (Watson 1976).

If $H_2CN^+$, which has the structure $HCNH^+$, is an intermediate in the formation of HCN, then the isomer HNC is likely to be produced at near the same rate as HCN. This may be true for other gas phase formation mechanisms, but will almost certainly not be the case for formation on grain surfaces. Thus the identification, as a result of laboratory measurements (Saykally et al. 1976; Cresswell et al. 1976; Blackman et al. 1976), of another strong emission line with the $J = 1-0$ transition of HNC supports gas phase formation mechanisms.

(C)  $H_2CO$

Formaldehyde has received special attention as the most complex molecule for which specific processes have been proposed for its formation. It thus becomes a test case for whether gas phase formation mechanisms that seem successful for the small molecules can be extended to the more complex species. Formaldehyde is ubiquitous; it is detected in both the dense and diffuse gas with radio telescopes and seems to have a relatively constant abundance $[H_2CO]/n_o \simeq 10^{-9}$. The discussion has centered around the following proposals.

$$i. \quad HCO^+ + H_2 \rightarrow H_3CO^+ + h\nu \ . \tag{35}$$

Then,

$$H_3CO^+ + e \rightarrow H_2CO + H \tag{36}$$

(Herbst and Klemperer 1973), though laboratory investigations are not encouraging due to the likelihood of small value for $k_{35}$ and a product distribution in equation (36) that favors $HCO + H_2$ or $CO + H + H_2$ (Fehsenfeld et al. 1974).

$$ii. \quad CH_3^+ + O \rightarrow H_2CO^+ + H \ . \tag{37}$$

Then,

$$H_2CO^+ + M \rightarrow H_2CO + M^+ \tag{38}$$

or,

$$H_2CO^+ + H_2 \rightarrow H_3CO^+ + H \tag{39}$$

followed by reaction (36). In equation (38), M is an atom with low ionization potential which can charge-exchange with $H_2CO^+$. Laboratory experiments now find that most of the reactions (37) lead to $HCO^+ + H_2$ (Fehsenfeld 1976) instead of those given in the above and that reaction (39) seems to proceed too slowly to be important (Huntress 1976).

$$\text{iii. } CH_3 + O \rightarrow H_2CO + H \tag{40}$$

Reaction (40) has been measured in the laboratory and proceeds rapidly at the low interstellar temperatures. The main problems with this proposal are producing adequate $CH_3$ by gas phase reactions and maintaining a high abundance of atomic oxygen in dense clouds. Proposals (ii) and (iii) are due to Dalgarno et al. (1973) and have been elaborated upon by Watson et al. (1975). The latter authors present arguments that the absence of HDCO at a detectable level is evidence against both proposals (ii) and (iii). With the recently observed ratio $[DCO^+]/[HCO^+] \simeq 1$, the absence of HDCO is now evidence against proposal (i) as well.

According to Langer (1976), the relative abundance of $H_2CO$ will be increased to the observed level under certain physical conditions if the gas cloud is collapsing at near its free-fall rate. The sensitivity of this conclusion to the assumed cross sections, role of grains, etc. is unclear. Of course, there exists the more basic question -- whether all clouds are collapsing at near the free-fall rate. In addition, formaldehyde apparently is observed in quite diffuse gas (Davies and Matthews 1972) where deviations from steady-state should normally be unimportant. Nevertheless, consideration of time-dependent effects represents a fresh approach.

V.  SUMMARY

The examples presented here indicate how investigations of gas phase reactions in the past three years have made possible quantitative analyses of molecule abundances in the ISM. Identification of interstellar $HCO^+$ has provided the strongest support for the general validity of this approach. Despite the considerable progress in the formation of small molecules, there has been little if any progress in the formation of molecules larger than $H_2CO$ through gas phase reactions. One possibility that has been examined is additional radiative association reactions of a molecular ion with $H_2$ (Herbst 1976). Surface reactions may be necessary, in which case it will be difficult to obtain a quantitative description of the molecule formation.

Fractionation of isotopes (D/H, $^{13}C/^{12}C$) in interstellar molecules is being discussed by the author in a parallel report (Watson 1977).

The author's research is supported in part by NSF Grant MPS 73-04781 and by an A. P. Sloan Fellowship for Basic Research.

REFERENCES

Bardsley, J.N. and Junker, B.R.: 1973, Astrophys. J. Letters 183, L135.

Bates, D.R. and Spitzer, L.: 1951, Astrophys. J. 113, 441.

Black, J., Dalgarno, A. and Oppenheimer, M.: 1975, Astrophys. J. 199, 633.

Blackman, G.L., Brown, R.D., Godfrey, P.D. and Gunn, H.I.: 1976, Nature 261, 395.

Blint, R.M., Marshall, R.F. and Watson, W.D.: 1976, Astrophys. J. 206, 627.

Cheung, A.C.: 1976, personal communication.

Cresswell, R.A., Pearson, E.F., Winnewisser, M. and Winnewisser, G.: 1976, Z. Naturforsch 31a, 221.

Crutcher, R.M. and Watson, W.D.: 1976a, **Astrophys**. J. Letters 203, L123.

Crutcher, R.M. and Watson, W.D.: 1976b, Astrophys. J. 209, 778.

Dalgarno, A.: 1976, Adv. Atomic Molecular Phys., in press.

Dalgarno, A. and Black, J.: 1976, Rep. Prog. Phys. 39, 573.

Dalgarno, A., Oppenheimer, M. and Black, J.: 1973, Nature Phys. Sci. 245, 100.

Davies, R.D. and Matthews, H.E.: 1972, Mon. Not. Roy. Astron. Soc. 156, 253.

Fehsenfeld, F.C.: 1976, Astrophys. J. 209, 638.

Fehsenfeld, F.C., Dunkin, D.B. and Ferguson, E.E.: 1974, Astrophys. J. 188, 43.

Giusti-Suzor, A. and Lefebvre-Brion, H.: 1977, "Comments on the Dissociative Recombination of $CH^+$ Ions", to be published.

Guisti-Suzor, A., Roueff, E. and van Regemorter, H.: 1976, J. Phys. B9, 1021.

Green, S., Montgomery, J.A. and Thaddeus, P.: 1974, Astrophys. J. Letters 193, L89.

Herbst, E.: 1976, Astrophys. J. 205, 94.

Herbst, E. and Klemperer, W.: 1973, Astrophys. J. 185, 505.

Herbst, E. and Klemperer, W.: 1976, Phys. Today 29, 32.

Herbst, E., Schubert, J.G. and Certain, P.: 1977, "The Radiative Association of $CH_2^+$", to be published.

Hollenbach, D. and Salpeter, E.E.: 1971, Astrophys. J. 163, 155.

Hollenbach, D., Werner, M. and Salpeter, E.E.: 1971, Astrophys. J. 163, 165.

Huntress, W.T.: 1976, personal communication.

Huntress, W.T. and Anicich, V.G.: 1976, Astrophys. J. 208, 237.

Jura, M.: 1975, Astrophys. J. 197, 575.

Kim, J.K., Theard, L.P. and Huntress, W.T.: 1974, J. Chem. Phys. 62, 45.

Krauss, M. and Julienne, P.: 1973, Astrophys. J. Letters 183, L139.

Langer, W.D.: 1976, Astrophys. J. 210, 328.

Langer, W.D. and Glassbold, A.E.: 1976, Astron. Astrophys. 48, 395.

O'Donnell, E.J. and Watson, W.D.: 1974, Astrophys. J. 191, 89.

Pearson, P. and Roueff, E.: 1976, J. Chem. Phys. 64, 1240.

Ray, S. and Kelly, H. P.: 1975, Astrophys. J. Letters 202, L57.

Saykally, R.J., Szanto, P.G., Anderson, T.G. and Woods, R.C.: 1976, Astrophys. J. Letters 204, L143.

Smith, W.H. and Zweibel, E.G.: 1976, Astrophys. J. 207, 758.

Snow, T.P.: 1975, Astrophys. J. Letters 202, L87.

Snow, T.P.: 1976, Astrophys. J. Letters 204, L127.

Snyder, L.E., Hollis, J. M., Lovas, F.J. and Ulich, B. L.: 1976, Astrophys. J. 209, 67.

Snyder, L.E., Watson, W.D. and Hollis, M.: 1977, Astrophys. J., in press (February 15).

Turner, B.E.: 1974, Astrophys. J. Letters 193, L83.

Watson, W.D.: 1973, Astrophys. J. Letters 183, L17.

Watson, W.D.: 1974, Astrophys. J. 188, 35.

Watson, W.D.: 1976, Rev. Mod. Phys. 48, 513.

Watson, W.D.: 1977, lecture presented at I.A.U. General Assembly, Grenoble, Fr. (1966), to be published in CNO Isotopes in Astrophysics, ed. J. Audouze (D. Reidel: Dordrecht).

Watson, W.D., Crutcher, R.M. and Dickel, J.R.: 1975, Astrophys. J. 201, 102.

Watson, W.D. and Salpeter, E.E.: 1972, Astrophys. J. 174, 321.

Woods, R.C., Dixon, T.A., Saykally, R.J. and Szanto, P.G.: 1975, Phys. Rev. Letters 35, 1269.

# THE NATURE OF DUST GRAINS

P.G. Martin
David Dunlap Observatory and Scarborough College,
University of Toronto, Toronto, Ontario, Canada.

This review is concerned not so much with the nature of the interstellar dust particles as with the nature of the investigations into the dust properties which have been going on in the three years preceding this General Assembly. Because space is limited, a subjective view of some areas of particular significance is presented. The topics of formation and destruction, dust in HII regions and molecule formation have been omitted intentionally, as they are discussed elsewhere. Two relevant reviews which have appeared (Aannestad and Purcell, 1973; Wesson, 1974) may be consulted for further details and omitted subjects. In addition it would be useful to refer to the proceedings of three conferences, on polarization studies (Gehrels, 1974a), on the dusty universe (Field and Cameron, 1975) and on solid state astrophysics (Wickramasinghe and Morgan, 1976).

A fundamental question is how much dust there is for a given amount of interstellar gas. Colour excess continues as a measure of the amount of dust. New methods of detecting the amount of gas have been used, directly with Lyman-$\alpha$ absorption (Jenkins and Savage, 1974; Bohlin, 1975), and indirectly through soft X-ray absorption (Gorenstein, 1975; Ryter et al., 1975). The ratio of gas to dust found by these techniques, $\sim$100, is in agreement with the best 21-cm line determinations, those towards globular clusters (Knapp and Kerr, 1974). This ratio is not too low to violate cosmic abundance constraints but is small enough to make gas depletion considerations interesting.

Of importance for many purposes, particularly distance determinations, is the value of $R$, the ratio of total to selective extinction. The variable extinction method (Herbst, 1975) and the colour-difference method (Hackwell and Gehrz, 1974; Schultz and Wiemer, 1975) give consistent results, the average being about 3.3. Anomalies in localized regions are still being reported and disputed however (e.g. Moffat and Schmidt-Kaler, 1976). The value of $R$ in Orion is now believed to be normal (Penston et al., 1975).

Variations in dust properties from place to place are nevertheless

*Hugo van Woerden (ed.), Topics in Interstellar Matter, 149-154. All Rights Reserved.*
*Copyright © 1977 by D. Reidel Publishing Company, Dordrecht-Holland.*

evident from a number of observations.  This is perhaps encouraging, since extreme uniformity would be difficult to explain.  For example, there are correlated changes in the polarization parameters $\lambda_{max}$ (Coyne et al., 1974; Serkowski et al., 1975), and $\lambda_c$ (Martin 1975a; Martin and Angel, 1976), and in the shape of the extinction curve.  These might be interpreted as the result of small changes in the mean grain size.  The dark cloud near $\rho$Oph has been favoured for studies of how the grains might grow by accretion and heavy element depletion (Carrasco et al., 1973; Whittet and Van Breda, 1975).  In other dense regions where both 3.1$\mu$ (ice) and 10$\mu$ (silicate) absorption are measured (Glass and Feast, 1973; Gillett et al., 1975; Cohen, 1976; Merril et al., 1976) the relative optical depths vary, perhaps indicating different ice-mantle thicknesses.  In addition there is evidence for differences in the amount of ultraviolet relative to optical extinction (Snow and York, 1975; Viotti and Lamers, 1975).  Apparently there is a shift in the relative importance of normal and smaller-sized particles.

As far as the albedo and phase function, or $g$, are concerned there is not much new to add.  More sophisticated radiative transfer techniques have been applied to diffuse galactic light (Mathis, 1973) and reflection nebula models (Rush, 1975), but because of the interdependence of albedo and $g$, neither is determined any more precisely.  It has been suggested that $g$ might be found independently from the brightness profiles of dark nebulae (Witt and Stephens, 1974).  The albedo inferred indirectly from interstellar circular polarization (Martin, 1974; Martin and Angel, 1975a) would exceed 0.7.  The ultraviolet region, including the dip in albedo at 2200Å and the suggested rise to shorter wavelengths has received further attention (Lillie and Witt, 1976).  Further progress in this area is surely needed, since already there are diverse applications for the radiation field in dusty regions (Brown, 1973; Whitworth, 1975; Krügel, 1975; Sandell and Mattila, 1975; Leung, 1975).

Grain alignment poses a problem because magnetic (Davis-Greenstein) alignment in the existing conditions in the interstellar medium is not efficient enough unless the grains possess special properties such as ferromagnetism.  A novel proposal is the 'pinwheel theory' (Purcell, 1975) in which surface 'rockets' spin the grains up to high speeds, so that more complete alignment is achieved in a weak magnetic field. However details concerning the effects of shape and of surface changes on the 'rocket' action are yet to be worked out.

There is now direct observational evidence from 10$\mu$ polarization measurements in Orion (Dyck and Beichman, 1974) that (ordinary) silicate grains can be aligned.  An interesting polarization measurement could be made at the 3.1$\mu$ ice band in the same source, to seek evidence on whether the ice exists as a coating on the silicates.  In the ultraviolet, polarization observations would be valuable too.  The few that exist (Gehrels, 1974) could be interpreted as  showing that the particles responsible for the 2200Å feature are spherical or unaligned

(Martin, 1975b).

Spectral features as a means of determining the grain composition
have attracted considerable interest.  Optical constants have been derived from laboratory measurements of amorphous silicates, hydrated
silicates and meteoritic material (Day, 1974; Zaikowski et al., 1975;
Zaikowski and Knacke, 1975) for application at the 10μ feature.  It has
also been shown that the observed 10μ polarization spectrum implies an
intrinsic silicate band strength significantly lower than was previously thought (Martin, 1975c).  Further identification of silicates
using the detected 18μ (Simon and Dyck, 1975) and 33μ (Hagen et al.,
1975) features is somewhat complicated by the temperature dependence of
the middle-infrared absorptivity (Day, 1976).  Some 10μ region emission
peaks are in fact suggestive of carbon compounds (Thomas et al., 1976);
in this connection, measurements of the optical constants of other
materials, such as carbonates and sulphates, are important (Penman,
1976).  Polyoxymethylene has been studied too (Wickramasinghe, 1975;
Cooke, 1976).

In the ultraviolet the 2200Å feature is of major interest.  The
long-standing identification as graphite has been the topic of continued investigation.  Direct laboratory measurements (Day and Huffman,
1973), and computations based on a small-particle modification to the
refractive index (Wickramasinghe et al., 1974) are not in complete
agreement with the astronomical observations.

In the optical the diffuse bands are still being studied, an
identification continuing to be elusive.  It has been established that
the lines are well-correlated one with another (Herbig, 1975).  Contrary to earlier interesting results, it has been shown that λ4430Å
does not possess an apparent blue emission wing (Danks and Lambert,
1975), and that polarization structure at λλ4430, 5780 and 6284 is absent (Martin and Angel, 1975b; Fahlman and Walker, 1975).  Both of
these results imply that normal-sized grains are not the carriers of
the diffuse features, and so a new working hypothesis is needed.  It is
possible that smaller unaligned grains are involved; this would be consistent with the measured asymmetry of the λ5780 profile (Savage, 1976).
Laboratory work has ruled out $H^-$, $C^-$ and $O^-$ as alternative sources
(Herbst et al., 1974).  Very broadband structure in the extinction
curve has received further attention (Hayes and Rex, 1976).  Structure
in the wavelength dependence of linear polarization was reported
(Mavko et al., 1974) but has apparently not been confirmed; similar observations are certainly worth pursuing.

A new frontier in the study of dust is the 50μ-1mm region where
both maps and spectra of thermal emission from numerous sources are
being produced (e.g. Werner et al., 1975; Werner et al., 1976; Harper,
1974; Ward and Harwit, 1974).  The background radiation from dust in
the galactic plane is being considered too (Baschek et al., 1974).
Consequently theories and measurements of the far-infrared absorptivity
of different grain materials will be needed (Andriesse, 1974).

Temperature fluctuations in small grains, resulting from absorption of ultraviolet photons, have been discussed extensively (Purcell, 1976) because of their importance to the questions of molecule formation and mantle accretion.

Scattering theory for interstellar dust has been based primarily on calculations for spheres and infinite circular cylinders.  Two advances should be noted.  Electromagnetic scattering by spheroids (prolate or oblate) has been treated by a boundary-value technique (Asano and Yamamoto, 1975).  An n-point interacting-dipole approximation has been devised for use with almost arbitrary shapes (Purcell and Pennypacker, 1973; Shapiro, 1975).  Both methods however require considerably more computational expense.

Even from this brief look it is apparent that there is a large and varied effort being made to understand interstellar dust.  There has been progress concerning many details, but the overall picture is far from complete.  The problem might be stated as one of too many alternatives!  Consider the question of uniqueness which arises in explaining even the shape of the extinction curve, where many different models have been proposed.  Clearly what is needed is a model which is capable of explaining a wide variety of different types of observation at the same time.  However even the observational constraints on this problem are not fully determined.  Therefore continued work on all fronts, as we have witnessed in the past three years, will have to be encouraged and supported, perhaps for many years to come.

This work was supported by the National Research Council of Canada.

REFERENCES

Aannestad, P.A., and Purcell, E.M.:  1973, Ann. Rev. Astron. Astrophys. 11, 309
Andriesse, C.:  1974, Astron. Astrophys. 37, 257
Asano, S., and Yamamoto, G.:  1975, Appl. Optics 14, 29
Baschek, B., Traving, G., and Wehrse, R.:  1974, Astron. Astrophys. 36, 147
Bohlin, R.C.:  1975, Astrophys. J. 200, 402
Brown, R.L.:  1973, Astrophys. J. 184, 693
Carrasco, L., Strom, S.E., and Strom, K.M.:  1973, Astrophys. J. 182, 95
Cohen, M.:  1976, Astrophys. J. 203, 169
Cooke, A.:  1976, Astrophys. Space Sci. 39, L13
Coyne, G.V., Gehrels, T., and Serkowski, K.:  1974, Astron. J. 79, 581
Danks, A.C., and Lambert, D.L.:  1975, Astron. Astrophys. 41, 455
Day, K.L.:  1974, Astrophys. J. Lett. 192, L15
Day, K.L.:  1976, Astrophys. J. Lett. 203, L99
Day, K.L., and Huffman, D.R.:  1973, Nature Phys. Sci. 243, 50
Dyck, H.M., and Beichman, C.A.:  1974, Astrophys. J. 194, 57
Fahlman, G.G., and Walker, G.A.H.:  1975, Astrophys. J. 200, 22

Field, G.B., and Cameron, A.G.W.: 1975, The Dusty Universe, Watson
    Academic Publications, New York
Gehrels, T.: 1974a, Planets, Stars and Nebulae Studied with Photopolari-
    metry, University of Arizona Press, Tucson
Gehrels, T.: 1974b, Astron. J. 79, 590
Gillett, F.C., Jones, T.W., Merrill, K.M., and Stein, W.: 1975, Astron.
    Astrophys. 45, 77
Glass, I.S., and Feast, M.W.: 1973, Astrophys. Lett. 13, 81
Gorenstein, P.: 1975, Astrophys. J. 198, 95
Hackwell, J.A., and Gehrz, R.D.: 1974, Astrophys. J. 194, 49
Hagen, W., Simon, T., and Dyck, H.M.: 1975, Astrophys. J. Lett. 201,
    L81
Hayes, D.S., and Rex, K.H.: 1976, preprint
Herbig, G.H.: 1975, Astrophys. J. 196, 129
Herbst, E., Patterson, T.A., Norcross, D.W., and Lineberger, W.C.:
    1974, Astrophys. J. Lett. 191, L143
Herbst, W.: 1975, Astron. J. 80, 498
Jenkins, E.B., and Savage, B.D.: 1974, Astrophys. J. 187, 243
Knapp, G.R., and Kerr, F.J.: 1974, Astron. Astrophys. 35, 361
Krügel, E.: 1975, Astron. Astrophys. 38, 129
Leung, C.M.: 1975, Astrophys. J. 199, 340
Lillie, C.F., and Witt, A.N.: 1976, Astrophys. J. 208, 64
Martin, P.G.: 1974, Astrophys. J. 187, 461
Martin, P.G.:1975a, Astrophys. J. 201, 373
Martin, P.G.:1975b, Astrophys. J. 202, 389
Martin, P.G.:1975c, Astrophys. J. 202, 393
Martin, P.G., and Angel, J.R.P.: 1975a, Astrophys. J. 193, 343
Martin, P.G., and Angel, J.R.P.: 1975b, Astrophys. J. 193, 379
Martin, P.G., and Angel, J.R.P.: 1976, Astrophys. J. 207, 126
Mathis, J.S.: 1973, Astrophys. J. 186, 815
Mavko, G.E., Hayes, D.S., Greenberg, J.M., and Hiltner, W.A.: 1974,
    Astrophys. J. Lett. 187, L117
Merril, K.M., Russell, R.W., and Soifer, B.T.: 1976, Astrophys. J. 207,
    763
Moffat, A.F.J., and Schmidt-Kaler, Th.: 1976, Astron. Astrophys. 48, 115
Penman, J.M.: 1976, Monthly Notices Roy. Astron. Soc. 176, 539
Penston, M.V., Hunter, J.K., and O'Neill, A.: 1975, Monthly Notices
    Roy. Astron. Soc. 171, 219
Purcell, E.M.: 1975, in The Dusty Universe (ed. G.B. Field and A.G.W.
    Cameron), Watson Academic Publications, New York, p. 155
Purcell, E.M.: 1976, Astrophys. J. 206, 685
Purcell, E.M., and Pennypacker, C.R.: 1973, Astrophys. J. 186, 705
Rush, W.F.: 1975, Astron. J. 80, 37
Ryter, C., Cesarsky, C.J., and Audouze, J.: 1975, Astrophys. J. 198,
    103
Sandell, G., and Mattila, K.: 1975, Astron. Astrophys. 42, 357
Savage, B.D.: 1976, Astrophys. J. 205, 122
Schultz, G.V., and Wiemer, W.: 1975, Astron. Astrophys. 43, 133
Serkowski, K., Mathewson, D.S., and Ford, V.L.: 1975, Astrophys. J.
    196, 261

Shapiro, P.R.:   1975, Astrophys. J. 201, 151
Simon, T., and Dyck, H.M.:   1975, Nature 253, 101
Snow, T.P. Jr., and York, D.G.:   1975, Astrophys. Space Sci. 34, 19
Thomas, J.A., Robinson, G., and Hyland, A.R.:   1976, Monthly Notices
   Roy. Astron. Soc. 174, 711
Viotti, R., and Lamers, H.:   1975, Astron. Astrophys. 39, 465
Ward, D.B., and Harwit, M.:   1974, Nature 252, 27
Werner, M.W., Elias, J.H., Gezari, D.Y., Hauser, M.G., and Westbrook,
   W.E.:   1975, Astrophys. J. Lett. 199, L185
Werner, M.W., Gatley, I., Becklin, E.E., Harper, D.A., Lowenstein, R.F.,
   Telesco, C.M., and Thronson, H.A.:   1976, Astrophys. J. 204, 420
Wesson, P.S.:   1974, Space Sci. Rev. 15, 469
Whittet, D.C.B., and Van Breda, I.G.:   1975, Astrophys. Space Sci. 38,
   L3
Whitworth, A.P.:   1975, Astrophys. Space Sci. 34, 155
Wickramasinghe, N.C.:   1975, Monthly Notices Roy. Astron. Soc. 170, 11P
Wickramasinghe, N.C., Lukes, T., and Dempsey, M.J.:   1974, Astrophys.
   Space Sci. 30, 315
Wickramasinghe, N.C., and Morgan, D.J.:   1976, Solid State Astrophysics,
   D. Reidel Publ. Co., Dordrecht, Holland
Witt, A.N., and Stephens, T.C.:   1974, Astron. J. 79, 948
Zaikowski, A., and Knacke, R.F.:   1975, Astrophys. Space Sci. 37, 3
Zaikowski, A., Knacke, R.F., and Porco, C.C.:   1975, Astrophys. Space
   Sci. 35, 97

# FORMATION AND DESTRUCTION OF GRAINS

N.C. Wickramasinghe,
Department of Applied Mathematics and Astronomy,
University College, Cardiff, Wales, U.K.

ABSTRACT

Refractory grains, consisting of graphite, SiC, silicate and iron particles, may form in mass flows from cool stars, novae and possibly supernovae. Tarry polymeric mantles (typified by polyoxymethylene copolymers) grow under conditions which prevail in massive molecular clouds. A large fraction of polymer-coated grains could be expelled into the general interstellar medium, and such grains could be responsible for the bulk of the observed interstellar extinction at optical wavelengths. Grain destruction occurs mainly by direct involvement in star formation.

## 1. INTRODUCTION

The existence of solid particles in interstellar space has been recognized for nearly five decades. Throughout this period, the search for their composition and origin has progressed on several fronts. Astronomical observations which have a bearing on these problems have been pursued in recent years using techniques of infrared and ultraviolet spectroscopy. Theories of grain optics, nucleation and growth have also been pursued with great vigour. Yet observational data and theoretical modelling have failed to converge upon a unique or generally acceptable solution.

Recent comprehensive reviews have dealt with many aspects of observational and theoretical work bearing on the formation and destruction of grains (Wickramasinghe and Nandy, 1972; Aannestad and Purcell, 1973). In the present article I shall mainly confine attention to developments which have not been covered in these recent reviews.

## 2. OBSERVATIONAL DATA

Important constraints on the nature of grains may be obtained

*Hugo van Woerden (ed.), Topics in Interstellar Matter, 155-162. All Rights Reserved.*
*Copyright © 1977 by D. Reidel Publishing Company, Dordrecht-Holland.*

directly from certain observational data. Studies of interstellar extinction, albedo (from diffuse galactic light data), and polarization (including linear and circular polarization) are consistent with a 3-component grain model (Wickramasinghe, 1976):

(i) Extinction and polarization observations in the waveband 10000 - 3000 Å demand a predominantly dielectric grain model; partially aligned dielectric needles of refractive index m = 1.5 with radii ∼ 1.5 x $10^{-5}$ cm are consistent with these data.

(ii) Extinction, phase function and albedo data in the 3000 - 1800 Å waveband require the dominance of a predominantly absorbing (low-albedo) grain population. This grain population must provide an explanation of the wellknown 2200-Å interstellar absorption band. If graphite is responsible for extinction in this waveband, the particles must be nearly spherical with radii ≤ 2 x $10^{-6}$ cm, and the 2200-Å band is a small-particle resonance in graphite.

(iii) Extinction and albedo data at wavelengths $\lambda \lesssim$ 1800 Å require a dielectric grain population with radii ≤ $10^{-6}$ cm.

A mixture of small spherical graphite particles, radii ≤ 2 x $10^{-6}$ cm, small dielectric spheres (radii < $10^{-6}$ cm) and larger elongated dielectric particles (radii ∼ 1.5 x $10^{-5}$ cm, for cylinders) could account for all the data. The mass densities of all three populations are roughly comparable. The question remains: what composition can be attributed to the dielectric particles? Two types of dielectric particle are currently en vogue: Silicate grains probably make up the smaller-sized population; icy grains, probably with silicate cores, account for the larger grains.

Silicate grains have been tentatively identified by a character-istic 10-μ absorption feature which appears in extinction against continuum infrared sources, as well as in emission in cool, oxygen-rich M-giant stars. $H_2O$ ice has also been identified by a 3.1-μ absorption feature, but the amount of ice involved inferred from the band strength is insufficient to explain the optical extinction data. An absorption feature at 3.3-μ has also been observed in stellar spectra, and this has been tentatively identified with polyoxymethylene polymers or co-polymers (Cooke and Wickramasinghe, 1977).

Other observational data which have a bearing on grain composition concern interstellar gas-phase abundances of various elements derived from ultraviolet-absorption-line studies (Spitzer and Jenkins, 1975). With respect to solar abundances (Ross and Aller, 1976) it is found that certain elements are grossly underabundant in the gaseous inter-stellar medium. Elements which can readily form refractory solids are most highly depleted, but the CNO elements are also depleted by signi-ficant factors. These data give general support to the idea that the bulk of the elements Mg, Si, Fe is condensed in grains or grain cores as iron, silicon carbide and silicate particles, and that the CNO ele-ments are also locked away in grains, presumably as grain mantles, but to a less complete extent.

## 3. GRAIN FORMATION IN MASS FLOWS FROM STARS

Dust formation under conditions of relatively high density as prevail in stellar atmospheres can be discussed with a reasonable degree of confidence. Thermodynamic equilibrium obtains, so that the theory is relatively simple. Refractory grains may readily condense in gaseous flows from cool stars. Carbon stars with a C/O ratio exceeding unity are ideal for the formation of graphite grains (Hoyle and Wickramasinghe, 1962), and SiC grains (Gilra, 1971) to a lesser extent. Oxygen-rich M-giant stars are expected to be sources of silicate and iron grains. Thermodynamic arguments as well as nucleation calculations have given strong support to these ideas (Donn et al., 1968; Salpeter, 1974a,b). Expulsion of grains occurs owing to the action of radiation pressure, particle sizes and fluxes being controlled by the appropriate value of stellar luminosity (Salpeter, 1974a,b) in any given case. Refractory grains are also expected to condense in mass flows from novae and super-novae (Hoyle and Wickramasinghe, 1970; Clayton and Hoyle, 1975; Clayton and Wickramasinghe, 1976).

## 4. GROWTH OF ICY GRAIN MANTLES

The question of grain mantle growth is fraught with more difficult-ies and uncertainty. Van de Hulst (1946, 1949) proposed that a hybrid mixture of volatile ices ($H_2O$, $NH_3$, $CH_4$) would condense under inter-stellar conditions, and this original suggestion has been pursued more recently in a modified form by Greenberg and Hong (1974). The relative weakness of the observed 3.1-$\mu$ absorption band would appear to militate against an ice-grain explanation for the bulk of optical interstellar extinction (Wickramasinghe and Nandy, 1972; Dempsey and Wickramasinghe, 1975).

My own preference is for "tarry" polymeric grain mantles comprised of CNO elements, which are thermodynamically more stable than icy mixtures. Organic polymers have typical melting temperatures $\sim 500^{\circ}K$, whereas ices have melting temperatures in the range $200 - 300^{\circ}K$. Polyformaldehyde polymers and co-polymers are a strong possibility, but other types of organic polymer could also occur. Interstellar polymeri-zation with particular reference to polyoxymethylene is discussed in the next section.

## 5. INTERSTELLAR POLYMERIZATION

The possibility of formaldehyde polymers occurring as mantles on cometary and interstellar grains has been discussed by several authors (Wickramasinghe, 1974, 1975; Vanysek and Wickramasinghe, 1974; Mendis and Wickramasinghe, 1975; Cooke, 1976; Cooke and Wickramasinghe, 1977). An identification of the 8-12 $\mu$ interstellar bands in at least some cases with polyformaldehyde-coated silicate grains seems tenable (Cooke, 1976). Moreover, the volatalization temperature of 10-$\mu$ emitting grains

inferred from cometary data (Mendis and Wickramasinghe, 1975) has been
shown to be consistent with tarry grains and probably inconsistent with
pure silicate grains.

Although a wide variety of organic and inorganic interstellar molecules have been discovered so far, CO and $H_2CO$ are the most ubiquitous,
being present in dense clouds as well as in the more tenuous regions of
the interstellar medium. In addition to millimeter-wave observations of
CO in denser regions, there are now also ultraviolet observations relating to CO in tenuous regions (Snow, 1975). Reactions leading to CO
formation have been described by Glassgold and Langer (1975) and by
Langer (1975). A substantial fraction of carbon in dense clouds
($n_H \gtrsim 10^4$ cm$^{-3}$) may be assumed to be in the form CO. Millimeter-wave
observations of CO may be interpreted to give $n_{CO}/n_H \simeq 10^{-5}$ in a typical
case (Leung and Liszt, 1976). The actual observed abundance of $H_2CO$ in
the gas phase gives a ratio $n_{H_2CO}/n_H$ which is three orders of magnitude
lower, $\sim 10^{-8}$ (Zuckerman and Turner, 1975). However, it is not inconceiv-
able that a much larger fraction of interstellar C is tied up as $H_2CO$,
in polymeric form, on grains. The gas phase observations may then merely
reflect a number density of molecules which is in equilibrium with
polymer/co-polymer phases under different interstellar conditions.

The processes which lead to the formation of $H_2CO$ (and also other
more complex molecules) in interstellar clouds are yet somewhat obscure.
They could involve ion-molecule reactions in the gas phase as well as
recombination reactions on grain surfaces. One possibility discussed by
Williams (1974) for $H_2CO$ formation involves direct additions of H atoms
to CO on grain surfaces. $H_2CO$ is expected to polymerize on grain surfaces,
whatever the mechanism of its formation (Wickramasinghe, 1974). Polymerization, in general, proceeds in three distinct stages (Bevington, 1961;
Blackadder, 1975):
     (i)   Initiation.  Initiation of polymerization, involving a bond-
breaking event, essentially converts a saturated molecule into a chemically active state with a free valence bond:

$$
\begin{array}{ccc}
H & & H \\
C = 0 & \rightarrow & C - 0 \cdot \\
H & & H
\end{array}
\qquad (1)
$$

This could be effected in a variety of possible ways – including interaction with UV photons, radicals or ions. Since UV photons will largely
be excluded from dense clouds, owing to dust opacity, reactions with
radicals and ions may provide the preferred initiation routes. For
example, if R· denotes a radical (which are among the most abundant
molecular species present in interstellar conditions), we may have an
initiation reaction of the form:

$$
\begin{array}{ccc}
& H & & H \\
R\cdot + & C = 0 & \rightarrow & R - C - 0 \cdot \\
& H & & H
\end{array}
\qquad (2)
$$

(ii) <u>Propagation and Co-polymerization.</u> Polymer-chain propagation could now proceed by simple addition reactions in the gas phase:

$$
\begin{array}{cccc}
\text{H} & \text{H} & \text{H} & \text{H}\\[-2pt]
\text{R} - \text{C} - \text{O·} \;+\; \text{C} = \text{O} \;\rightarrow\; \text{R} - \text{C} - \text{O} - \text{C} - \text{O·}\\[-2pt]
\text{H} & \text{H} & \text{H} & \text{H}
\end{array} \tag{3}
$$

The chain thus propagates spontaneously. $H_2CO$ is known to form highly stable co-polymers with many suitable co-monomers. These include acetaldehyde ($CH_3CHO$), isocyanic acid (HNCO) and cyanoacetylene ($HC_3N$), all of which are known to co-exist with $H_2CO$ in dense interstellar clouds. The presence of such molecules, with which co-polymerization can occur, would lead to the formation of polyoxymethylene co-polymers.

(iii) <u>Termination.</u> Closure or termination of a growing polymer chain could also occur in a variety of ways. In the case of gas-phase polymerization in the interstellar medium a likely mechanism is the approach of a radical or atom which could combine with the chain, rendering its growing tip effectively inert. Closure with radicals other than OH, which seems likely, will block depolymerization and produce polymers endowed with a high degree of thermal stability (Fawcett, 1975; Wickramasinghe and Santhanan, 1975).

So far, we have confined our attention to gas-phase polymerization reactions. If dirty-ice mantles initially form, it is still highly probable that interaction of these mantles with ultraviolet light, soft X-rays and cosmic rays leads to polymerization reactions in the solid state. The net result will be a hybrid organic polymer.

## 6. RATE OF MANTLE GROWTH

The deposition of successive layers of $H_2CO$ co-polymers, which in many cases may be cross-linked, is expected to form highly stable refractory grain mantles. The rate of growth of a formaldehyde co-polymer mantle in a dense molecular cloud of hydrogen (nucleon) density n, temperature T, is given by

$$
dr/dt = \alpha n_{[C]}/s \; (kTm/2\pi)^{\frac{1}{2}} \tag{4}
$$

where r is the mantle radius, $\alpha$ the sticking coefficient, s is the density of the solid polymer, and $n_{[C]}$ is the number density of C-atoms containing species whose interaction with grains leads directly to the production of $H_2CO$. We assume here that $n_{[C]} = n_{CO}$ so that $n_{[C]}/n_H = 10^{-5}$. $H_2CO$ formation could occur, for example, by the exothermic addition of H atoms as proposed by Williams (1974). With $\alpha/s \simeq 1$ we have from (4)

$$
dr/dt \simeq 1.05 \times 10^{-11} \; (n_{[C]}/n) \; n \; (T/100 \text{ K})^{\frac{1}{2}} \text{ cm yr}^{-1} \tag{5}
$$

It has been noted that dense massive molecular clouds are not in a state of free-fall collapse (Zuckerman and Palmer, 1974). Collapse could be slowed down by several processes, including effects of magnetic pressure, rotation and turbulence. Turbulence might be generated as well as maintained by the effect of continuing star-formation within molecular cloud complexes of the type considered here. We may suppose that typical contraction timescales for an entire cloud complex as well as for separate fragments within it are ~ $10^6$ yr; such a time, together with the estimated total mass of protostellar clouds, gives a star-formation rate which is in good agreement with observational data (Hoyle and Wickramasinghe, 1976).

Setting $t \simeq 10^6$ yr as the appropriate timescale, together with $n_{[C]}/n \simeq 10^{-5}$, $T \simeq 100$ K, we obtain a mantle radius ~ $10^{-5}$cm in regions of hydrogen density $n_{H_2} \simeq 5 \times 10^4$ cm$^{-3}$. These densities are typical of extended regions in many dense molecular clouds in the Galaxy. Although clumping of such co-polymer-coated grains could occur in subregions of the cloud which are directly associated with protostellar contraction (Hoyle and Wickramasinghe, 1976), a significant fraction of individual co-polymer-coated grains may be expected to be expelled with systematic gas flows into the general interstellar medium. Such dielectric grains with typical radii ~ $10^{-5}$ cm may be the main source of interstellar extinction at optical wavelengths (Wickramasinghe, 1975).

## 7. DESTRUCTION OF GRAINS

Ideas concerning the destruction of grains have not evolved or advanced to any significant extent since the time of the earlier reviews. The main destructive mechanisms in order of importance are:

(i)   Thermal evaporation.  Grains evaporate whenever their temperatures exceed a critital temperature which is a function of the grain material. This occurs only for grains directly involved in star formation and for those in the immediate vicinity of hot stars.

(ii)   Sputtering.  Interaction of grains with high-speed atoms or ions leads to ejection of lattice atoms (Wickramasinghe, 1965; Stuart and Wehner, 1962). Under interstellar conditions He is the most effective sputtering agency, and relative speeds ~ 200 km s$^{-1}$ are required to produce significant sputtering yields. This process could be important for grains in the vicinity of supernova explosions, and also for grains which are being accelerated due to radiation pressure of parent stars. Sputtering could limit the radius of the injected grains.

(iii) Grain-Grain Collisions.  Experimental data on this type of destructive process are virtually non-existent. For grain-grain collisions at relative velocities $\leq$ 10 km s$^{-1}$, destruction (either by shattering or evaporation) is unlikely to occur for refractory grains. In any event, since grains carry an electric charge and are coupled to magnetic field lines, this destructive process which is probably restricted to cloud-cloud collisions is likely to be of negligible importance.

(iv)  Interaction with UV photons.  Although UV photons could produce photodesorption of adsorbed molecules, they are unlikely to destroy grains.

The typical lifetime of a grain against destruction (mainly by process (i)) may be estimated as ~ $10^9$ yr.

REFERENCES

Aannestad, P.A. and Purcell, E.M.  1973, Ann. Rev. Astron. Astrophys. 11, 309.
Bevington, J.C.  1961, Radical Polymerization (London, Academic Press).
Blackadder, D.A.  1975, Some Aspects of Basic Polymer Science (Londen, The Chemical Society).
Clayton, D.D. and Hoyle, F.  1976, Astrophys. J. 203, 490.
Clayton, D.D. and Wickramasinghe, N.C.  1976, Astrophys. Space Sci. 42, 463.
Cooke, A.  1976, Astrophys. Space Sci. 39, L13.
Cooke, A. and Wickramasinghe, N.C.  1977, Astrophys. Space Sci., in press.
Dempsey, M.J. and Wickramasinghe, N.C.  1975, Astrophys. Space Sci. 34, 185.
Donn, B., Wickramasinghe, N.C., Hudson, J. and Stecher, T.  1968, Astrophys. J. 153, 451.
Fawcett, A.H.  1975, Nature 257, 159.
Gilra, D.P.  1971, Nature 229, 237.
Glassgold, A.E. and Langer, W.D.  1975, Astrophys. J. 197, 347.
Greenberg, J.M. and Hong, S.S.  1974, IAU Symposium No.60, eds. Kerr, F.J. and Simonson, S.C. (D. Reidel Co.)
Hoyle, F. and Wickramasinghe, N.C.  1962, Mon.Not.Roy.Astr.Soc. 124, 417.
Hoyle, F. and Wickramasinghe, N.C.  1970, Nature 226, 62.
Hoyle, F. and Wickramasinghe, N.C.  1976, Nature 264, 45.
Langer, W.D.  1976, Astrophys. J. 206, 699.
Leung, C.M. and Liszt, H.S.  1976, Astrophys. J. 208, 732.
Mendis, D.A. and Wickramasinghe, N.C.  1975, Astrophys. Space Sci. 37, L13.
Onishi, T.  1976, Prog. Theor. Phys. 55, 1669.
Ross, J.E. and Aller, L.H.  1976, Science 191, 1223.
Salpeter, E.E.  1974a, Astrophys. J. 193, 579.
Salpeter, E.E.  1974b, Astrophys. J. 193, 585.
Snow, T.P.  1975, Astrophys. J. 201, L21.
Spitzer, L. and Jenkins, E.B.  1975, Ann.Rev. Astron. Astrophys. 13, 133.
Stuart, R.V. and Wehner, G.K.  1962, J. Appl. Phys. 33, 2345.
Van de Hulst, H.C.  1946, Rech. Astron. Obs. Utrecht, XI, Part 1.
Van de Hulst, H.C.  1949, Rech. Astron. Obs. Utrecht, XI, Part 2.
Vanysek, V. and Wickramasinghe, N.C.  1975, Astrophys. Space Sci. 33, L19.
Wickramasinghe, N.C.  1965, Mon.Not.Roy.Astr.Soc. 131, 177.

Wickramasinghe, N.C.    1974, Nature 252, 462.
Wickramasinghe, N.C.    1975, Mon.Not.Roy.Astr.Soc. 170, 11.
Wickramasinghe, N.C. and Nandy, K.    1972, Rep. Progr. Phys. 35, 157.
Wickramasinghe, N.C. and Santhanan, K.S.V.    1975, Nature 257, 159.
Wickramasinghe, N.C.    1975, in Solid State Astrophysics, eds.
       Wickramasinghe and Morgan (D. Reidel Ltd., Dordrecht, Holland).
Williams, D.A.    1974, Observatory 94, 66.
Zuckerman, B. and Palmer, P.    1974, Ann. Rev. Astr. Astrophys. 12, 279.
Zuckerman, B. and Turner, B.E.    1975, Astrophys. J. 197, 123.

Chapter 4


LARGE-SCALE DISTRIBUTION OF
INTERSTELLAR MATTER IN THE GALAXY


Papers presented in a joint session of IAU Commissions
33 (Galactic Structure) and 34 (Interstellar Matter),
Grenoble, 30 August 1976.


Session Chairman:  Frank J. Kerr

# COMPARATIVE MORPHOLOGY OF GALACTIC CARBON MONOXIDE AND HYDROGEN

W. B. Burton and M. A. Gordon
National Radio Astronomy Observatory[*]
Green Bank, West Virginia

ABSTRACT

   We compare here overall galactic characteristics derived from observations of the $\lambda 2.6$-mm J=1→0 rotational transition of $^{12}$CO with
those derived from observations of the $\lambda 21$-cm hyperfine transition of
atomic hydrogen.  The CO observations show that the cold, compressed
component of the interstellar medium is confined to larger extent than
the more diffuse HI gas to the inner Galaxy, is confined to a thinner
layer, and is more clumpily distributed.  The two tracers share the
same overall kinematics, and both are relatively depleted at R < 4 kpc
except in the galactic nucleus itself.  Quantitative measures of the
relative distributions have been derived.  Several assumptions underlying these derivations have been verified by numerical experiments.

---

   This review compares some of the salient results of the large-
scale exploration of our Galaxy using emission observations of carbon
monoxide and atomic hydrogen.  At this writing, still-current contributions to the CO exploration have been made by Scoville and Solomon
(1975), Burton et al. (1975), Bash and Peters (1976), Gordon and Burton
(1976), Scoville et al. (1976), Burton and Gordon (1976), Cohen and
Thaddeus (1977), Roberts and Burton (1977), and Burton and Gordon (1977).
Our review consists for the most part of figures assembled from our own
recent work, connected by a minimum of text.  For details the reader
must consult the original papers.

   In several respects CO is an advantageous galactic probe.  The
J=1→0 spectral line at 115.27 GHz occurs in an accessible part of the
spectrum, is relatively intense and unobstructed by interstellar ex-

---

* Operated by Associated Universities, Inc., under contract with the
  National Science Foundation.

*Hugo van Woerden (ed.), Topics in Interstellar Matter, 165-177. All Rights Reserved.*
*Copyright © 1977 by D. Reidel Publishing Company, Dordrecht-Holland.*

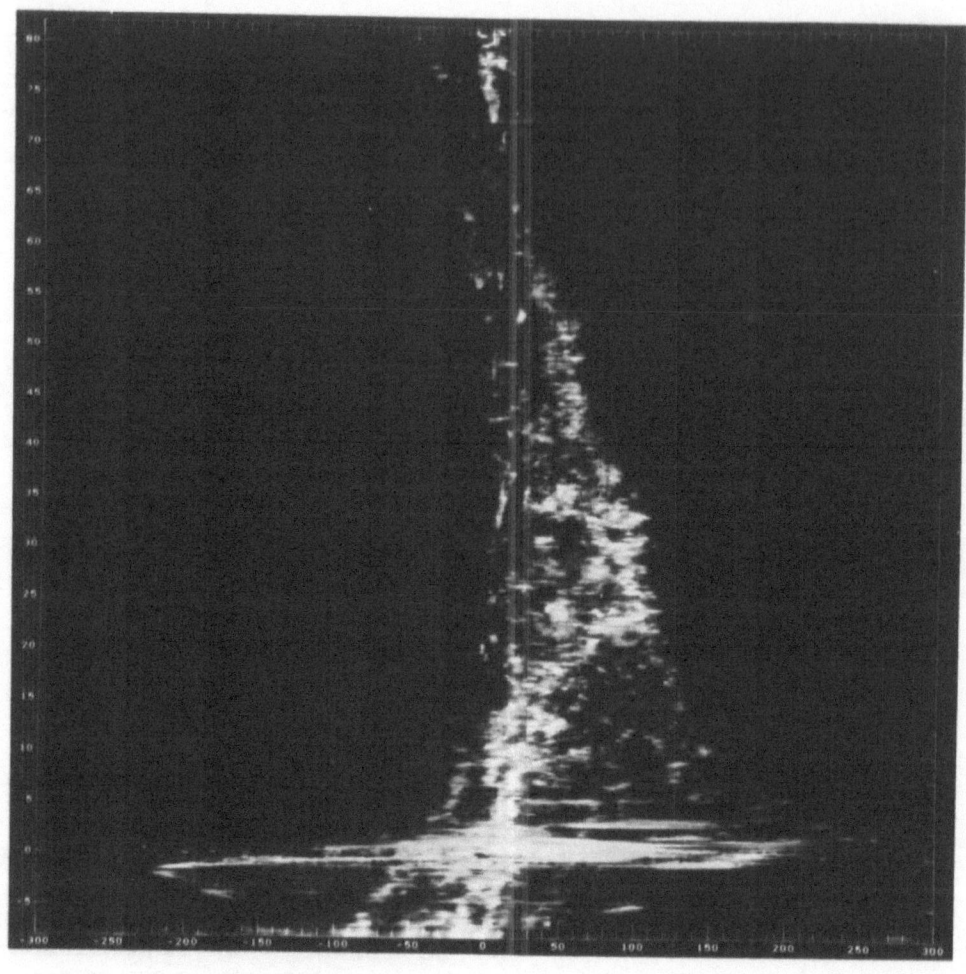

Figure 1. Grey-scale representation of $^{12}C^{16}O$ emission intensities in
longitude-velocity coordinates at the galactic equator. Little CO emission occurs in the portions of this figure corresponding to R > 9 kpc
(except for the exceptional Cygnus region) or to R < 4 kpc (except for
the exceptional 3-kpc arm and the intense nuclear sources). The observations at $\ell$ < 10° are due to Bania (1977), those at 10° < $\ell$ < 36° to
Gordon and Burton (1976), and those at $\ell$ > 36° to Burton and Gordon
(1977). The sampling interval of 0°.2 sets the angular resolution; the
velocity resolution is 1.3 km s$^{-1}$ at $\ell \geq$ 10° and 2.6 km s$^{-1}$ at $\ell$ < 10°.

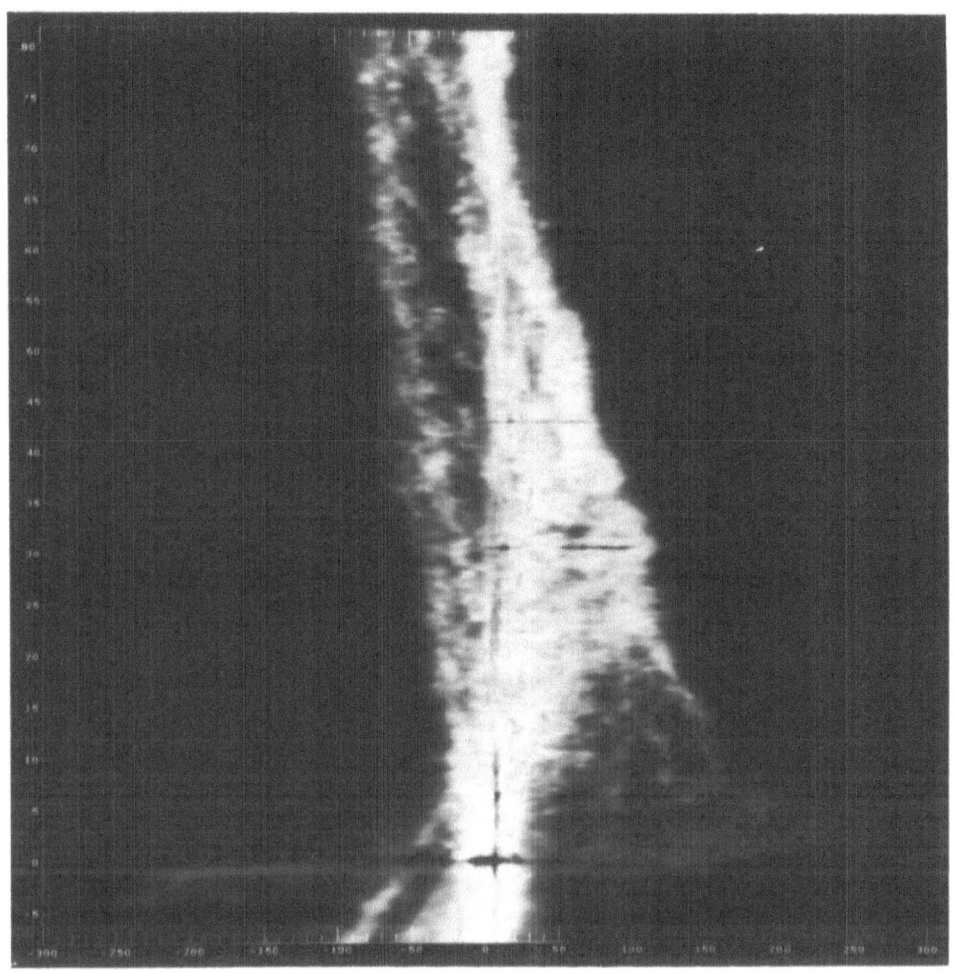

Figure 2. Grey-scale representation of HI emission intensities in longitude-velocity coordinates at the galactic equator. Westerhout (1976)
made the observations at $\ell \geq 2°$; Bania (1977) made those at $\ell < 2°$.
For comparison with the CO situation, the scale is the same as in Figure 1. The HI observations similarly sample every $0°2$ and their velocity resolution is 1.3 km s$^{-1}$. The HI galactic disk has a diameter
approximately twice that of the CO disk. In the region where the CO
and HI distributions overlap, the two tracers show the same kinematics.
The CO distribution is evidently clumpier than that of the HI.

tinction, and is only moderately broadened by mechanisms other than
those governing the overall galactic kinematics. Furthermore, CO is
found in intimate association with the cold, dense component of the
interstellar medium. For the first time, astronomers have a tool which
can locate throughout the Galaxy the compressed component of interstellar gas in which stars form. Information inherent in observations
of galactic CO complements that available from observations of the more
diffusely distributed, and warmer, atomic hydrogen gas. Like the HI
situation, observations of CO give velocity information from transgalactic paths and can provide distances to the emitting regions in
terms of a kinematic model of the Galaxy. Unlike the HI situation, individual clouds of CO are opaque to emission from the $J=1 \to 0$ line, making
difficult conversion of measured intensities into physical quantities
such as column density. Fortunately no severe opacity problems hinder
derivation of the overall distribution of CO. A number of arguments
support this point, including arguments based on numerical simulations
and on boundary conditions imposed by other kinds of observations. Im-
proved technology in the millimeter-wave range will allow in the future
extensive observations of the $J=2 \to 1$ transition of CO at 230 GHz. The
CO gas at this transition (which lies in a favorable atmospheric window)
has a small optical depth.

Figures 1 and 2 show the velocity-longitude distribution along the
galactic equator of CO and HI emission, respectively. Comparison of
these figures shows directly that the CO is more confined to the inner
Galaxy than is the HI: little CO emission is observed (at $\ell > 0°$) with
negative velocities. The CO emission is more clumpily distributed
than that of HI. Thus the peak-to-valley ratio in CO profiles is typically 10 or more; for HI profiles this ratio is typically 2 or less,
even for those observed with angular resolution comparable to that of
the CO data (Baker and Burton, see Burton 1976a). A more detailed comparison of the CO and HI data shows that absorption features in the HI
data are often emission features in the corresponding CO profiles, confirming the association of CO with cold material.

The galactic kinematics of the CO and HI distributions are, however, quite similar in the region of overlap. Investigation of the
galactic rotation function is a direct use of these observations. This
involves measuring in a suitable way the terminal velocity for each
line of sight, corresponding (at $\ell > 0°$) to the extreme positive velocity and thus, under assumed conditions, to gas at a known galactocentric distance. The major assumed conditions are that the gas is
quasi-uniformly distributed within the galactic disk and that radial
motions are not important in this context. These assumptions are
amenable to observational verification; the terminal velocities then
provide $\Theta(R)$, which gives the dependence of the linear velocity of
circular differential galactic rotation on galactocentric radius.
Symbols in Figure 3 correspond to the HI and CO terminal velocities.
The agreement between the CO and HI terminal velocities shows that the
two constitutents share the same overall kinematics. The abrupt variation of the observed terminal velocities near the direction of the
galactic center supports the contention that the terminal velocities

represent rotational, not radial, motions.  The larger scatter in the
CO data is not of kinematic origin, but is due to the discrete-
cloud nature of the molecular distribution.

Also in Figure 3 is a newly-derived rotation curve, fit in the
sense of least squares to the data.  The ordered perturbations to this
curve probably represent large-scale motions induced by the gravitational torque of spiral arms (see references in the review by Burton
1976b).  Indeed these ordered variations provide the best (if not only)
evidence for the spiral nature of the overall gas distribution.

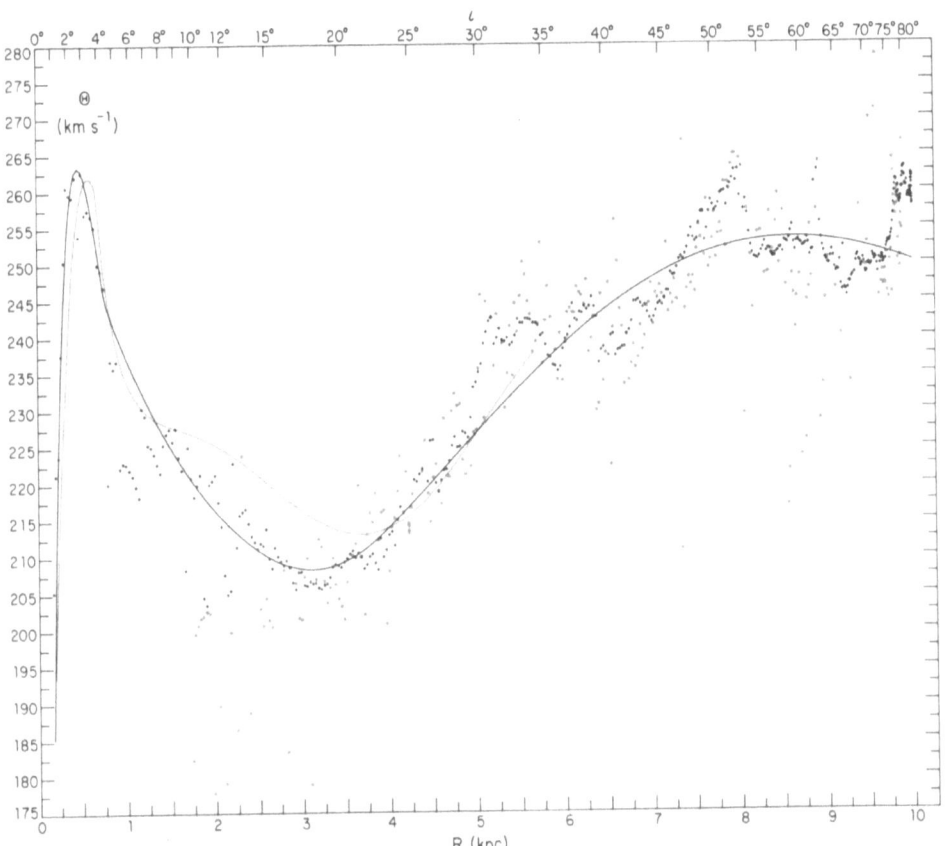

Figure 3.  Variation of the linear velocity of differential circular
galactic rotation as a function of distance from the galactic center
(lower scale) and of longitude (upper scale) (Burton and Gordon 1977).
The foreground symbols correspond to the HI terminal velocities; the
grey symbols correspond to the CO terminal velocities.  The CO and HI
gas share the same overall kinematics.  The black line represents a
smooth approximation to the HI points and applies to the inner Galaxy
at $2° < \ell < 90°$.  The grey line is the galactic rotation curve found
by Simonson and Mader (1973) as a compromise between conflicting data
from $\ell > 0°$ and from $\ell < 0°$.

Having the galactic rotation curve allows assignment of each intensity, measured at a particular velocity and longitude, to a galactocentric distance. (Unlike distance from the Sun, this distance is unambiguous.) The left side of Figure 4 shows the accumulated intensity
of CO emission as a function of galactic radius. Smoothing in azimuth
is inherent in the derivation. It is crucial to establish that the
integrated intensity measure is proportional to abundance. Burton and
Gordon (1977) deal with this problem at length, and show that the
measure is proportional to abundance essentially because the mean-free-
path between the discrete CO clouds is sufficiently large that more
than one cloud rarely occurs at the same velocity on the same line of
sight. For the diffusely distributed HI gas, optical-depth effects
near the galactic plane are such that column densities do not follow
directly from profile integrals. The galactic HI abundance plotted on
the right-hand side of Figure 4 incorporates a correction for partial
saturation. It is ironic that the CO gas viewed as a whole is trans-
parent, although the gas is an assemblage of opaque elements, whereas
the HI gas viewed as a whole is partially saturated, although the sub-
elements are generally optically thin in emission profiles.

The comparative radial distributions given in Figure 4 show CO
more confined to the inner Galaxy than HI. The CO distribution is by
implication that of dark material in general, because, for reasons of
formation and survival, CO cannot long exist in quantity outside dark
environments (see Stief 1973). Within the observational uncertainties
(which are in some cases large), the CO distribution is roughly equivalent to those found independently for ionized hydrogen (e.g., Gordon
and Cato 1972; Lockman 1976; Hart and Pedlar 1976), for giant HII regions (e.g., Mezger 1970; Burton et al. 1975), for supernova remnants
(e.g., Ilovaisky and Lequeux 1972; Kodaira 1974), for pulsars

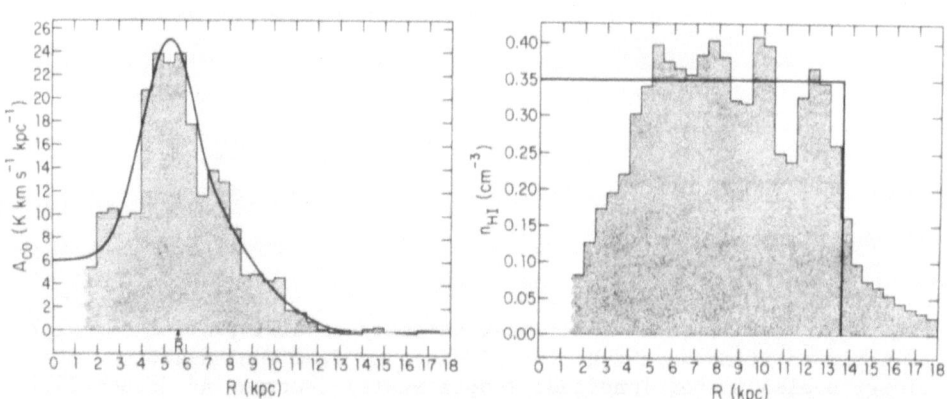

Figure 4. (left) Radial distribution of CO emission at b = 0° expressed as the accumulated total emission in galactic annuli. The
galactic-plane distributions of most Population I constituents follow
the form of the CO distribution. The line is a smoothed approximation
to the measured histogram. (right) Radial distribution of HI column
densities at b = 0° (Burton and Gordon 1977).

(Seiradakis 1976; Taylor and Manchester 1977), for γ-radiation (e.g., Strong 1975; Stecker et al. 1974), and for galactic continuum radiation (e.g., Westerhout 1958; Price 1974). It is clear that among the constituents of the interstellar medium, atomic hydrogen has a unique fundamental distribution.

Although 21-cm observations of atomic hydrogen show it to be an ubiquitous tracer of a number of galactic characteristics, they do not provide the true distribution of interstellar hydrogen. Hydrogen in the molecular form undoubtedly predominates over all other material in opaque, compressed regions where it is shielded against photo-dissociation after formation on grain surfaces. $H_2$ is, however, not directly observable on a galactic scale. Observations of CO are especially important because CO is excited by collisions with $H_2$ (see e.g., Zuckerman and Palmer 1974); the intensity of CO is therefore probably proportional to the abundance of $H_2$. Defining the proportionality in quantitative terms is difficult. The problem has been approached in

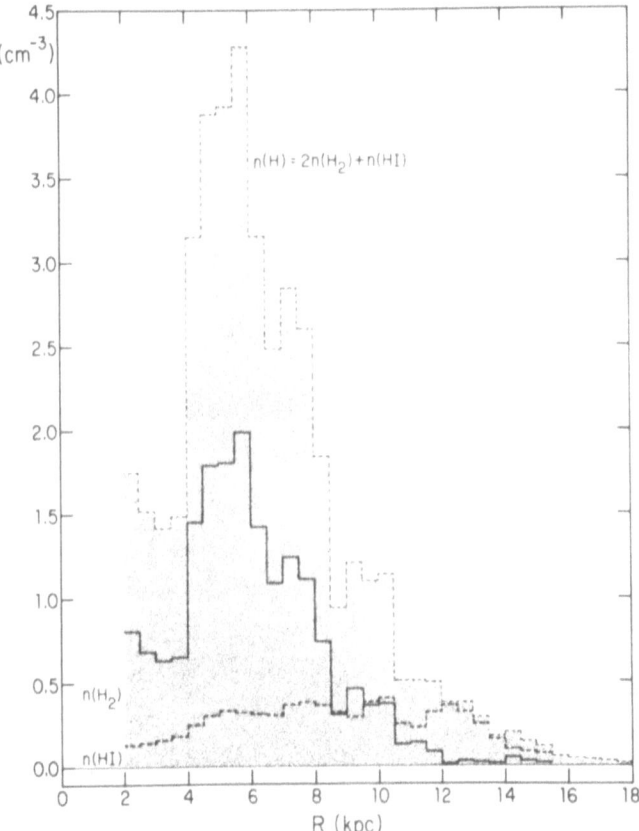

Figure 5. Radial distribution in the plane of the Galaxy of volume densities of atomic and (tentatively) of molecular hydrogen (Gordon and Burton 1976). The distribution of the sum $2\,n(H_2) + n(HI)$ shows the morphology of interstellar nucleons, because hydrogen predominates.

a quite tentative way by Gordon and Burton (1976), who derive from the
CO data a radial abundance distribution of $H_2$ which provides good agreement with $H_2$ column densities found by other methods by Stecker <u>et al.</u>
(1975).  The assumptions remain troublesome, but the systematic nature
of the possible errors suggest that the comparative $H_2$ and HI distributions shown in Figure 5 are probably correct.  At greater galactocentric
distances than that of the Sun, the interstellar gas is mostly atomic;
at smaller distances, the gas becomes increasingly molecular.  Almost
no molecular gas is found at $R > R_0$.  Little gas of either form is
found between 4 kpc and the galactic nucleus.

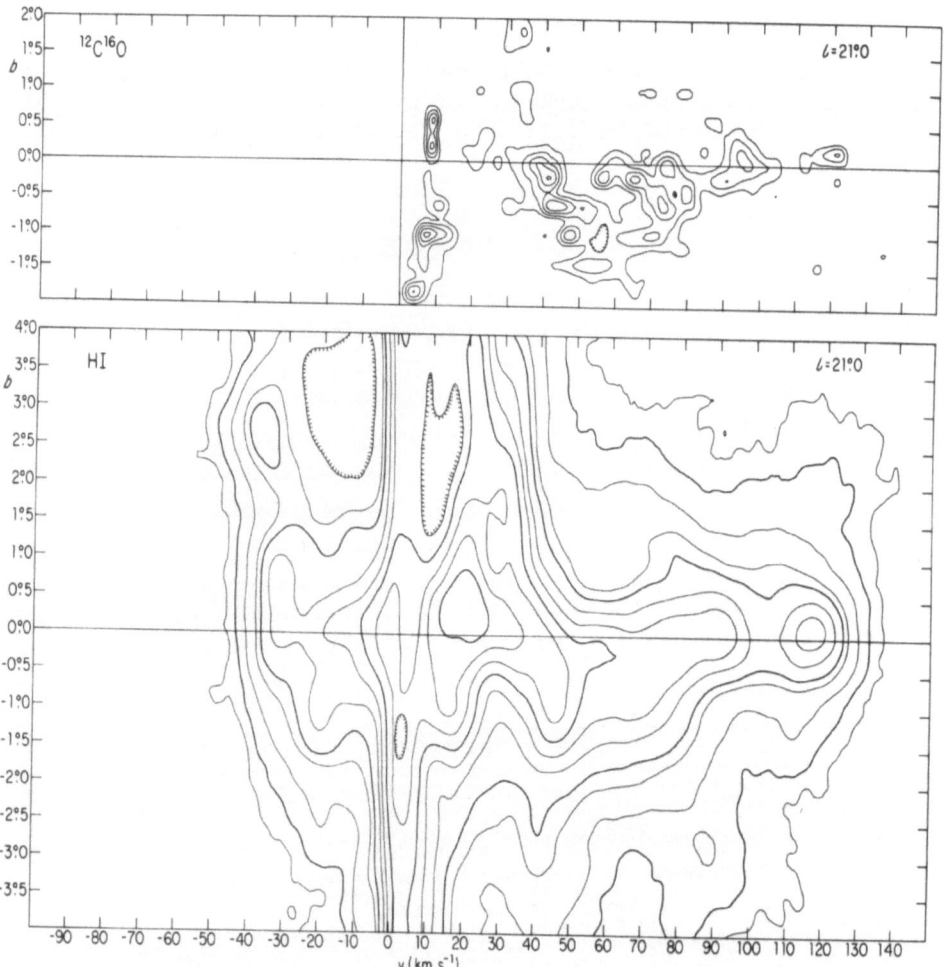

Figure 6.  Comparison, on the same scale, of CO and HI latitude-velocity
distributions in the direction $\ell = 21°$, showing that the z-thickness of
the HI layer is more than twice that of the CO layer (Burton and Gordon
1976).  The HI observations are due to Weaver and Williams (1973).

Derivation of the total distribution of gas in the Galaxy requires knowledge of the thickness of the galactic layer. Figure 6 shows the comparative latitude-velocity distribution of HI and CO for the representative longitude 21°. The line of sight at $\ell$ = 21°, b $\approx$ 0°, passes through the galactic annulus of maximum CO abundance. HI is much less confined to the plane z = 0 pc than CO. Burton and Gordon (1976) modelled the Figure 6 distributions and found a scale height for the CO of 50 pc, compared to 120 pc for the HI in the same galactic region. The width of the CO layer is comparable to that observed for O and B stars in the solar neighborhood, and for other tracers. The $H_2$ extent must also be that measured for CO. More extensive relevant CO observations are reported by Cohen and Thaddeus (1977) and by Scoville et al. (1976). The centroid of the CO layer deviates in a systematic way from the plane z = 0 pc, as do the centroids of other Population I constituents (see Lockman 1977). (Note that our results here pertain mostly to b = 0°.) There is some evidence that the CO layer becomes increasingly thick as R increases. This is certainly the case for the extensive HI layer (e.g., Jackson and Kellman 1974; Baker and Burton 1975).

Figure 7 shows the surface density and differential mass of $H_2$ and HI calculated using the measured layer thickness. Also plotted is the distribution of total surface density (of course mainly contributed by stars) deduced from dynamical considerations. The ratio $\sigma_{gas}/\sigma_{total}$ is approximately 4% over a large range of galactic radius. The inner-Galaxy zone of gas depletion is exceptional. The constancy of the

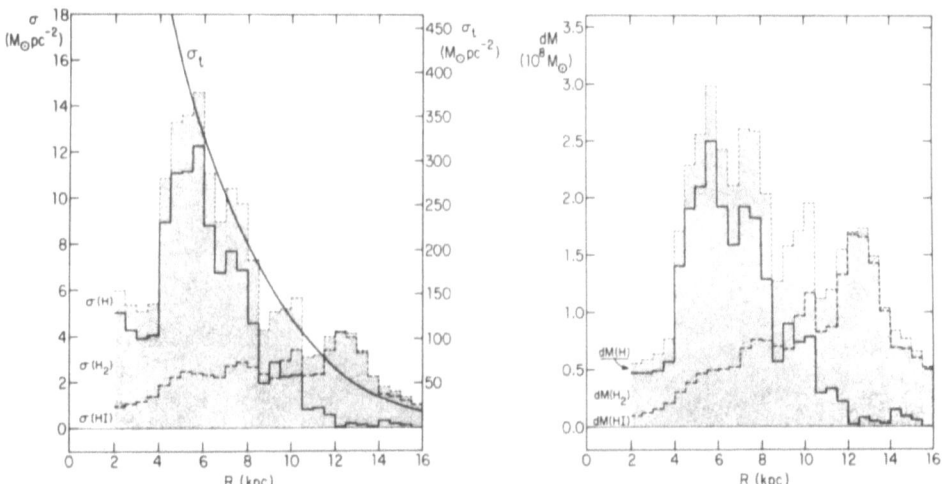

Figure 7. Radial distribution of projected surface densities (left) and of differential masses (right) of atomic and molecular hydrogen (Gordon and Burton 1976). At smaller distances hydrogen in molecular form dominates the mass distribution of the interstellar gas; at larger distances most of the interstellar gas is in the form of atomic hydrogen. Also shown is the total surface density $\sigma_t$ predicted by Innanen (1973).

ratio suggests that long-term star-formation efficiency does not strongly vary with galactic radius, contrary to conclusions based on the HI
distribution alone (c̲f̲. Talbot and Arnett 1975).  While the gas and
total mass distributions must be inferred indirectly, and accordingly
are subject to errors, they give the only evidence available on efficiency of star formation on a galactic scale.  The total galactic
gas masses derived following the assumptions of the original papers
are $2 \times 10^6$ $M_\odot$ of CO, $2 \times 10^9$ $M_\odot$ of $H_2$, and $2.3 \times 10^9$ $M_\odot$ of HI.

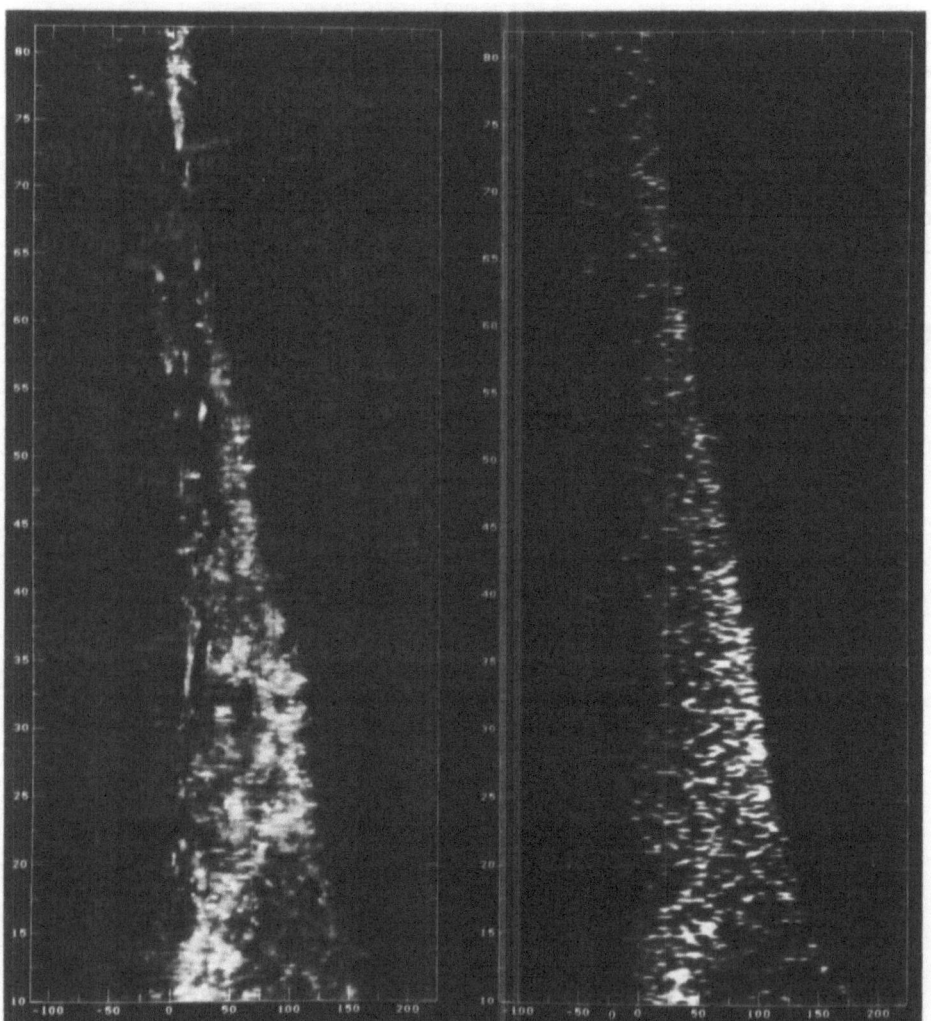

Figure 8.  (left)  Grey-scale representation of the longitude-velocity
arrangement of $^{12}C^{16}O$ emission observed at $0°.2$ intervals along the
galactic equator.  (right)  Grey-scale representation of the longitude-
velocity arrangement of emission inherent in synthetic spectra representing stochastically distributed discrete CO clouds (Burton and Gordon
1977).

    An important difference between the Figure 1 CO observations and
the Figure 2 HI observations concerns the amount of small-scale structure apparent.  HI features extending over many beam widths are common,
CO features are rarely extended.  The characteristic appearance in the
CO spectra of isolated features allows estimates of the number and separation of emitting clouds, and of limits on their characteristic size.
We have modelled the CO observations by generating synthetic profiles
corresponding to a stochastic assemblage of dark clouds.  Figure 8 shows
for comparison the longitude-velocity distributions of the observed and
modelled CO intensities.  The stochastic distribution is governed by
the probability of finding a cloud in a particular interval on the line
of sight.  This probability depends crucially on the mean free path
between clouds, which we describe by $\lambda(R) = \lambda_0/A_{CO}(R)$, where $\lambda_0$ is the
mean free path at the peak of the (normalized) smoothed CO radial
abundance distribution drawn in Figure 4.  The kinematics of an individual model cloud is set by the smooth rotation function in Figure 3
and by a random perturbation characterized by a dispersion $\sigma_c$.

    There are several quantitative measures which constrain the model.
The most important are the observed runs with longitude of the terminal
velocities (see Bash and Peters 1976) and of the profile integrals.
The stochastic formulation together with values for the major parameters
$\lambda_0$ = 1000 pc and $\sigma_c$ = 4 km s$^{-1}$ provides profiles which well represent
the observations.  We discuss the modelling in detail elsewhere (Burton
and Gordon 1977).  Here we show in Figure 9 the total-profile integrals,
observed (top) and synthesized (bottom).  The scatter in the variations
is not measurement noise (the synthetic profiles are noise-free), but
is due to the irregular nature of the gas distribution. Because of the
controlled conditions inherent in the procedure, we know that shadowing
of one cloud by another at the same velocity is not important for the
best-fit model.  Thus, the radial abundance distribution derived using
the model profiles yields the input distribution.  This would not be
the case if macroscopic saturation were important.  Numerical experiments of this sort convince us that the CO observations accurately reflect the total amount of CO in the Galaxy.

    We stress that the Figure 8 model is azimuthially symmetric, and
smoothly varying in radius. Although the model gives agreement with
many aspects of the observations, it does not show the clustering into
several major bands seen in Figure 1 or on the left in Figure 8.  Clearly, the cloud distribution is not entirely random; its further study
may well reveal fundamental azimuthal structure, spiral arms, in our
Galaxy.

Figure 9.  (top)  Longitude variation of the total-velocity integrated
intensities calculated from the CO observations of Figure 1.  (bottom)
Longitude variation of the integrated emission calculated from noise-
free synthetic profiles (Burton and Gordon 1977).  The lines show the
data every 0°.5 of ℓ, smoothed over ±0°.5.  The model CO distribution
consists of discrete clouds distributed stochastically.  The clouds
are generally opaque, but nevertheless almost all clouds contribute
to the profiles.

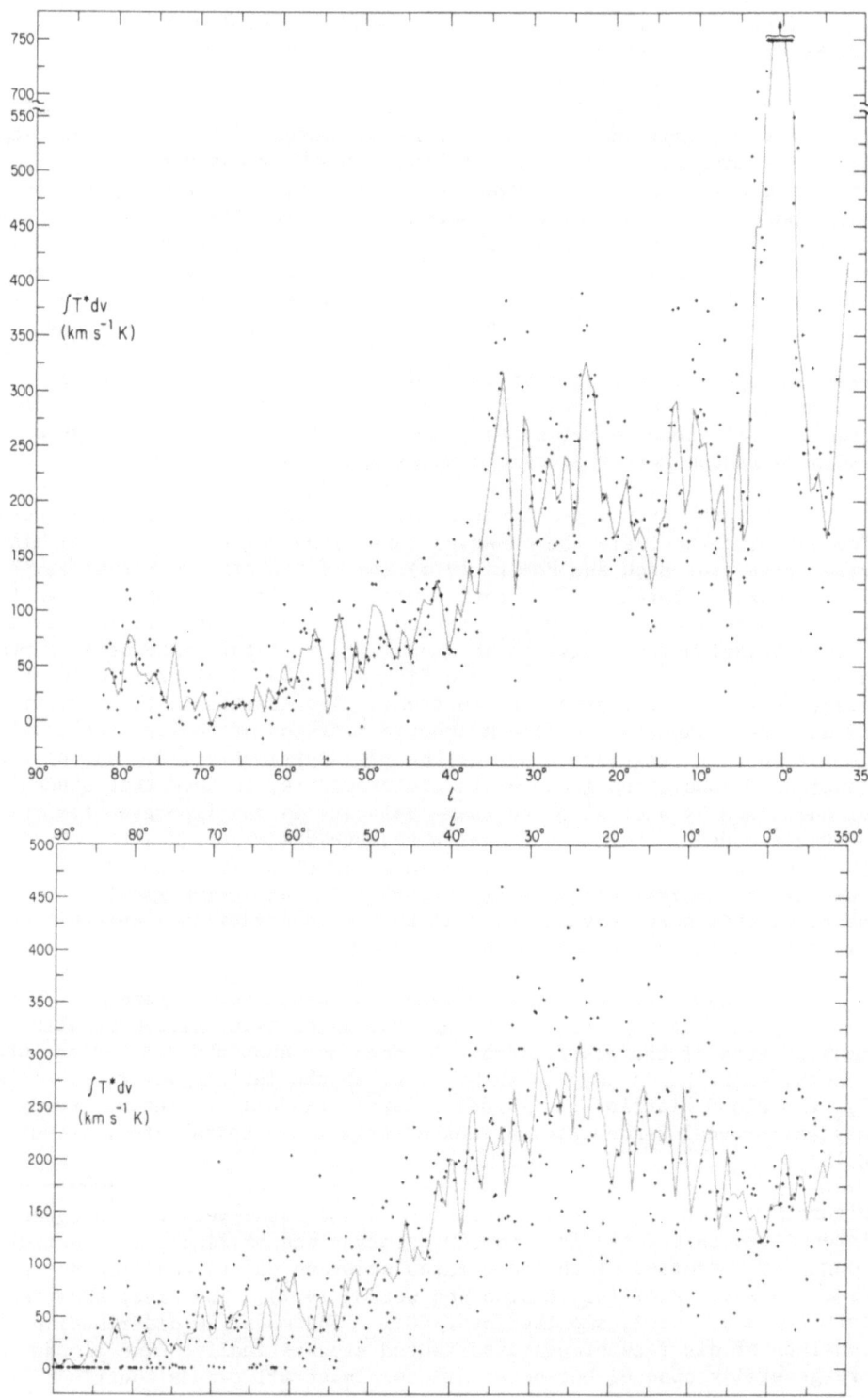

RERERENCES

Baker, P. L., and Burton, W. B. 1975, Ap. J., 198, 281.
Bania, T. M. 1977, Ap. J., submitted.
Bash, F. N., and Peters, W. L. 1976, Ap. J., 206, 786.
Burton, W. B. 1976a, in C. E. Fichtel and F. W. Stecker (eds.), "The
    Structure and Content of the Galaxy and Galactic Gamma Rays",
    177, Goddard Space Flight Center, Greenbelt.
Burton, W. B. 1976b, Ann. Rev. Astr. and Ap., 14, 275.
Burton, W. B., Gordon, M. A., Bania, T. M., and Lockman, F. J. 1975,
    Ap. J., 202, 30.
Burton, W. B., and Gordon, M. A. 1976, Ap. J. (Letters), 207, L189.
Burton, W. B., and Gordon, M. A. 1977, Astr. and Ap., submitted.
Cohen, R. S., and Thaddeus, P. 1977, Ap. J. (Letters), submitted.
Gordon, M. A., and Cato, T. 1972, Ap. J., 176, 96.
Gordon, M. A., and Burton, W. B. 1976, Ap. J., 208, 346.
Hart, L., and Pedlar, A. 1976, Mon. Not. R. astr. Soc., 176, 547.
Ilovaisky, S. A., and Lequeux, J. 1972, Astr. and Ap., 18, 169.
Innanen, K. A. 1973, Ap. and Space Sci., 22, 343.
Jackson, P. D., and Kellman, S. A. 1974, Ap. J., 190, 53.
Kodaira, K. 1974, Publ. Astron. Soc. Japan, 26, 255.
Lockman, F. J. 1976, Ap. J., 209, 429.
Lockman, F. J. 1977, Astr. J., submitted.
Mezger, P. G. 1970, Proc. I.A.U. Symp. 38, 107.
Price, R. M. 1974, Astr. and Ap., 33, 33.
Roberts, W. W., and Burton, W. B. 1977, in H. van Woerden (ed.),
    "Topics in Interstellar Matter", Reidel, Dordrecht.
Seiradakis, J. H. 1976, in C. E. Fichtel and F. W. Stecker (eds.),
    "The Structure and Content of the Galaxy and Galactic Gamma Rays",
    299, Goddard Space Flight Center, Greenbelt.
Scoville, N. Z., and Solomon, P. M. 1975, Ap. J. (Letters), 199, L105.
Scoville, N. Z., Solomon, P. M., and Sanders, D. B. 1976, in C. E.
    Fichtel and F. W. Stecker (eds.), "The Structure and Content of
    the Galaxy and Galactic Gamma Rays", 163, Goddard Space Flight
    Center, Greenbelt.
Simonson, S. C., and Mader, G. L. 1973, Astr. and Ap., 27, 337.
Stecker, F. W., Puget, J. L., Strong, A. W., and Bredekamp, J. H.
    1974, Ap. J. (Letters), 188, L59.
Stecker, F. W., Solomon, P. M., Scoville, N. Z., and Ryter, C. E.
    1975, Ap. J., 201, 90.
Stief, L. J. 1973, in M. A. Gordon and L. E. Snyder (eds.) "Molecules
    in the Galactic Environment", Wiley, New York.
Strong, A. W. 1975, J. Phys. A, 8, 617.
Talbot, R. J., and Arnett, W. D. 1975, Ap. J., 197, 551.
Taylor, J. H., and Manchester, R. N. 1977, Ap. J., in press.
Weaver, H., and Williams, D.R.W. 1973, Astr. and Ap. Suppl., 8, 1.
Westerhout, G. 1958, Bull. Astron. Inst. Neth., 14, 215.
Westerhout, G. 1976, Maryland-Bonn Galactic 21-cm Line Survey, Uni-
    versity of Maryland.
Zuckerman, B., and Palmer, P. 1974, Ann. Rev. Astr. and Ap., 12, 279.

# GAMMA-RAY ASTROPHYSICS AND GALACTIC STRUCTURE

J.L. Puget,
Observatoire de Meudon, D.A.P.H.E., 92190 Meudon, France.

## 1. INTRODUCTION

Gamma-ray data (together with far-infrared and radio observations) are a powerful tool to study large-scale galactic structure, because of the absence of absorption even for column densities up to $\sim 10^{26}$ cm$^{-2}$. Two satellite experiments have brought us gamma-ray intensities for $E_\gamma > 35$ MeV (Fichtel et al., 1975, COS B Caravane Collaboration 1976). These gamma-rays are concentrated in the galactic plane and even at high latitude most of them are probably of galactic origin (Puget et al., 1976; Stecker, 1977; Schlickeiser, 1976). The extragalactic contribution at high latitude is less than $5 \times 10^{-6}$ cm$^{-2}$ s$^{-1}$ sr$^{-1}$.

It has been shown from spectral (Stecker et al., 1974) as well as production-rate considerations (Kniffen et al., 1976b; Stecker, 1976; Piccinotti and Bignami, 1976; Shukla and Paul, 1976) that for $E_\gamma > 100$ MeV, $\pi^0$ production is dominant as the source of gamma-rays, compton production being negligible and bremsstrahlung contributing for about 25%. The production rate in the solar neighbourhood for $E_\gamma > 100$ MeV per hydrogen atom is:

$$q = 1.3 \times 10^{-25} \text{ s}^{-1} \tag{1}$$

The similar rigidity of cosmic-ray electrons and protons together with the similar longitude profiles of gamma-ray intensity and 150-MHz intensity (mostly synchrotron emission) led Paul et al. (1976) to argue that:

$$\rho_{mat} \propto \rho_{C.R.e} \propto \rho_{C.R.p} \propto B^2 \tag{2}$$

at least on a large scale ($\gtrsim 1$ kpc). On scales for which the relationships (2) hold, the relative contribution of $\pi^0$ production and bremsstrahlung will stay the same, as long as secondary leptons are negligible compared to the primaries, which is true everywhere except in the Galactic-Center region (R < 500 pc). We have to deal with the large-scale distribution of two quantities, if we want to investigate more precisely

*Hugo van Woerden (ed.), Topics in Interstellar Matter, 179-185. All Rights Reserved.*
*Copyright © 1977 by D. Reidel Publishing Company, Dordrecht-Holland.*

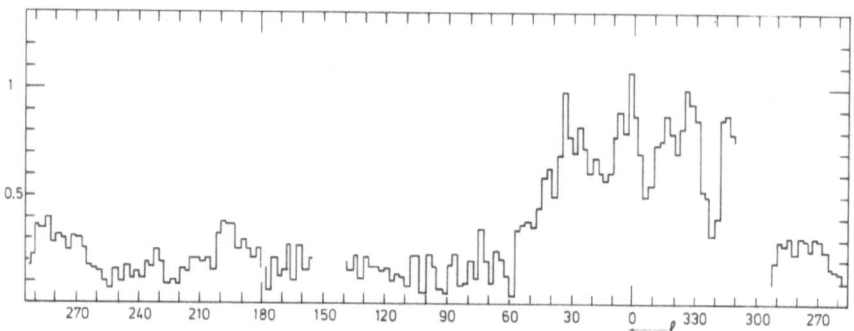

Figure 1.   Gamma-ray longitude profile, after subtraction of local contributions (< 1 kpc) and known point sources.

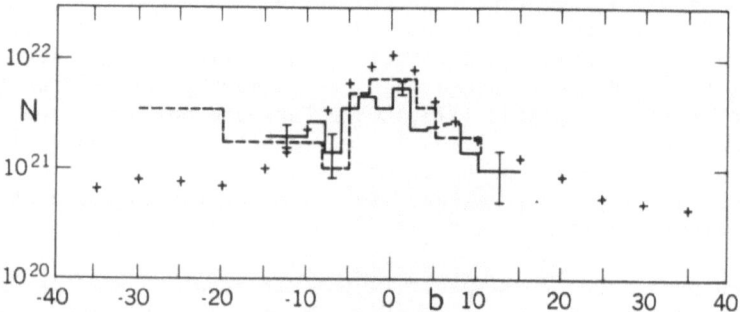

Figure 2.   Gamma-ray latitude profiles expressed in terms of column density, compared with HI (crosses) and absorption (dashed lines).

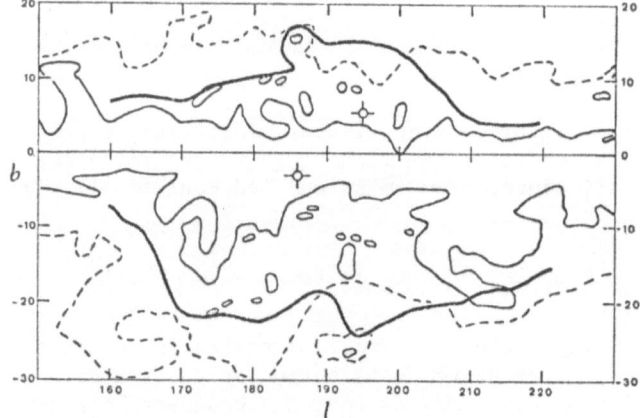

Figure 3.   Comparison of gamma-ray intensity and reddening in the Anti-centre region. Heavy line:  gamma-ray isophote $3 \times 10^{-5}$ $cm^{-2}$ $s^{-1}$ $sr^{-1}$. Solid line:  $E_{B-V} = 0.7$, dashed line:  $E_{B-V} = 0.3$.

on what scale relationships (2) hold: $\rho_{CR}$ and $\rho_{mat}$. Most of the interstellar medium in the inner Galaxy is known to be contained in massive
molecular clouds. It should be remembered that the angular resolution
of gamma-ray detectors is a few degrees, which implies that for clouds
a few hundred parsecs away, the gamma-ray maps always average the matter
distribution over several tens of parsecs. Such dimensions correspond to
"CO envelopes" (in Zuckerman and Palmer, 1974, terminology) and not with
the dense cores of molecular clouds (1 pc or so). Furthermore, this poor
angular resolution implies that the longitude profile integrates the
whole thickness of the galactic plane for regions more than 1 kpc away
from the Sun, and that the same amount of matter placed nearby or far
away will not contribute equally to the γ-ray flux and to a flux measured
with a beam smaller than the thickness of the plane even 25 kpc away from
the Sun (as in CO data). This implies that no direct comparison can be
made of the longitude profiles of γ-rays and for example CO-line intensity. The CO data can be averaged and interpreted assuming cylindrical
symmetry and standard differential rotation, to obtain an average large-
scale radial distribution of molecular material in the Galaxy (e.g.
Burton's contribution in this book, and references therein). Several
hypotheses have to be made to convert $^{12}$CO-line intensities into mole-
cular-hydrogen column densities. Nevertheless, other methods relying on
absorption by dust (near infra-red) or heavy elements (X-rays) allow one
to derive the column density between the Sun and the Galactic Center
(Ryter et al., 1975; Ryter, 1975). The best value: $N_H = 6.5 \times 10^{22}$ cm$^{-2}$
is in agreement with the one implied by the radial distribution of Gordon
and Burton (1976), and confirms that the amount of molecular material in
the Galaxy is probably known with an accuracy better than 25%.


2. LOCAL FEATURES

    In order to use gamma-ray data to deduce the large-scale radial
distribution of cosmic rays, one should take into account the comparatively large contribution of the first kiloparsec. On such a scale,
interstellar material is distributed in a very non-uniform manner. Puget
et al. (1976a) have shown that reddening data give the best estimate of
the local (< 1 kpc) column density. Using

$$I_\gamma = 7.25 \times 10^{-5} \, E_{B-V} \, \text{cm}^{-2} \, \text{s}^{-1} \, \text{sr}^{-1} \qquad (3)$$

and Fitzgerald's (1968) reddening data, the subtraction of the locally
produced gamma-rays leaves the longitude profile shown in Figure 1.

    Local features can be important in another way. Looking at gamma-
ray intensities at intermediate latitudes ($10^{\circ} < b < 40^{\circ}$) and comparing
them with absorption data (star and galaxy counts) or reddening data,
one can get experimental evidence on the behaviour of cosmic-ray intensity in molecular clouds. One can use either latitude profiles as in Figure
2, or isophotes as can be done in the Anticentre direction (Figure 3).

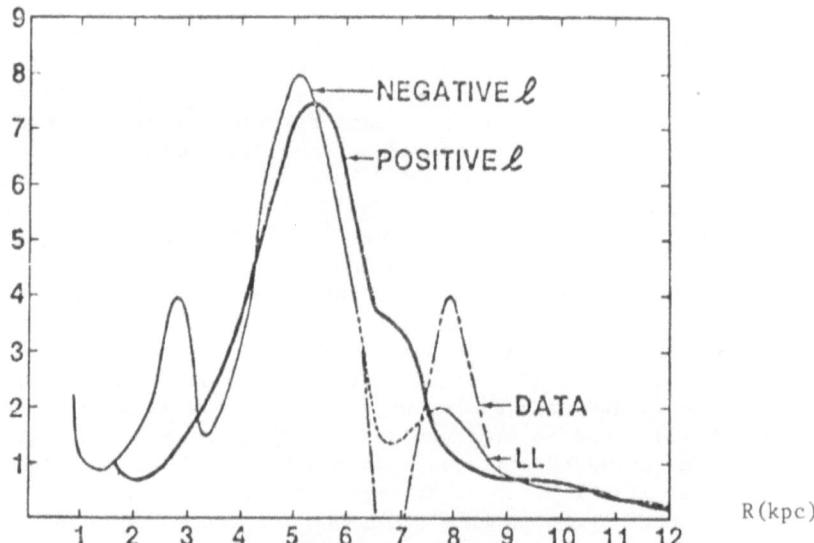

Figure 4.  Unfolded gamma-ray production rate in units of $10^{-4}$ $cm^{-2}$ $s^{-1}$, plotted against distance R(kpc) from the Galactic Centre.

In the recent SAS-2 data for the Anticentre region (Thomson et al., 1976), in all extended regions with $A_V \geq 3$ at $|b| > 5^o$ and far enough from known point sources (Galaxy counts below 1 per square degree, Khavtasi Atlas) a gamma-ray intensity of about $10^{-4}$ $cm^{-2}$ $s^{-1}$ $sr^{-1}$ ($E_\gamma >$ 100 Mev) is observed. This implies that in those clouds the cosmic-ray intensity is equal to the intensity in the solar vicinity within 30%.

## 3.  LARGE-SCALE FEATURES

The general shape of the profile shown in Figure 1 is quite symmetrical with respect to longitude 0, as expected for the profile of an intensity due only to large-scale structure.

The central peak is very likely to be associated with the Galactic-Centre region. The intensity of this peak can be interpreted as $\Pi^o$ production and bremsstrahlung emission in the massive molecular clouds within 200 pc from the Galactic Centre. Wolfendale and Worall (1976) have shown that, taking into account secondary electrons and positrons, the flux observed can be explained if one assumes an intensity for cosmic-ray protons 6 times larger than the value in the solar vicinity.

Aside from this component two prominent, almost symmetrical maxima can be seen at $|\ell| \approx 30^o$, with an intensity of about 1 $m^{-2}$ $s^{-1}$ $rad^{-1}$, falling quite sharply at smaller longitudes to a value of about 0.6 $m^{-2}$ $s^{-1}$ $rad^{-1}$. For longitudes $|\ell| > 60^o$ the flux is about 0.15 $m^{-2}$ $s^{-1}$ $rad^{-1}$. As shown by Puget and Stecker (1974) from earlier data, this longitude

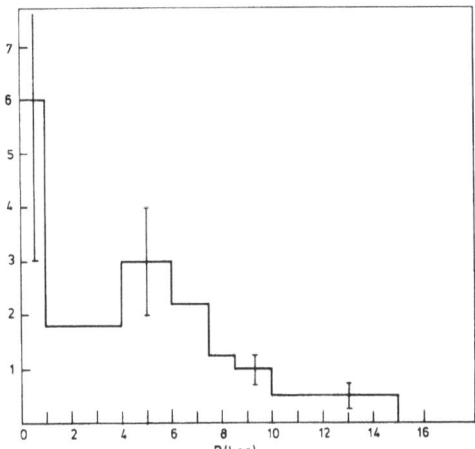

Figure 5. Cosmic-ray intensity as a function of galactic radius, normalized to the intensity in the solar neighbourhood.

distribution may be represented by a simple two-component model: a fairly small and uniform production rate for R > 8 kpc, and an inner component which can be unfolded if one assumes circular symmetry. The result of an unfolding of the most recent data (Stecker, 1977) is shown in Figure 4.

The gamma-ray production rate in the 5-kpc region is 7 or 8 times larger than in the solar vicinity. If one just divides this production rate per unit area of the Galaxy by the surface density of interstellar matter (Gordon and Burton, 1976), one gets the ratio of the predicted cosmic-ray intensity to the one measured in the solar vicinity as a function of galactic radius. This ratio is shown in Figure 5. The average value shown for 10 < R < 15 kpc is deduced from the average flux observed in the Anticentre direction. The fact that cosmic-ray intensity decreases as one goes toward the Anticentre can be seen directly from comparison of latitude profiles, as first pointed out by Dodds et al. (1975) and shown in Figure 2 (Puget et al., 1976a).

The result obtained on the cosmic-ray distribution in the Galaxy is in good agreement with Paul et al. (1975), who argued that $\rho_{mat} \propto \rho_{C.R.p}$. This radial distribution is also quite similar to those of HII regions (Stecker et al., 1974; Puget et al., 1976a), supernova remnants (Stecker, 1976), and pulsars, suggesting a strong connection between cosmic-ray sources and extreme population I.

## 4. ARM FEATURES AND COSMIC-RAY MODULATION

Several groups have considered spiral-structure models for our Galaxy to try to explain the peaks of the gamma-ray longitude profile. The two more recent and detailed studies (Paul, Cassé and Cesarsky, 1976; Kniffen et al., 1976) assume that the proportionality between cosmic-ray

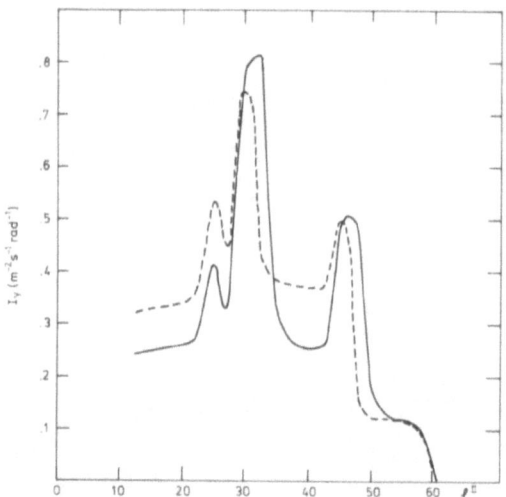

Figure 6.   Synthetic longitude profiles for a structure with 3 rings
(arms). Dashed line: arm width 500 pc, arm-interarm contrast 3; solid
line: arm width 300 pc, contrast 2.


intensity and matter density holds on the scale of spiral-arm width.
There is so far no direct evidence for this assumption. Obviously for a
given amount of matter the flux predicted will depend directly on the
width chosen for the spiral arms, which modulate the average distribution of matter.

     In Figure 6, two simple models with three arms are shown, with
different width and arm-interarm contrast. The profiles obtained should
be compared with the central excess of gamma rays between longitudes
$300°$ and $60°$. It is clear that the very bright peak observed in the
longitude profile can be reproduced only with an arm width of less than
500 pc and a contrast in the production rate of 3 or more. In the density-wave model of spiral structure (Roberts, 1976) the density contrast
is of this order, and the width of the increased density is less than
500 pc, so there is no need for a strong modulation of cosmic rays on
such a scale which is unlikely because of cosmic-ray diffusion.

     Matter is not distributed uniformly along the arms but in massive
clouds or groups of clouds. This implies that no detailed fit of the
data can be obtained from large-scale smooth models. The next step
should be a comparison per line of sight, using the radial-velocity
information to take account of the gradient in the average cosmic-ray
intensity and check if any cosmic-ray modulation is needed. It may be
noted that a model with wide (1 kpc) spiral arms (Kniffen et al., 1976)
does not reproduce the fluctuations of the gamma-ray intensity.

REFERENCES

C.O.S. B Caravane Collaboration: 1976, in The structure and content of
     the Galaxy and galactic gamma-rays, ed. C.E. Fichtel and F.W.
     Stecker, N.A.S.A., G.S.F.C., X-662-76-154, p. 39
Dodds, D., Strong, A.W. and Wolfendale, W.W.: 1975, Monthly Notices Roy.
     Astron. Soc. 171, 569
Fichtel, C.E., Kniffen, D.A., Thomson, D.J., Bignami, G.F., Ögelman, H.,
     Özel, M.A. and Turner, T.: 1975, Ap. J. 198, 163
Fitzgerald, M.P.: 1968, Astron. J. 73, 983
Kniffen, D.A., Fichtel, C.E. and Thomson, D.J.: 1976a, preprint
Kniffen, D.A., Fichtel, C.E. and Thomson, D.J.: 1976b, in The structure
     and content of the Galaxy and galactic gamma-rays, ed. C.E. Fichtel
     and F.W. Stecker, N.A.S.A., G.C.F.C., X-662-76-154, p. 341
Paul, J., Cassé, M. and Cesarsky, C.J.: 1976, Ap. J., in press
Puget, J.L. and Stecker, F.W.: 1974, Ap. J. 191, 323
Puget, J.L., Ryter, C., Serra, G. and Bignami, G.: 1976a, Astron. Astro-
     phys. 50, 247
Puget, J.L., Ryter, C. and Serra, G.: 1976b, in The structure and content
     of the Galaxy and galactic gamma-rays, ed. C.E. Fichtel and F.W.
     Stecker, N.A.S.A., G.S.F.C., X-662-76-154, p. 255
Piccinotti, G. and Bignami, G.F.: 1976, preprint
Roberts, W.: 1976, in The structure and content of the Galaxy and
     galactic gamma-rays, ed. C.E. Fichtel and F.W. Stecker, N.A.S.A.,
     G.S.F.C., X-662-76-154, p. 128
Ryter, C., Cesarsky, C. and Audouze, J.: 1975a, Ap. J. 198, 103
Ryter, C.: 1975b, Meeting of the American Physical Society, Washington
     D.C., 28 April - 1 May
Shukla, P.G. and Paul, J.: 1976, preprint
Schlickeisser, R. and Thielheim, K.O.: 1976, preprint
Stecker, F.W., Puget, J.L., Strong, A.W. and Bredekamp, J.H.: 1974,
     Ap. J. Lett. 188, L59
Stecker, F.W.: 1976, in The structure and content of the Galaxy and
     galactic gamma-rays, ed. C.E. Fichtel and F.W. Stecker, N.A.S.A.,
     G.S.F.C., X-662-76-154, p. 357
Thomson, D.J., Fichtel, C.E., Hartman, R.C., Kniffen, D.A. and Lamb,
     R.C.: 1976, preprint
Wolfendale, A.W. and Worall, D.M.: 1976, preprint
Zuckerman, B. and Palmer, P.: 1974, Annual Review of Astron. Ap. 12, 279.

# RECOMBINATION LINE OBSERVATIONS OF IONIZED HYDROGEN

L. Hart
N.R.A.L. University of Manchester
Jodrell Bank, Cheshire, England.

ABSTRACT

    Recombination line surveys of the large scale distribution of
ionized hydrogen in the Galaxy indicate a concentration of material
between 4 and 6 kpc from the galactic centre. This is very similar to
the distribution derived for molecular hydrogen from CO observations
but is unlike that found for atomic hydrogen.

## I.   RECOMBINATION LINE SURVEYS

    Surveys of H109$\alpha$ recombination line emission from the higher elec-
tron density HII regions in our Galaxy have been made by Reifenstein et
al. (1970) and Wilson et al. (1970). These surveys have naturally followed the distribution of such sources in our Galaxy. Two surveys of H166$\alpha$
emission with observations spaced uniformly along the galactic plane
have now also been made by Hart and Pedlar (1976a) and Lockman (1976).
These lower frequency surveys are relatively more sensitive to emission
from the low density ionized regions in our Galaxy. Such regions may be
extended low density envelopes around giant HII regions (see for example
Hart and Pedlar, 1976b) or the result of ionization by numerous smaller
weak sources (Jackson and Kerr, 1975),

    From the recombination line observations of HII regions and also
observations of emission from directions apparently free of discrete
sources (Matthews, Pedlar and Davies, 1973; Gordon and Cato, 1972) it
was known that the bulk of the line emission originated in the inner
regions, R < 9 kpc, of the Galaxy. Hart and Pedlar therefore concentrated
their observations to the galactic longitude range $5^{\circ} < \ell < 70^{\circ}$ while
Lockman also included the very interesting galactic centre region in the
range, $358^{\circ} < \ell < 51^{\circ}$, he observed. Because of the weakness of the line
emission from low density regions, typically $T_L \sim 0.05$ K, integration
times of 5-7 hours per position were employed. After such a time the
limiting factor in detecting very weak emission is the stability and
shape of the instrumental zero level. These two independent surveys

*Hugo van Woerden (ed.), Topics in Interstellar Matter, 187-193. All Rights Reserved.*
*Copyright © 1977 by D. Reidel Publishing Company, Dordrecht-Holland.*

therefore provide us with a useful comparison on the very weak emission detected in each.

The intensity, $\int T_L dv$, of the H166α emission observed every $1°$ along the galactic plane is shown in Fig. 1. The emission for $\ell < 5°$ is not shown. The distribution is characterized by strong emission, $T_L \gtrsim 0.1$ K, $\Delta V \sim 40$ km s$^{-1}$, from the direction of well studied HII regions with weak distributed emission, $T_L \sim 0.05$ K, $\Delta V \sim 100$ km s$^{-1}$ from adjacent directions. The distributed emission accounts for about 30% of the total recombination line power detected and originates predominantly in low density, $n_e \leq 10$ cm$^{-3}$, high temperature, $T_e \sim 5000$ K, gas (Jackson and Kerr, 1975; Matthews, Pedlar and Davies 1973; Hart and Pedlar, 1976a,b). Some emission from a cold component in the interstellar medium also occurs. Lockman notes that approximately 25% of his spectra show possible C166α emission at levels close to the noise.

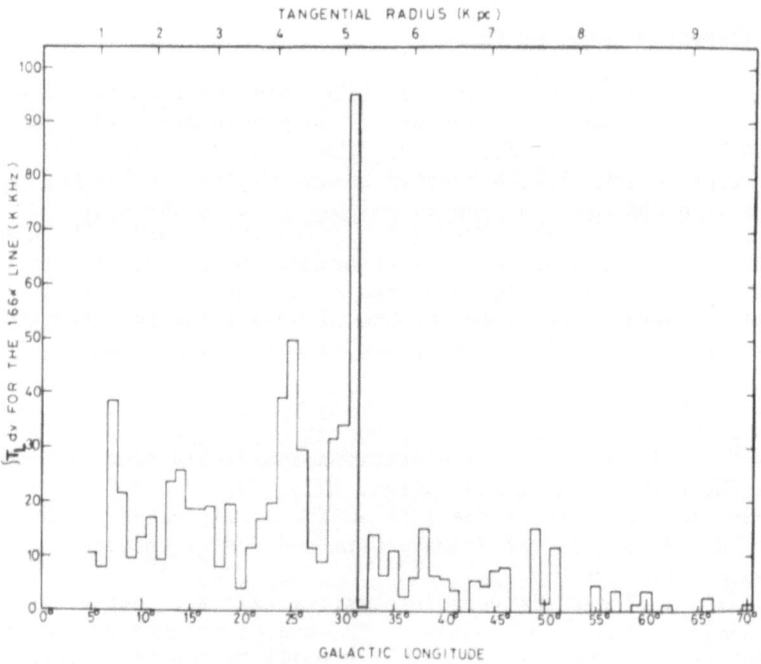

Figure 1.   Intensity of H166α emission as a function of galactic longitude
            (Hart and Pedlar, 1976a)

II.   COMPARISON OF IONIZED HYDROGEN WITH HI AND CO

    The H166  observations contributing to Fig. 1 are shown as a
velocity-longitude (v-ℓ) diagram in Fig. 2. Only contour levels greater
than four times the rms noise level are given. The dashed lines indicate
the run of HI peak emission with longitude in the Scutum and Sagittarius
arm regions. A comparison of Fig. 2 with a similar figure for HI reveals
how restricted the ionized gas is in velocity and longitude. Perhaps the
simplest comparison that can be made is between the centroid velocities
of the H166α and HI emission. This is shown in Fig. 3. The horizontal
bars denote ±2 sd about the centroid value obtained by fitting Gaussian
profiles. Some spectra have been decomposed into two main components. It
appears from this figure that the centroid velocity of the ionized gas
is similar to that for the higher velocity component of an HI spiral
feature. This is just the region where density wave shocks are stronger
and more frequent.

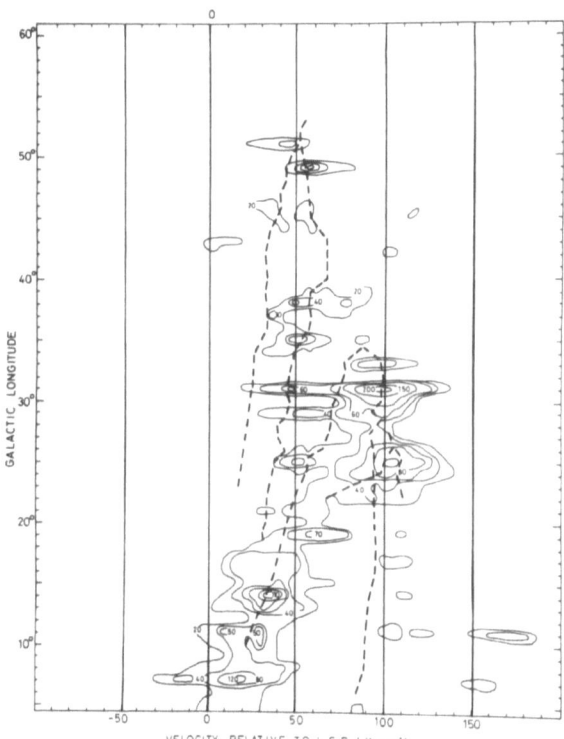

Figure 2.   Velocity-longitude diagram of H166α emission obtained every
            1° along the galactic plane.

A comparison between the v-ℓ diagram for ionized hydrogen and CO emission is given in Fig. 4. The CO observations are taken from an early survey by Scoville and Solomon (1975). Here the v-ℓ diagrams are very alike presumably because the dense neutral material has a similar spatial distribution in the Galaxy.

## III.   RADIAL DISTRIBUTION OF IONIZED HYDROGEN

The radial distribution can be derived from the recombination line observations by using a rotation curve for the material in our galaxy and assuming, in the first instance, circular motions about the galactic centre. The emission over a small velocity range $\Delta V$ can then be attributed to gas along the line of sight between R and R + $\Delta R$ from the galactic centre. Initially the Schmidt (1965) rotation curve, which has been approximated analytically by Burton (1971), is used. Lockman has also carried out the calculations using the Simonson and Mader (1973) form of the rotation curve for R < 5 kpc which changes the final distribution only slightly. A comparison between the two surveys and the two rotation curves used is shown in Fig. 5.

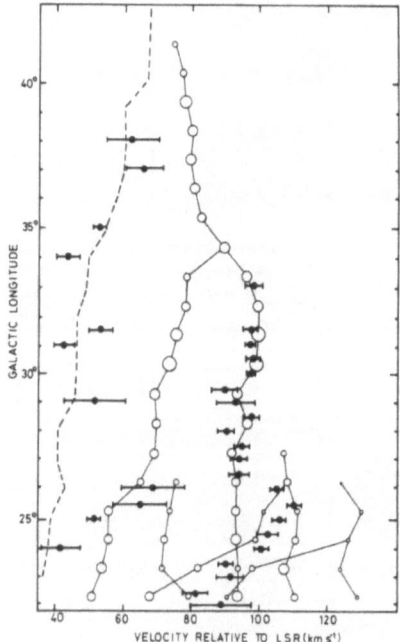

Figure 3.   Comparison of HII and HI centroid velocities. The locus of HI points has been adapted from Shane (1972) and Burton and Shane (1970).

It is clear that the ionized hydrogen is concentrated in the 4-6 kpc region of our Galaxy. This is not an entirely new result since the continuum observations of the Galaxy by Westerhout (1958) were inter- preted in a similar way. The recombination line observations however, refer specifically to hydrogen and a simple kinematic model used to derive the radial distribution as opposed to continuum optical depth considerations. Interior to 4 kpc the abundance of the ionized hydrogen decreases much more rapidly than it does for radii larger than 6 kpc. Outside 6 kpc the abundance falls by $e^{-1}$ over a scale length of 2 kpc. Lockman points out also that the low density gas interior to 4 kpc is relatively more abundant than the high density gas when compared to the region outside 6 kpc. The neutral hydrogen abundance remains roughly constant for radii larger than 5 kpc and decreases less abruptly than the ionized gas interior to 5 kpc. The CO radial abundance on the other hand (Gordon and Burton, 1976) shows a very similar kind of distribution to the ionized hydrogen in Fig. 5. Although interior to 4 kpc the CO abundance also decreases less rapidly than the ionized gas.

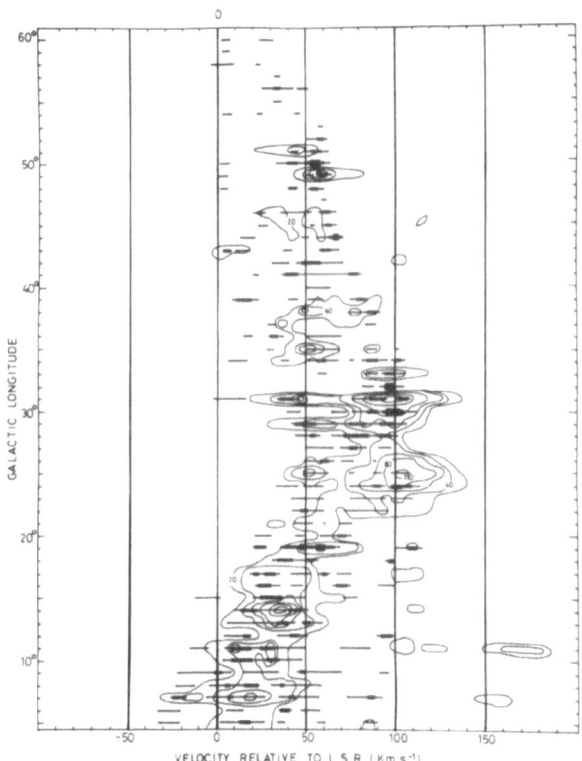

Figure 4.    Comparison of v-ℓ diagrams for ionized hydrogen and CO
             emission.

## IV.  LATITUDE EXTENT

As yet there are very few recombination line observations relating to the z extent of the low density ionized hydrogen. Those observations that are available suggest that the half-thickness, $z_{\frac{1}{2}}$, of the ionized layer increases for radii less than about 5 kpc from the galactic centre. The observations are at

R $\sim$ 4.3 kpc    $z_{\frac{1}{2}}$ $\sim$ 80 pc    (Hart and Pedlar, 1976a; Lockman, 1976)

R $\sim$ 5.5 kpc    $z_{\frac{1}{2}}$ $\sim$ 35 pc    (Gordon et al., 1972)

R $\sim$ 6.5 kpc    $z_{\frac{1}{2}}$ $\sim$ 35 pc    (Hart and Pedlar, 1976a)

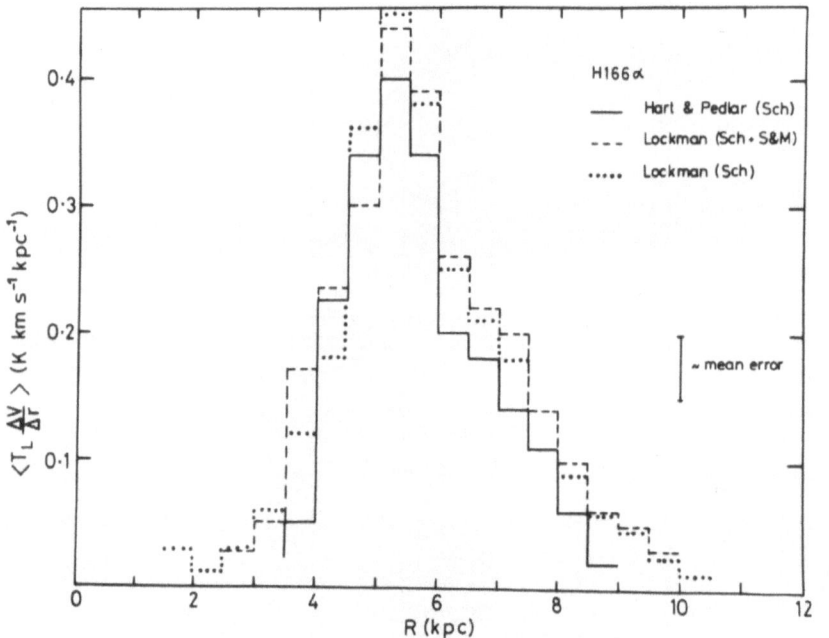

Figure 5.  Radial distribution of H166α line temperature abundance. The rotation curves used are indicated by (Sch) for Schmidt and (S & M) for Simonson and Mader.

REFERENCES

Burton, W.B.: 1971, Astron. Astrophys. 10, 76
Burton, W.B. and Shane, W.W.: 1970, IAU Symposium No. 38, 397
Gordon, M.A. and Cato, T.: 1972, Astrophys.J. 176, 587
Gordon, M.A. and Burton, W.B.: 1976, Astrophys.J. 208, 346
Gordon, M.A., Brown, R.L., and Gottesman, S.T.: 1972, Astrophys.J. 178, 119.
Hart, L. and Pedlar, A.: 1976a, Monthly Not.Roy.Astron.Soc. 176, 547
Hart, L. and Pedlar, A.: 1976b, Monthly Not.Roy.Astron.Soc. 176, 135
Jackson, P.D. and Kerr, F.J.: 1975, Astrophys.J. 196, 723
Lockman, F.J.: 1976, Astrophys.J. 209, 429
Matthews, H.E., Pedlar, A., and Davies, R.D.: 1973, Monthly Not.Roy. Astron.Soc. 165, 149
Reifenstein, E.C., Wilson, T.L., Burke, B.F., Mezger, P.G., and Altenhoff, W.J.: 1970, Astron. Astrophys. 4, 357
Schmidt, M.: 1965, Stars and Stellar Systems 5, 513
Scoville, N.Z. and Solomon, P.M.: 1975, Astrophys. J. Lett. 199, L105
Shane, W.W.: 1972, Astron. Astrophys. 16, 118
Simonson, S.C. and Mader, G.L.: 1973, Astron. Astrophys. 27, 337
Westerhout, G.: 1958, Bull. Astron. Inst. Neth. 14, 215
Wilson, T.L., Mezger, P.G., Gardner, F.F., Milne, D.K.: 1970, Astron. Astrophys. 6, 364

# THE LARGE-SCALE DISTRIBUTION OF INTERSTELLAR MATTER IN THE CONTEXT OF THE DENSITY-WAVE THEORY

William W. Roberts, Jr.
University of Virginia
Charlottesville, Virginia    and

W. B. Burton
National Radio Astronomy Observatory[*]
Green Bank, West Virginia

## ABSTRACT

The theoretically viable prospect that density waves and the associated galactic shock fronts are present in disk-shaped galaxies has received support in recent years from a variety of observational studies. Large-scale shocks in the interstellar gas may play an important role in determining the kinematics and the relative distribution of various galactic tracers. This is particularly apparent in some external spirals, because of the advantageous perspective, and for the tracers HI and CO in our own Galaxy. Simulation of CO observations according to the precepts of the density-wave theory shows that these precepts are supported by several observational results.

## INTRODUCTION

In this review we focus on the large-scale distribution and dynamics of interstellar matter in spiral galaxies. Our main interest here is our own Galaxy, although--because our Galaxy apparently differs in no fundamental way from many other spirals--we do make use of the advantageous perspective offered by external galaxies. It was observations of external galaxies which for the most part motivated the density-wave interpretation of spiral structure. According to this interpretation, which originated with Lindblad (1963) and which has been developed toward a coherent theory by Lin and his colleagues and others (see Lin 1971), many observed characteristics of a spiral arm are attributed to the recent passage through the interstellar medium of the crest of a spiral density wave.

## GALACTIC SHOCK WAVES

In such a passage a large-scale galactic shock wave can develop in the ambient interstellar medium and the resulting compression can

---

[*] Operated by Associated Universities, Inc., under contract with the National Science Foundation.

*Hugo van Woerden (ed.), Topics in Interstellar Matter, 195-205. All Rights Reserved.*
*Copyright © 1977 by D. Reidel Publishing Company, Dordrecht-Holland.*

trigger the formation of high-density clouds, and, in turn, of stars,
molecules, and other tracers along a spiral arm (Roberts 1969, Roberts
and Yuan 1970, also see Fujimoto 1966). Figure 1 illustrates such a
shock and the nonlinear response of the gas density distribution along
a typical streamline.

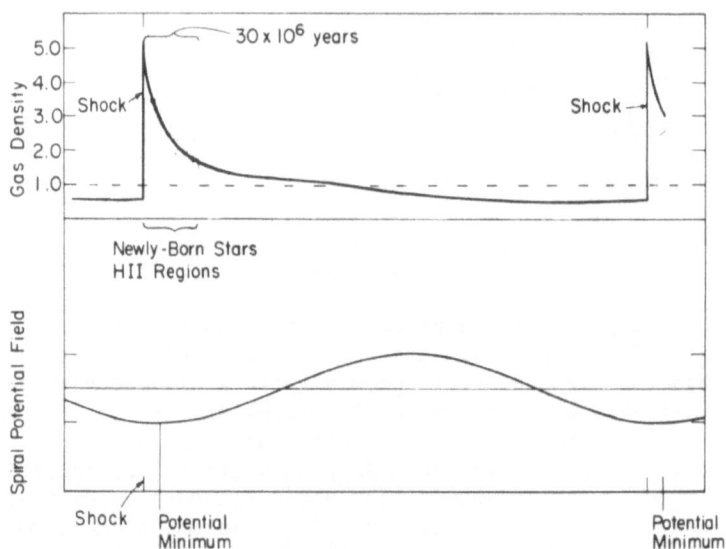

Figure 1. Distribution of gas density, relative to the underlying
(stellar) gravitational potential, along a typical streamline (Roberts
1969). The gaseous response has a narrow peak induced by the shock.
The regions of strong compression resulting in molecule and star formation lie within and just outside the shock. The shaded area shows
the extent of these objects corresponding to a formation and evolution
time of 30 million years.

        Galactic shocks form if the driving force of the interstellar gas,
which is contributed by the spiral gravitational field of the stars, is
sufficiently strong to force the velocity component normal to a spiral
arm $W_\perp$, to oscillate at transonic speeds about its unperturbed value,
$W_{\perp 0}$. Figure 2 shows that there are actually two different regimes of
shocked gas flow (Shu et al. 1973). For regime (1), with $W_{\perp 0}$ > effective acoustic speed (left panel), the shocks are strong and produce narrow zones of high gas compression (as in Figure 1). For regime (2),
with $W_{\perp 0}$ < effective acoustic speed (right panel), the shocks are weak
or absent and yield broad zones of low compression. This difference
accounts for the observed differentiation between the narrow, filamentary spiral arms, observed in some galaxies and the broad, massive
spiral arms observed in others.

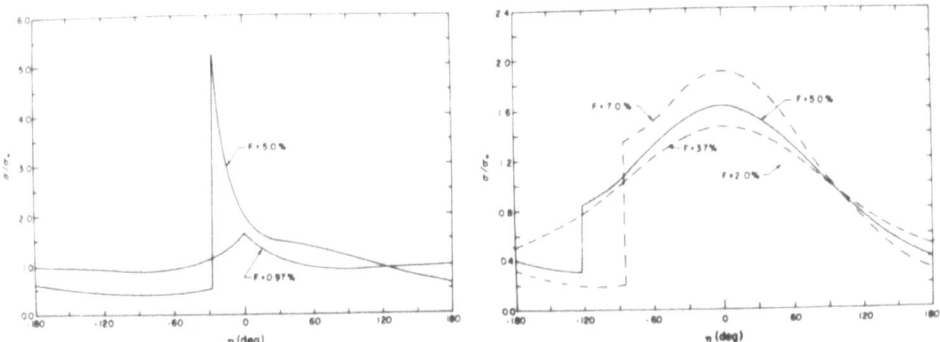

Figure 2. Variation of the density enhancement, $\sigma/\sigma_0$, as a function of phase, $\eta$, between adjacent arms, for various strengths F of the driving field, showing the distinction between the nonlinear and linear regimes (Shu et al. 1973). Regime (1), with $W_{\perp 0}$ > effective acoustic speed, is characterized by strong shocks and narrow regions of high gas compression (left panel). Regime (2), with $W_{\perp 0}$ < effective acoustic speed, is characterized by weak shocks (or no shocks) and broad regions of weak gas compression (right panel). The morphological characteristics of a given galaxy would depend on which regime is predominant.

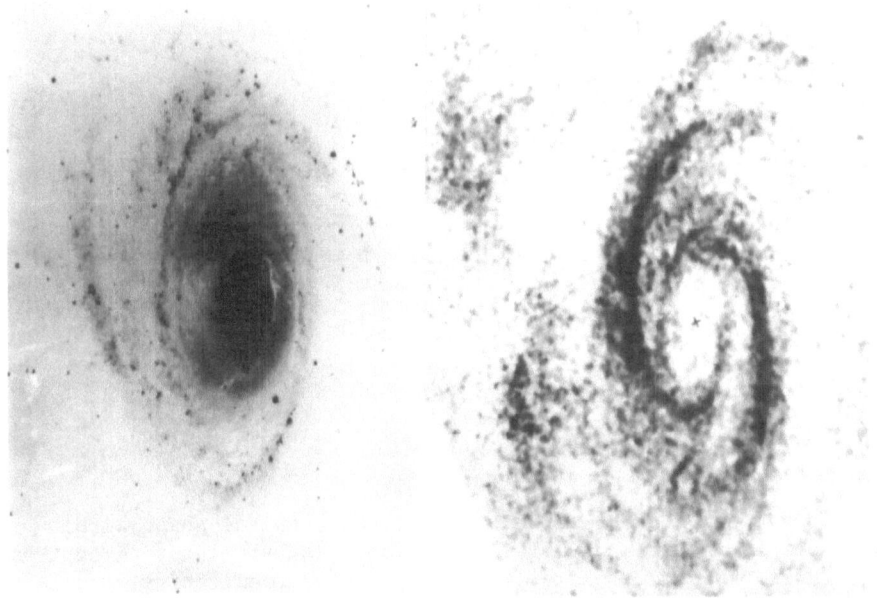

Figure 3. Negative, blue-light photograph of the sample galaxy M81, showing the well-developed narrow luminous spiral arms characteristic of strong galactic shocks (Kitt Peak National Observatory).

Figure 4. Radiograph of the HI surface density distribution in M81, showing that the gaseous arms are also narrow and well-defined (Rots and Shane 1975). The scale of this figure is about half that of Figure 3.

## M81--A REPRESENTATIVE GALAXY

A representative galaxy characterized by regime (1) is M81. We view the well-developed, narrow spiral arms of the luminous-star distribution to be a consequence of strong shocks. This structure is apparent in the photograph of M81 shown in Figure 3. The observed surface density distribution of neutral hydrogen in M81, shown in Figure 4, also exhibits similarly well-developed structure, extending to large distances from the center of the galaxy (Rots and Shane 1975).

Figure 5 shows the observed HI surface density distribution in M81 plotted against spiral phase in three separate galactic annuli (Rots 1975). The HI arm peaks are narrower than the interarm troughs, and some peaks are skewed. Both of these characteristics are suggestive of strong shocks (cf. Figure 1).

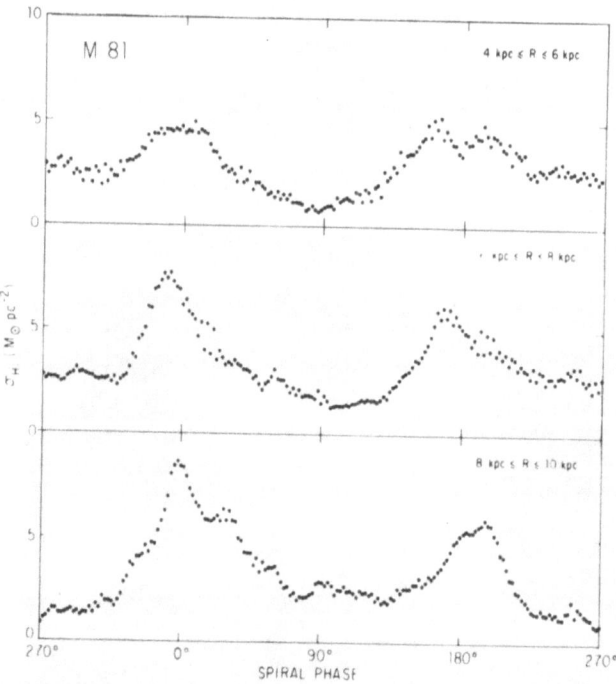

Figure 5. Variation of HI surface density with spiral phase within three annuli in the disk of M81 (Rots 1975). The arm-interarm contrast and the skewed shape of some of the arm peaks are indicative of non-linear phenomena with shocks.


## OUR GALAXY IN THE PERSPECTIVE OF EXTERNAL SPIRALS

Figure 6 shows for 24 representative galaxies the relationship between $W_{10}$ and potential shock strength on the one hand, and luminosity classification and degree of development of spiral structure on the other (Roberts et al. 1975). Those galaxies with potentially strong shocks show well-developed spiral structure, while those for which weak or no shocks are predicted via the $W_{\perp o}$ parameter show poorly developed structure.

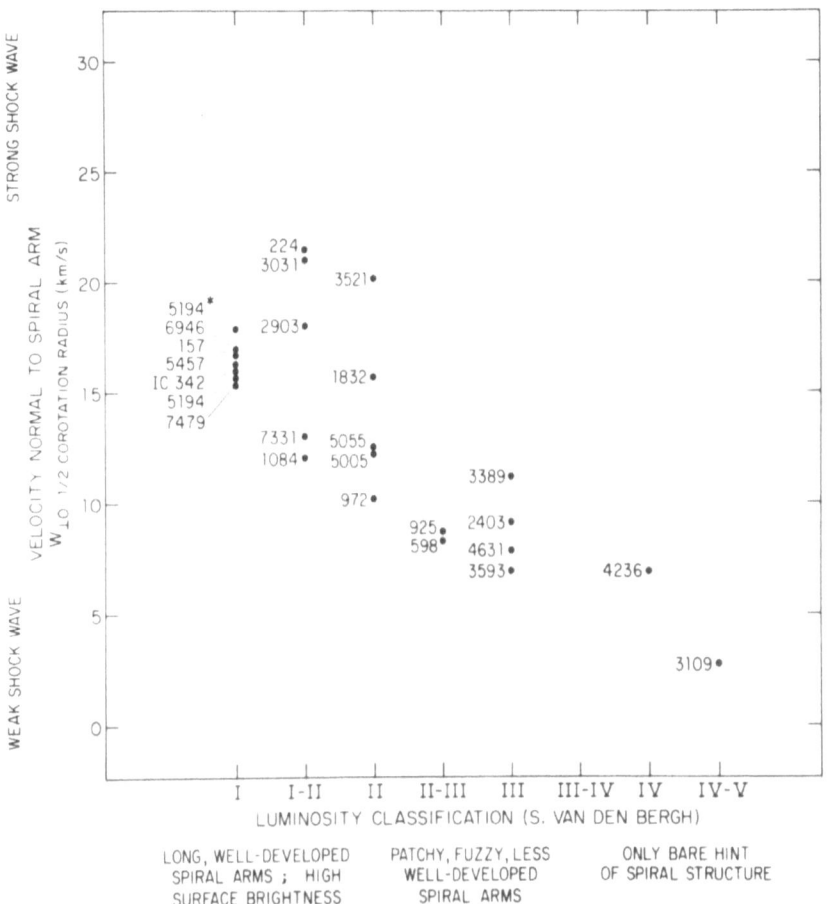

Figure 6. Variation of the kinematic parameter $W_{\perp o}$ with luminosity classification for 24 representative spirals (Roberts et al. 1975). The correlation suggests that strong shocks are associated with well-developed spiral structure; weak shocks are associated with poorly-developed spiral structure.

The parameter $W_{\perp o}$ and the shock strength in turn depend on two even more fundamental quantities: (1) the total mass of the galaxy divided by a characteristic dimension, $M/\varpi_c$, and (2) the degree of concentration of mass toward the galactic center, measured by $\varpi_{.5M}/\varpi_C$. Figure 7 shows this dependence (Roberts <u>et al</u>. 1975). A galaxy as high on the $W_{\perp O}$ ridge as M81 (NGC 3031) can form strong shocks and thereby exhibit well-developed, narrow spiral arms. Our own Galaxy lies in the same general region of the parameter space as M81, suggesting similarly well-developed spiral structure.

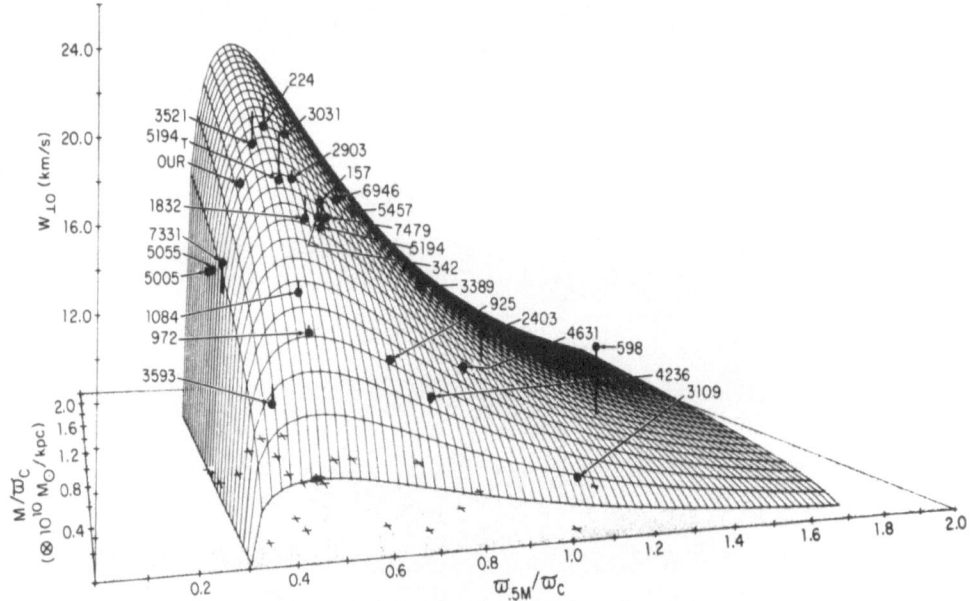

Figure 7. Theoretical categorization of disk-shaped galaxies in terms of the two-dimensional parameter space of the fundamental quantities $M/\varpi_C$ and $\varpi_{.5M}/\varpi_C$ (Roberts <u>et al</u>. 1975). The distances $\varpi_C$ and $\varpi_{.5M}$ refer to the corotation radius and to the radius inward of which lies half the total galactic mass, respectively. Location on the $W_{\perp O}$ surface determines the potential strength of galactic shocks. Galaxies with relatively large central concentration of mass (low $\varpi_{.5M}/\varpi_C$), <u>e.g</u>., NGC 3031 (M81), have the strongest shocks; those galaxies with low central mass concentration (high $\varpi_{.5M}/\varpi_C$), <u>e.g</u>., NGC 2403, have the weakest shocks.

## RADIAL ABUNDANCE DISTRIBUTION OF INTERSTELLAR MATTER

Several observational aspects of the structure of our Galaxy are amenable to interpretation in the context of the density wave theory. One of the most impressive of such aspects concerns the different morphological characteristics of atomic hydrogen compared to those of all other constituents of the galactic disk which can be observed along transgalactic paths (see the review by Burton 1976). The diameter of the galactic disk as defined by atomic hydrogen is fully twice as large as that of the galactic disk as defined by the ionized and molecular states of hydrogen, as well as by other molecules, supernova remnants, pulsars, $\gamma$-radiation, and synchrotron radiation. Analogously for external galaxies, the HI disk is larger than the luminous disk (M. S. Roberts 1974). This separation of the peak concentrations of these tracers, representing recent and current activity, from that of the HI implies that there exists a factor other than the average HI density which controls the formation rate of molecules and stars. Furthermore, the rough equivalence of the inner Galaxy distributions suggests that a single mechanism is responsible.

Compression due to the passage of a galactic density wave and the associated shock has been invoked specifically to account for the relative HII regions/HI distributions (e.g., Mark 1971, 1974; Lin 1971; Shu 1973; Roberts 1975) and for the relative CO/HI and other distributions (Scoville and Solomon 1975, Burton et al. 1975, Gordon and Burton 1976). Figure 8 indicates schematically for our own Galaxy the regimes of strong and of weak compression (Burton 1976, following Roberts et al. 1975). Strong shocks are possible in the inner Galaxy because $W_{\perp o}$ > the effective sound speed. Here galactic shock fronts trigger molecular and stellar formation. In the outer region compression is weak and conditions for molecular and stellar formation are unfavorable except in unusual local environments. The inner Galaxy is all the more susceptible to galactic shock influences because the frequency at which the gas meets the spiral-wave pattern, $2[\Omega(\varpi)-\Omega_p]$, increases with decreasing distance from the galactic center. Other aspects of the problem, including the dependence of compression efficiency on ambient gas density, are dealt with by Oort (1973), Shu (1975), and Segalovitz (1975).

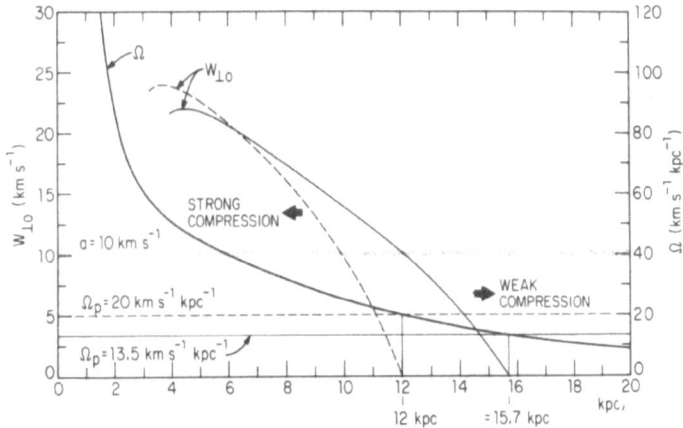

Figure 8. Schematic representation for our Galaxy of the variation of the density-wave parameter $W_{\perp o}$ calculated for two plausible values of the wave pattern speed, $\Omega_p$. Relatively strong compression is expected in the inner Galaxy where $W_{\perp o} > a$, the effective acoustic speed in the interstellar medium. The frequency at which the shock wave acts on the gas, $2(\Omega-\Omega_p)$, also enhances the compression effects in the inner Galaxy.

CO AS A TRACER OF THE GALACTIC SHOCK IN OUR GALAXY

    The overall kinematics of CO are generally similar to those of HI
(see Burton and Gordon 1977a,b).  The ordered perturbations in the galactic rotation curve derived from these tracers provide the most direct
observational evidence for spiral structure on the large-scale in our

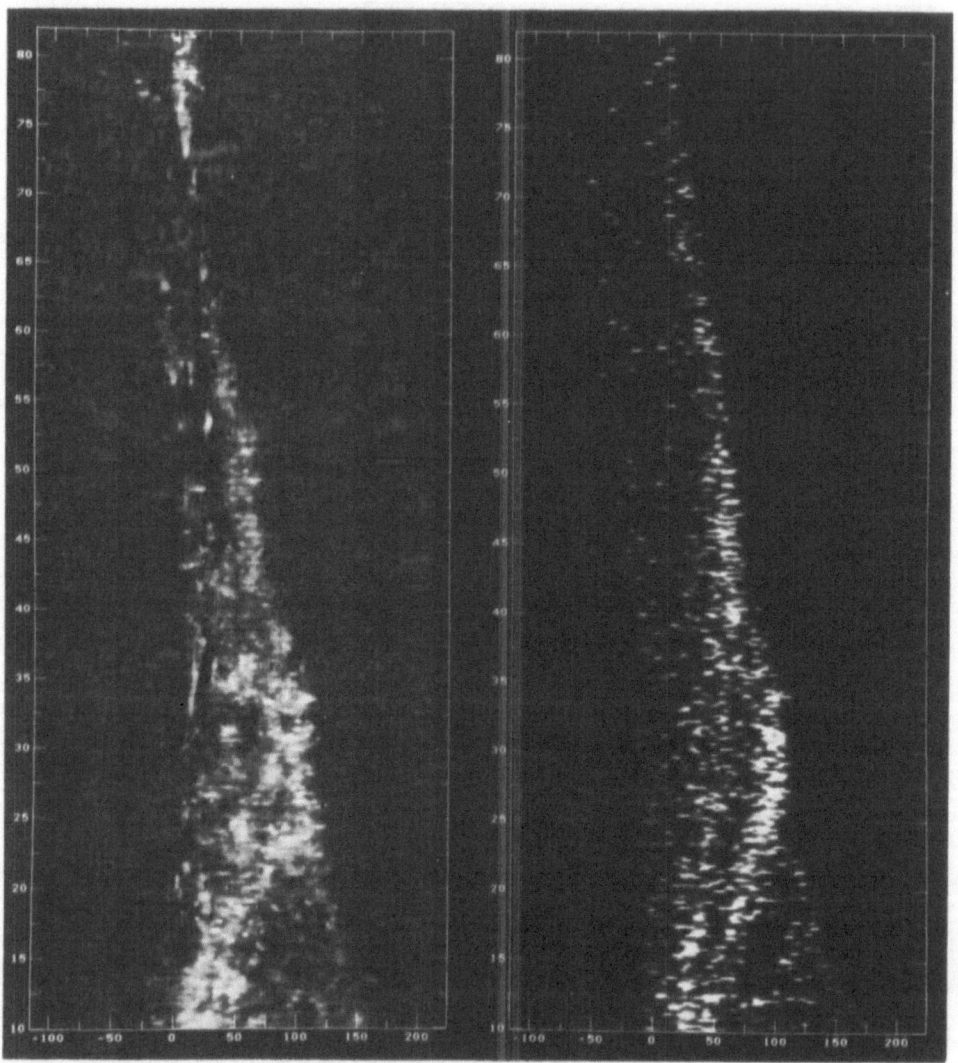

Figure 9.  (left)  Grey-scale representation of the longitude, velocity
arrangement of $^{12}C^{16}O$ emission observed at b = 0° (Burton and Gordon
1977a,b).  (right)  Longitude, velocity arrangement of emission inherent in the synthetic line profiles, representing CO distributed
stochastically in discrete clouds in the spiral model.  The clouds's
kinematics and probability of occurrence follow the predictions of the
density-wave theory.

Galaxy (for the case of HI, see e.g., Yuan 1969, Burton and Shane 1970, Burton 1971). However, because these ordered variations, particularly in the CO data, are not more sharply confined in angle and of larger amplitude in velocity, it is natural to ask if the CO kinematics are consistent with the presence of large-scale shocks.

We address this question by following the dynamics of the cloud component in a spiral density-wave model with a two-phase interstellar medium similar to that of Shu et al. (1972). In this model the clouds are viewed as embedded bodies which expand or contract to adjust to changes of the ambient pressure in the intercloud medium. The increase in pressure across a galactic shock occurring in the hot intercloud phase is in turn transmitted to the cold clouds. We consider the cloud-intercloud medium interaction due to turbulent viscosity following Sawa (1975). The clouds themselves plow supersonically, as ballistic particles, through the shock and beyond until the drag force due to turbulent viscosity slows them down. Consequently a smoothed peak

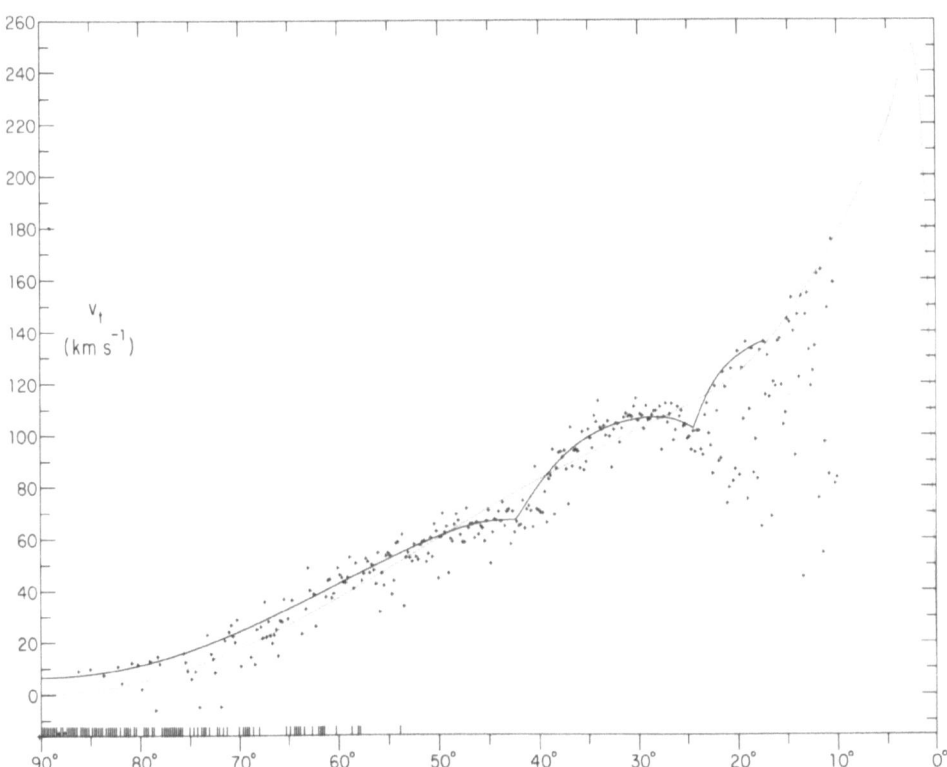

Figure 10. Distribution with longitude of the terminal velocities measured on the synthetic CO profiles shown on the right in Figure 9. The black line shows how the terminal-velocity locus predicted by the spiral model differs from that of the axisymmetric model (grey line). The scatter in the points is due to the stochastic nature of the distribution. The ordered variations are due to the density-wave perturbation, and reproduce the general trends observed.

forms in the overall density distribution of the cloud component, in
contrast to the sharply peaked intercloud component.

We apply the CO profile simulation method of Burton and Gordon
(1976, 1977a,b), and the approach which they applied to the axisymmetric
case, to the cloud component in this spiral model.  The right-hand side
of Figure 9 contains the resulting longitude, velocity arrangement of
the modelled CO emission.  Comparison with the left-hand side of Figure
9, which shows the observed arrangement from Burton and Gordon (1977a,b),
indicates rough agreement in a number of respects.

One quantitative test of the reality displayed by the model involves comparison of the observed and modelled terminal velocities
which represent the high-velocity cutoffs of the profiles comprising
Figure 9.  This test has also been used by Bash and Peters (1976), who
also consider (from a somewhat different point of view) ballistic cloud
trajectories.  Figure 10 shows the longitude distribution of the terminal velocities measured on the synthetic profiles.  The amplitude
and locations of the ordered variations resemble those observed (Burton
and Gordon 1977a,b).  The ordered variations vary smoothly, despite
the presence of large-scale shocks in the model.

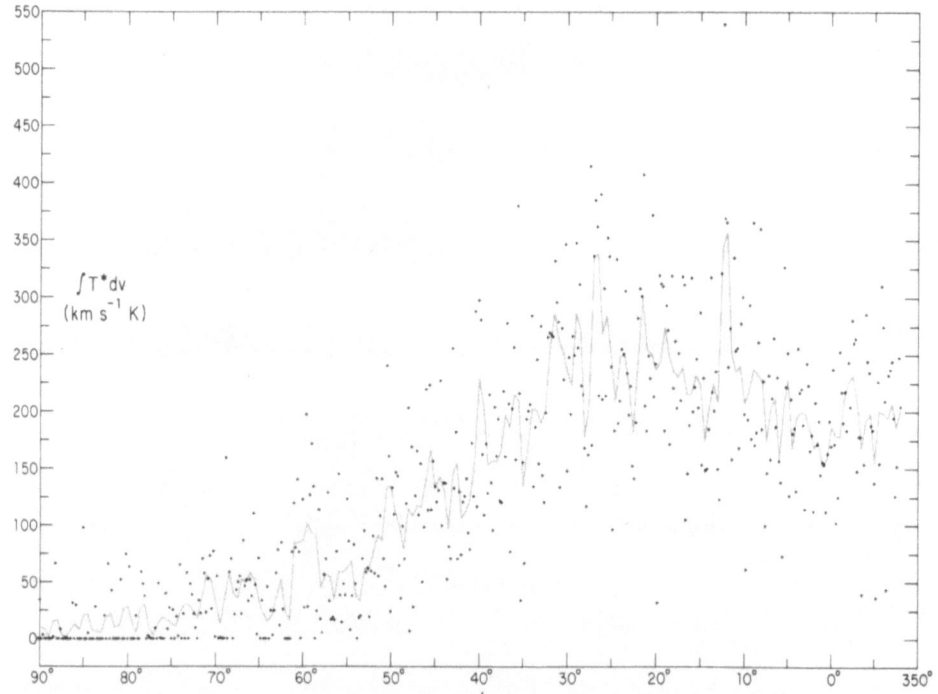

Figure 11.  Variation with longitude of the integrated emission of the
synthetic CO profiles for the spiral model shown in the right panel of
Figure 9.  The grey line shows every 0°5 the data smoothed over ±0°5
of longitude.  The scatter of the points is due to the stochastic
nature of the distribution of clouds in the model.  This variation is
consistent with that observed.

A second quantitative measure available from the CO observations is the total-velocity integrated emission at each direction. The total-line powers derived from the simulated profiles are plotted in Figure 11. These show general agreement with the integrals for the observed emission (Burton and Gordon 1977a,b). We remark that no steps occur in the variation of integrated emission in the spiral model despite the concentration of material to spiral arms. This is because information is lost in integration over velocity. Note in Figure 9, however, that the large-scale clustering of the observed CO emission to major bands extending over some 15° of longitude is reproduced by the spiral model. We conclude that the CO observations do not rule out concentration of molecular material to spiral arms, and indeed provide some (rather indirect) evidence for such concentration. We will discuss these points in detail elsewhere.

The work of one of us (WWR) was supported in part by the National Science Foundation under Grant AST72-05124 A03.

REFERENCES

Bash, F. N. and Peters, W. L. 1976, Ap. J., 205, 786.
Burton, W. B. 1971, Astron. & Astrophys., 10, 76.
Burton, W. B. 1976, Ann. Rev. Astron. & Astrophys., 14, 275.
Burton, W. B. and Gordon, M. A. 1976, Ap. J. (Letters), 207, L189.
Burton, W. B. and Gordon, M. A. 1977a, Astron. & Astrophys., submitted.
Burton, W. B. and Gordon, M. A. 1977b, in "Topics in Interstellar Mat-
    ter", H. van Woerden (ed.), Reidel, Dordrecht.
Burton, W. B. and Shane, W. W. 1970, Proc. I.A.U. Symp. No. 38, 397.
Fujimoto, M. 1966, Proc. I.A.U. Symp. No. 29, 453.
Gordon, M. A. and Burton, W. B. 1976, Ap. J., 208, 346.
Lin, C. C. 1971, in Highlights of Astr., 2, 88, Reidel, Dordrecht.
Lindblad, B. 1963, Stockholm Obs. Ann., 22, 3.
Mark, J.W.-K. 1971, Proc. Natl. Acad. Sci., 68, 2095.
Mark, J.W.-K. 1974, Ap. J., 193, 539.
Oort, J. H. 1973, Proc. I.A.U. Symp. No. 58, 375.
Roberts, M. S. 1974, Science, 183, 371.
Roberts, W. W. 1969, Ap. J., 158, 123.
Roberts, W. W. 1975, Vistas in Astron., 19, 91.
Roberts, W. W., Roberts, M. S., and Shu, F. H. 1975, Ap. J., 196, 381.
Roberts, W. W. and Yuan, C. 1970, Ap. J., 161, 877.
Rots, A. H. 1975, Astron. & Astrophys., 45, 43.
Rots, A. H., and Shane, W. W. 1975, Astron. & Astrophys., 45, 25.
Sawa, T. 1975, Sci. Reports Tohoku Univ. Series I, LVIII, 13.
Scoville, N. Z. and Solomon, P. M. 1975, Ap. J. (Letters), 199, L105.
Segalovitz, A. 1975, Astron. & Astrophys., 40, 401.
Shu, F. H. 1973, Am. Sci., 61, 524.
Shu, F. H. 1975, Proc. Colloquium on Spiral Galaxies, Bures-sur-Yvette,
    309.
Shu, F. H., Milione, V., Gebel, W., Yuan, C., Goldsmith, D. W., and
    Roberts, W. W. 1972, Ap. J., 173, 557.
Shu, F. H., Milione, V., and Roberts, W. W. 1973, Ap. J., 183, 819.
Yuan, C. 1969, Ap. J., 158, 871.

Chapter 5


INTERSTELLAR MATTER IN EXTERNAL GALAXIES




Papers presented in a joint session of IAU Commissions
34 (Interstellar Matter) and 28 (Galaxies), Grenoble,
26 August 1976.


Session Chairmen: Hugo van Woerden, Erik B. Holmberg

# HII REGIONS IN GALAXIES OF THE LOCAL GROUP *

G. Courtès,
Observatoire de Marseille,
Laboratoire d'Astronomie Spatiale du C.N.R.S.

ABSTRACT

The galaxies of the Local Group showing HII regions have the great advantage of being close enough to provide, sometimes more easily than in our Galaxy: the fine morphology and distribution of those HII regions; their unambiguous positions compared with those of stars and HI contours; the precise shape of spiral patterns and their true galactocentric distances.
Recent observations using high-sensitivity monochromatic detection and new optical designs have given very efficient means for understanding of galactic and extragalactic structures.
Efforts have been made to obtain standard sizes of HII regions, in order to resolve extragalactic distance-scale problems.
Studies of star formation at the front of spiral features rich in HII regions have been discussed in relation with the kinematics of the gas as well as with stellar distribution and evolution.
Several new kinds of HII regions have been distinguished, up to the very extended diffuse emissions of spiral arms, disk and central regions.
Results from high-sensitivity and high-resolution spectrography, and from far-UV space astronomy, give a first general explanation of the different modes of excitation of these emission phenomena.

TABLE OF CONTENTS

---

* Part of this paper is an extension of a previous communication to the Symposium in Mittelberg on HII regions (Wilson and Downes, 1975) which unfortunately did not reach the publisher.

*Hugo van Woerden (ed.), Topics in Interstellar Matter, 209-242. All Rights Reserved.*
*Copyright © 1977 by D. Reidel Publishing Company, Dordrecht-Holland.*

INTRODUCTION

The main advantage of the HII regions in the Local Group is that they allow us the best observing conditions because of their relatively small distances, which are very well determined thanks to Cepheĩds.

Diameter, classification, shape, circular loops, filaments, diffuse emission, etc., relation with the positions of exciting stars are, in the Local Group, easy to observe. Exact positions of radio maxima are obtained with good resolution for comparison with optical results.

In the constant astrophysical problem of penetrating the universe more and more deeply, HII regions are (owing to their monochromatic emission and their character of extended sources) (Courtès, 1960, 1972) one of the best optical signals in more distant galaxies. One continues to observe them when the brightest stars are already under the limit of detection of the largest telescopes. This advantage is due: 1) to the possibility of selecting the emission lines with narrow interference filters (exposure time not limited by sky background), and 2) to the opportunity of using, for extended sources, a very high focal ratio (high illumination of the detector).

The usual logic of astrophysics consists of a precise study of the nearest galaxies, in which the role and properties of HII regions can be more easily understood among the other components of the galaxy. When all these relations and criteria are well established, a safe extrapolation to more distant galaxies is possible. This is true not only for problems of morphology and distance, but also for spectrographic and photometric observations.

## THE LOCAL GROUP

Among the 20 members of the Local Group described by van den Bergh (1968), we have selected some galaxies showing the HII region phenomenon to best advantage:

| Spirals | Irregulars |
|---|---|
| M 31 | SMC |
| M 33 | LMC |
| The Galaxy | IC 1613 |
| | NGC 6822 |

We point out, first, a selection effect due to the instrumentation and the angular size. Paradoxically, there is relatively little information on the largest galaxies, because of the need for several dozen spectra or direct photographs, which one cannot obtain in the amounts of observing time usually allocated on the largest optical telescopes.[*] Wide-field optical techniques have been used to improve these observational limitations (Courtès, 1972).

As the first means of detection, Hα emission has been mapped in these galaxies with several filter techniques. The most selective one, using narrow-band interference filters, has in fact been applied to all these galaxies, but with its extreme refinements to M31, M33 and our Galaxy (Courtès, 1964). We shall describe the main results obtained on these galaxies. The Galaxy itself will be considered as an edge-on spiral, as it is shown on wide-field camera photographs. On these, the resolution is of the same order as in the Andromeda Nebula observed with a medium-size telescope. We shall also make some remarks and comparisons with the four irregular galaxies cited above.

## THE ANDROMEDA NEBULA (M31)

The Andromeda Nebula was the first extragalactic nebula in which HII regions were discovered by W.Baade, more than twenty-five years ago (Baade and Mayall, 1950). These HII regions distributed in long chains gave the best evidence of the spiral arms, and this fact of observation permitted Morgan to test the hypothesis of spiral structure of our own Galaxy. The first optical drawing of galactic spiral arms was obtained immediately by plotting the photometric distances of the exciting stars of HII regions only. We will appreciate later from other examples the quality of this intuition and its real meaning related to the new-born stars' location.

Baade's result, obtained with coloured filters and an F/5 telescope (100", Mount Wilson), was one of great merit. The full reduction of Baade's Hα survey was recently completed by Arp, owing to some comple-

---

[*]Conditions are quite different with the radiotelescopes, which are sometimes devoted for very long runs to this type of observation.

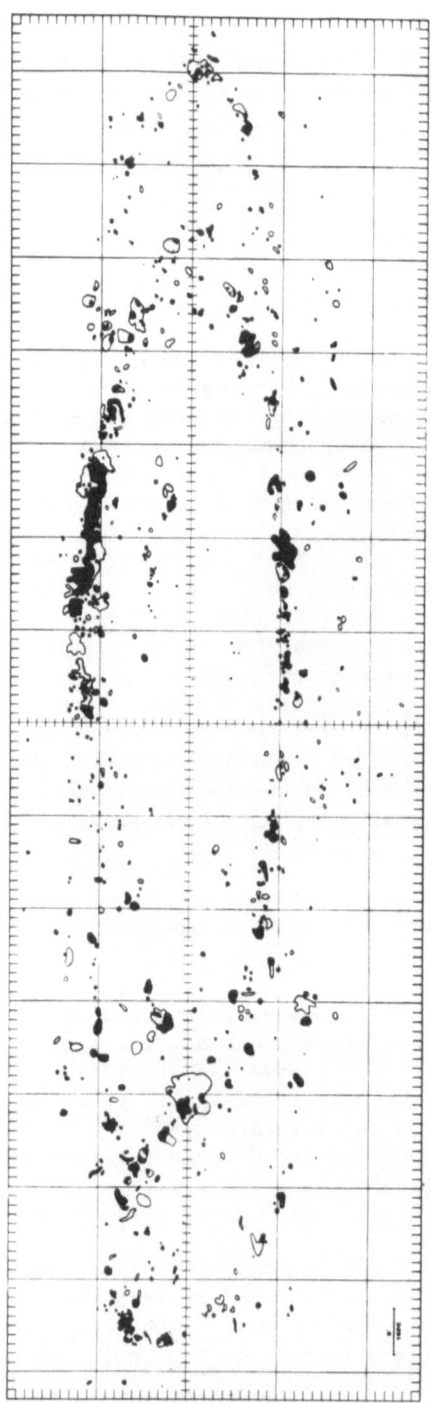

Figure 1 – Distribution of HII regions in M31

From the HII region Atlas of M31, prepared by Maucherat, Pellet and Simien (Marseilles and Haute Provence Observatories), with technical assistance of Mrs. Astier, Mrs. Viale and J.C. Gonin.

mentary plates obtained at the 48" Schmidt telescope (Arp and Brueckel, 1973). This important catalogue will be followed by a new Atlas of the HII regions, prepared by Hodge (1976). Another Atlas is prepared by the Marseille group (Courtès et al., 1975a), after a survey made by Maucherat, Pellet and Simien with interference filters about 25 to 60 times more selective than the coloured filters (Courtès, 1972), and a focal ratio of F/1 (focal reducers) increasing the illumination of the detector by a factor 25 (Fig. 1). In some cases, an image tube reduces the exposure time by a factor 10, but the larger field ($1^{\circ}12'$) is obtained on photographic emulsion (Boulesteix et al., 1975).

Spectrographic studies by Rubin and Ford (1970) and by Deharveng and Pellet (1975), and Fabry-Pérot observations by Courtès, Pellet and Deharveng with the 200" Hale telescope (Courtès, 1972; Deharveng and Pellet, 1975), have extended the detection of faint $H_\alpha$. Between the conventional HII regions, a general "diffuse arm emission" was observed in 1969 by Courtès, Deharveng and Pellet (200" telescope) and described by Monnet (1974). This "diffuse arm emission" is very similar to the one in the Milky Way first pointed out by Courtès (1951) and measured now with high spectroscopic resolution by the remarkable Fabry-Pérot experiment of Reynolds, Roesler and Scherb (1973) and the wide-field observations of Courtès and Sivan (1972) and Sivan (1974). Its emission measure is of the order of 100 $cm^{-6}$ pc (Monnet, 1974) (Plate 1).

## Sizes of HII regions

As a distance indicator and for physical studies, the sizes of HII regions have been intensely observed. A general remark concerns the instrumental resolving power. It is obvious that surveys made with the 200" telescope will be rich in small-size HII regions; hence the maximum in the statistics of sizes may be shifted to smaller diameters. The same effect appears in the Galaxy because of the small distances of the HII regions (Georgelin, 1971). In nonselective statistics, there are some other ambiguities in the diameter evaluation, for example the "bubble bath" effect for HII regions located in dust-rich spiral arms (Brand and Zealey, 1975) seems to appear in the NW outer arm of M31. The same difficulty could appear in the Galaxy itself if all emission regions were put in the statistics, without careful grouping (Georgelin and Georgelin, 1976). A possible solution is, as Gum and de Vaucouleurs (1953) suggested a long time ago, to select only the largest, circular ring-shaped HII regions; these are of greatest interest for the distance scale, they are true Strömgren spheres, and possibly ionization-bounded.Filamentary HII regions have little meaning in size determinations. The statistics made recently (Maucherat, 1976) on Strömgren spheres give 37 HII regions in M31 between 80 and 180 pc. In our Galaxy the 7 largest HII regions are between 150 and 200 pc (Georgelin, 1971; Churchwell, 1975; Sivan, 1976). Sandage and Tammann (1974) have shown that the diameter of the largest HII regions is indeed a function of morphological type (Georgelin, 1971), but also of luminosity class. The absolute size of HII regions for Sc galaxies has been discussed by Hodge (1966), but very few Sb galaxies have been considered.

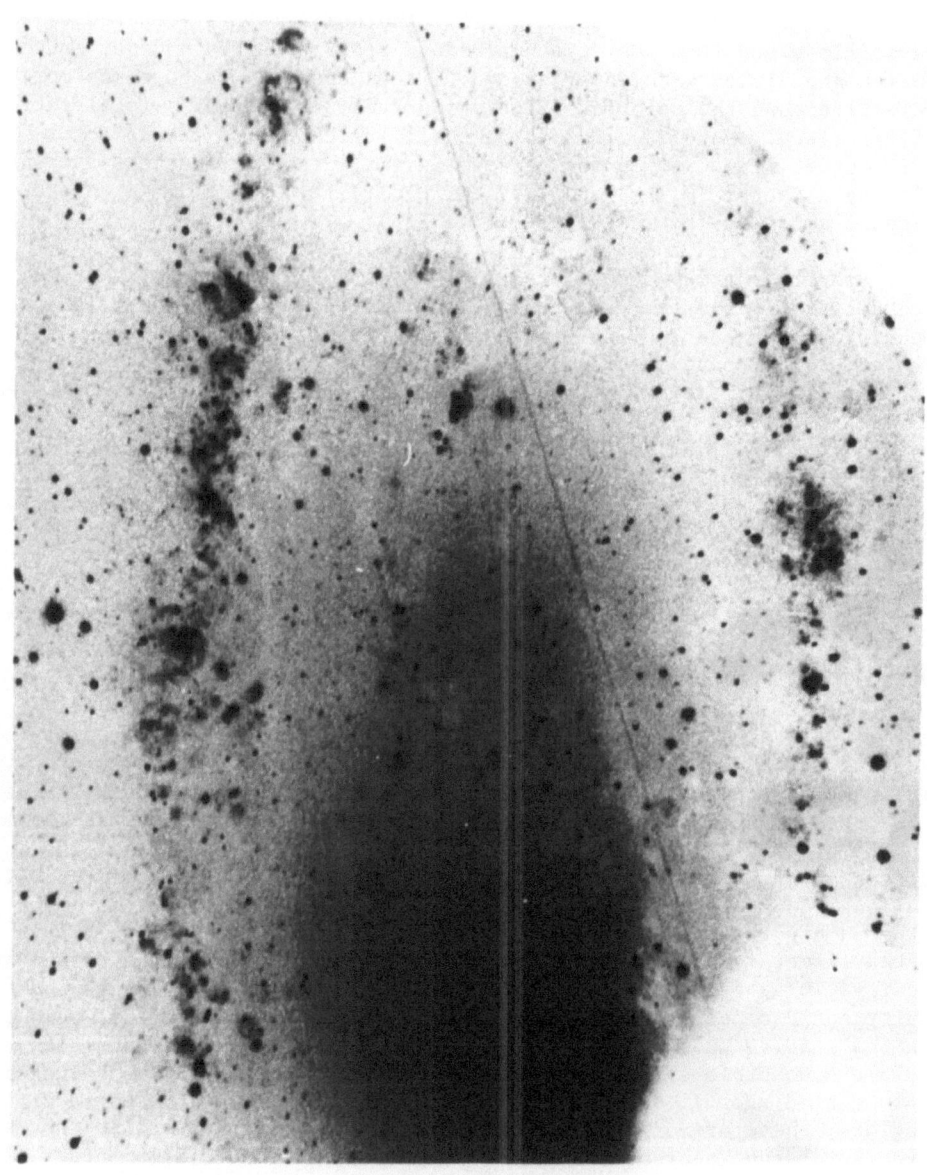

Plate 1

Diffuse arm emission in the N4 spiral arm of M31. Part of the $1°12'$
field of the F/1 "reducteur focal" of the 193-cm telescope at
Observatoire de Haute Provence; interference filter $\Delta\lambda = 20$ Å (Hα);
film Kodak 103 aE. Observation by J. Maucherat and H. Petit.

In the general statistics for M31, the most frequent diameter (Arp and Brueckel, 1973) is 80 pc, with no clear cut-off in the larger diameters. Thus it cannot be a very good distance indicator. If one excludes the largest peculiar HII regions, the mean size of the general statistics is different in M33 and M31 by a factor 2: it is 40 pc for M33 (Courtès and Cruvellier, 1965; Arp and Brueckel, 1973; Boulesteix et al., 1974). However, the largest ring-shaped HII regions have diameters of the same order in both systems (cf. Figures 2 and 4).

## New results in M31

The latest F/1 survey of M31 with the 193-cm telescope of the Haute Provence  Observatory equipped with the F/1 "reducteur focal" shows mainly:

1) The diffuse arm emission with some filamentary structure in the N4 spiral arm (Baade and Arp notation) (Plate 1).

2) The true structure of some already catalogued HII regions. A certain selection effect due to the lack of contrast of the coloured-filter techniques gave, in the first catalogues, groups of several small HII regions corresponding in fact to the brightest maxima of a large HII region. Thus, as has already been noticed (Arp and Brueckel, 1973), the more recent catalogues have fewer HII regions than the first one of Baade and Arp.

3) The conventional HII regions seem to be located around the peaks of HI emission (Emerson, 1974) in the HI gradient areas (Boulesteix et al., 1974), but very bright spots of HI seem to coincide with condensed HII regions (Baldwin, 1976). The overall distributions of HI and HII are very similar (Berkhuijsen, 1976).

4) Some very large and faint rings in the outer parts of M31, often with an exciting star at the periphery in a possible compression zone.

5) A "clumpy" structure of large areas in the inner N and S arms. A better resolution than that of the 193-cm telescope of the Haute Provence Observatory is probably necessary (Plate 2).

6) Until now, no disk interarm emission.

7) A central emission ($R < 5' \approx 1$ kpc) (Rubin and Ford, 1970 ; Deharveng and Pellet, 1975; Burbidge and Burbidge, 1975).

## The central HII regions

A spectrographic survey, when the resolution is well adapted (Courtès and Cruvellier, 1965; Baranne et al., 1975), say, about 40 $\overset{\circ}{A}$/mm, can be a good means of detection, especially in the central parts, where the continuum background is strong. The monochromatic contrast coefficient

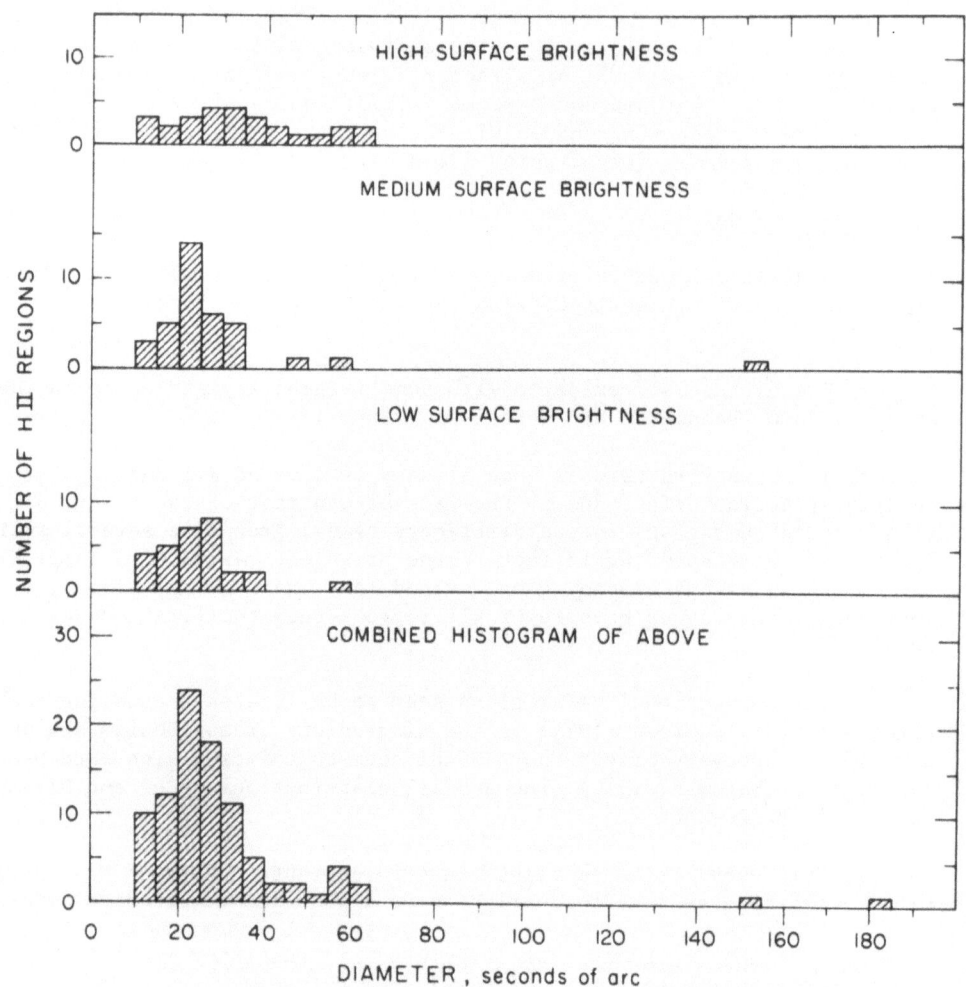

Figure 2

Histograms showing frequency distributions for diameters of HII regions
in M31. The three upper histograms show the numbers of regions in the
three different classes of surface brightness, while the bottom histogram combines the three above, showing the final frequency distribution
of HII for all classes of surface brightness. See on Fig. 4 the scale
in pc; M31 and M33 are approximately at the same distance. From Arp and
Brueckel, 1973 (Courtesy, Astrophys. J.).

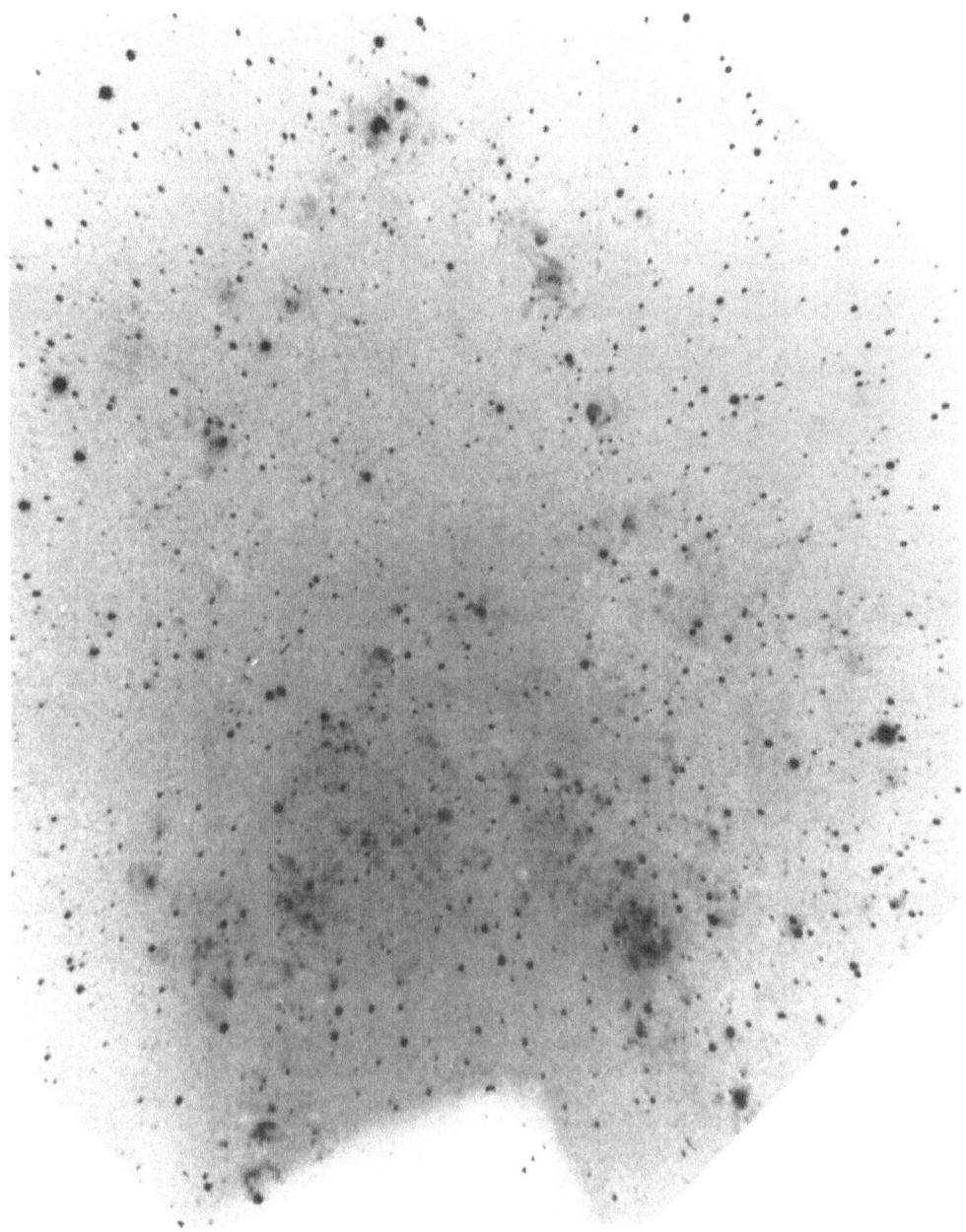

Plate 2

Clumpy structure in the inner arms of M31. For technical details see
Plate 1.

Γ is 210 (Courtès, 1972, Table 2, p. 104). For comparison: the Palomar
sky survey is defined to have Γ = 1.

Rubin and Ford (1970) and Deharveng and Pellet (1975) have observed
Hα, [N II] 6584 Å and 6548 Å, and the [S II] doublet in the central part
of M31 up to 1 kpc from the nucleus. At a certain distance from the
nucleus, the decrease of intensity of the emission lines permits only to
observe 6584 Å of [N II] owing to the usual high [N II]/Hα ratio in the
central part of galaxies (Burbidge and Burbidge, 1975). This [N II] line
(6584 Å) is unfortunately in coincidence, because of the radial velocity
along the north half of the major axis of M31, with the 6577.2 Å band of
OH from the night sky. Reliable determination of the diameter of the
central emission would require a better spectral resolution and a better
spectrophotometric analysis. A new generation of spectrographs and Fabry-
Pérots equipped with photon counting devices will bring a better solution
to this problem.

The central emission of M31 has a radius of about 1 kpc, but its
true morphology is not yet known, in spite of the refined methods used
by Deharveng and Pellet. According to them, extension beyond 1 kpc is
unlikely. Near the nucleus, observation with a mask to cancel the continuum light of the nucleus, like the observation of M51 (Warner, 1973),
should permit to reach the very central part of the emission phenomenon.
One remembers that Johnson (1972) detected dust filaments down to 6"
from the centre. In spite of many spectroscopic studies since the Münch
(1960) observations of [O II] lines, no good morphology has been given
of this central part, because of the strong stellar continuum. A Lyot
combination filter and coronograph-like device studied for the 2.4 m
Space Telescope are now being prepared in Marseille to improve the
detection of the emission features close to the nucleus.

## Mode of excitation of the central part

The central part of M31 does not exhibit blue supergiant stars.
Rubin and Ford's (1971) evaluation of the Hα flux inside 400 pc radius,
$\Phi = 0.2 \times 10^{-5}$ erg cm$^{-2}$ s$^{-1}$ ster$^{-1}$, permits obtaining an upper limit to
the Lyman continuum and to the possible star population under the limit
of individual detection producing it. In connection with this central
emission, it is of interest to recall space observations by OAO (Code,
1969) which show a turnup of the far-UV spectrum below 2000 Å. The space
resolution of OAO was 10', and that of a first observation with the
French rocket experiment Persée (Cruvellier et al., 1970) was only 60'.
An UV image of the central part was required. A new Wynne rocket telescope, FAUST (F/1.12; 16 cm diameter; 7.5° flat field; Thomson TH 9301
image intensifier equipped with C-I photocathode and P20 phosphor; resolution 1') has been launched from Kourou, French Guyana (Deharveng et
al., 1976). In the 1400–1800 Å bandpass ($\Delta\lambda = 470$ Å), the central part
of M31 appears as a roughly spherical bright core with an apparent diameter of 70 arc sec. The integrated far-UV flux corresponds to an AOV
star of visual magnitude 9.3.

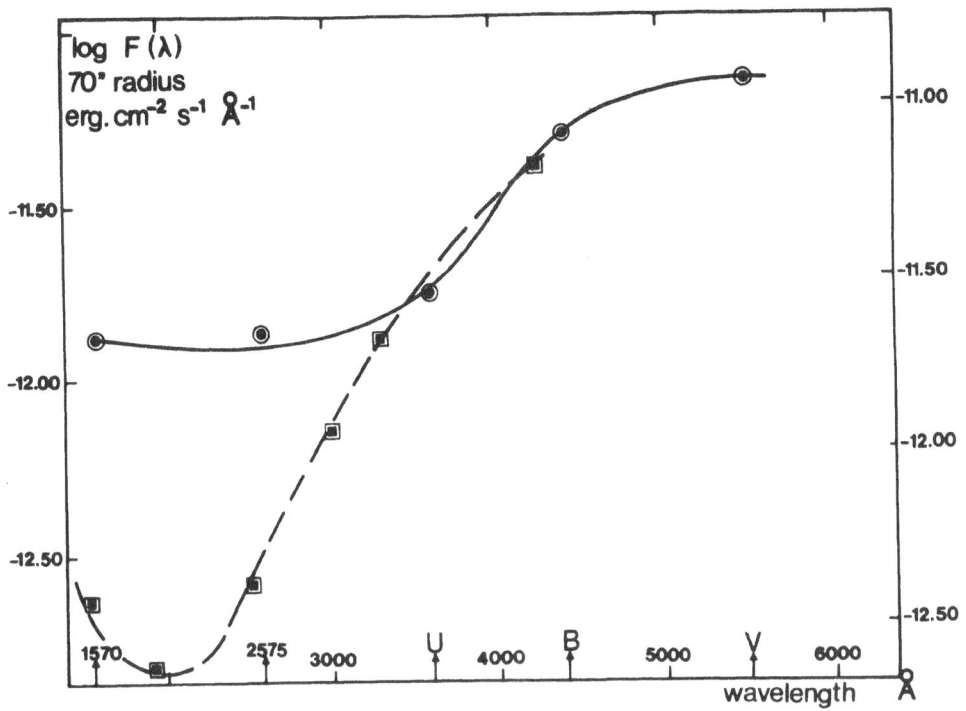

Figure 3.   Comparison of spectral distributions in the central part of
M31.   Squares: Code (1969);   Circles: Deharveng, Laget, Vuillemin and
Monnet (1976).


        Another observation at 2574 Å ($\Delta\lambda$ = 358 Å) was made twice with the
S 183 French experiment (field 6x9°) aboard Skylab (Courtès et al., 1975)
and gives a diameter of 100 arc sec. Images of field stars obtained at
the same time permit to conclude again for a central, pseudo-extended
source and a flux equivalent to an AOV star of V = 8.3. These two observations at 1570 and 2574 Å correspond to a flat distribution of the UV
spectrum[*] (Fig. 3), with a far-UV flux about 0.3 of the blue flux, in
good agreement with relatively low temperatures ($10^4$ to $2\times10^4$ K) for the
blue horizontal branch of Population II stars. Conclusions similar to
those of Deharveng et al. (1976) are obtained from other considerations
(Ulrich, 1975) for M31 and for the Sb and elliptical galaxies with central emission.


[*] Similar to the M31 observation of Burton (1974) with the TD1 satellite,
   and the globular-cluster observations of the Astronomical Netherlands
   Satellite (Wesselius, 1975).

THE TRIANGULUM NEBULA (M33)

The M33 galaxy shows certainly the best conditions of observation: a small distance permitting good comparisons of stars and HII regions, clear spiral structure in the southern arm (Courtès and Cruvellier, 1965; Courtès and Dubout-Crillon, 1971), and a relatively small tilt permitting good geometrical positions. As was pointed out by Carranza et al. (1968) and recently by Monnet (1974), M33 is also the best example of a galaxy showing the main manifestations of extended Hα emission:

1) The conventional HII regions distributed at the front of the spiral arms. The HII regions of the southern spiral arms follow a quasi-perfect logarithmic spiral (Plates 3 and 4). The general statistics on 369 HII regions give a most frequent diameter of 40 pc (Boulesteix et al., 1974) (Fig. 4).

2) The diffuse arm emission, often of filamentary structure ($\approx 100$ pc), in close relation with the conventional HII regions.

3) The diffuse emission of the disk (Mayall and Aller, 1939), discovered in its largest extent in Hα by Carranza et al. (1968) up to 2 or 3 kpc from the centre, confirmed by Palomar 200" observations of Courtès, Georgelin, and Monnet in 1968 owing to a special optical arrangement computed by Baranne (Courtès, 1972). This general disk emission has been the key to complete mapping of the gas kinematics.

4) Ring-like HII regions in the outer parts, with diameters of about 200 pc (Boulesteix et al., 1974), and a linear increase of diameter with distance to the nucleus (Plate 5).

5) Emission of the nucleus itself (Benvenuti et al., 1973).


Distribution of HII regions

In general, the distribution of the HII regions follows the gradient of HI intensities (Emerson, 1974; Boulesteix et al., 1974), more than the peak intensities. This is most obvious in the southwestern area, in which a large HI cloud lies from 3 to 6 kpc from the centre (Wright et al., 1972). The HII regions are situated at the periphery of this HI cloud. One will find a deep analysis of this phenomenon in Boulesteix et al. (1974) and in Dubout-Crillon (1976), as well as a precise morphology of the nearest (720 kpc, 1' = 209 pc) and best example of spiral structure, the southern arm. Its inner front is exceptionally sharp and follows a pure logarithmic spiral, with a pitch angle of $27^{\circ}$:
$$\theta = 130^{\circ} + 260^{\circ} \log_{10} \tilde{\omega}.$$

The situation is summarized in Figure 5 and in the following tabulation:

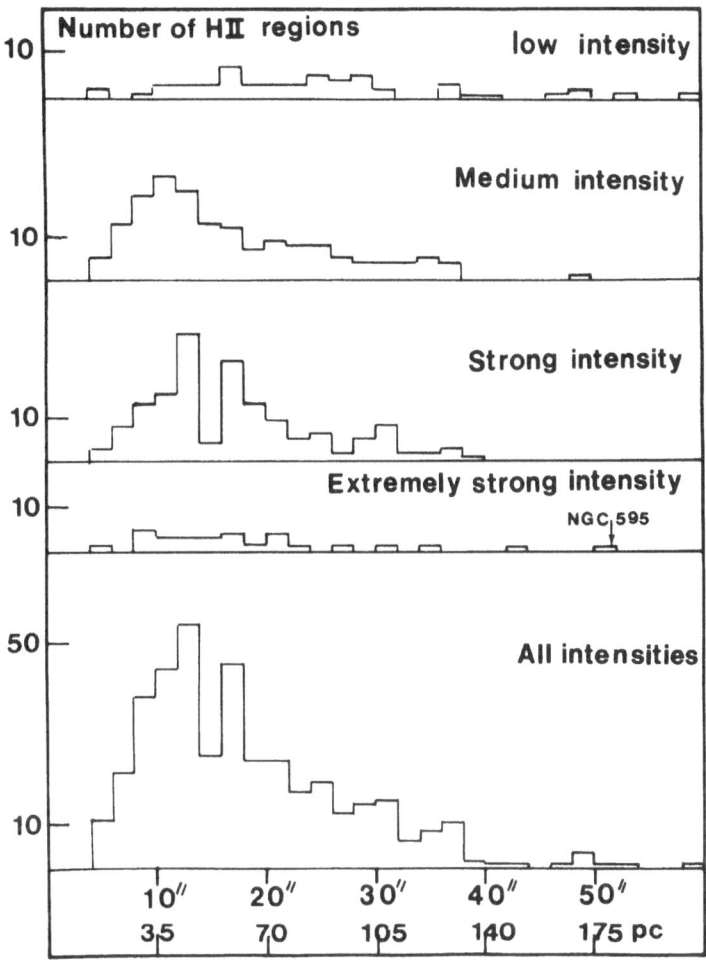

Figure 4.   Histogram of frequency distributions of diameters of HII regions in M33.   From Boulesteix, Courtès, Laval, Monnet and Petit, 1974.

| Distance from the nucleus | Structure of southern spiral arm in M33. |
|---|---|
| $0.4 < \tilde{\omega} < 2.5$ kpc | Narrow, 4 kpc long chain of HII regions, dust, stars and young clusters, 100 pc wide. |
| $2.5 < \tilde{\omega} < 4.5$ kpc | Drop of the density of all components except neutral hydrogen. Large HI complex (Wright et al., 1972). |
| $4.5 < \tilde{\omega} < 7.6$ kpc | Arm again traced by a few large HII regions. |

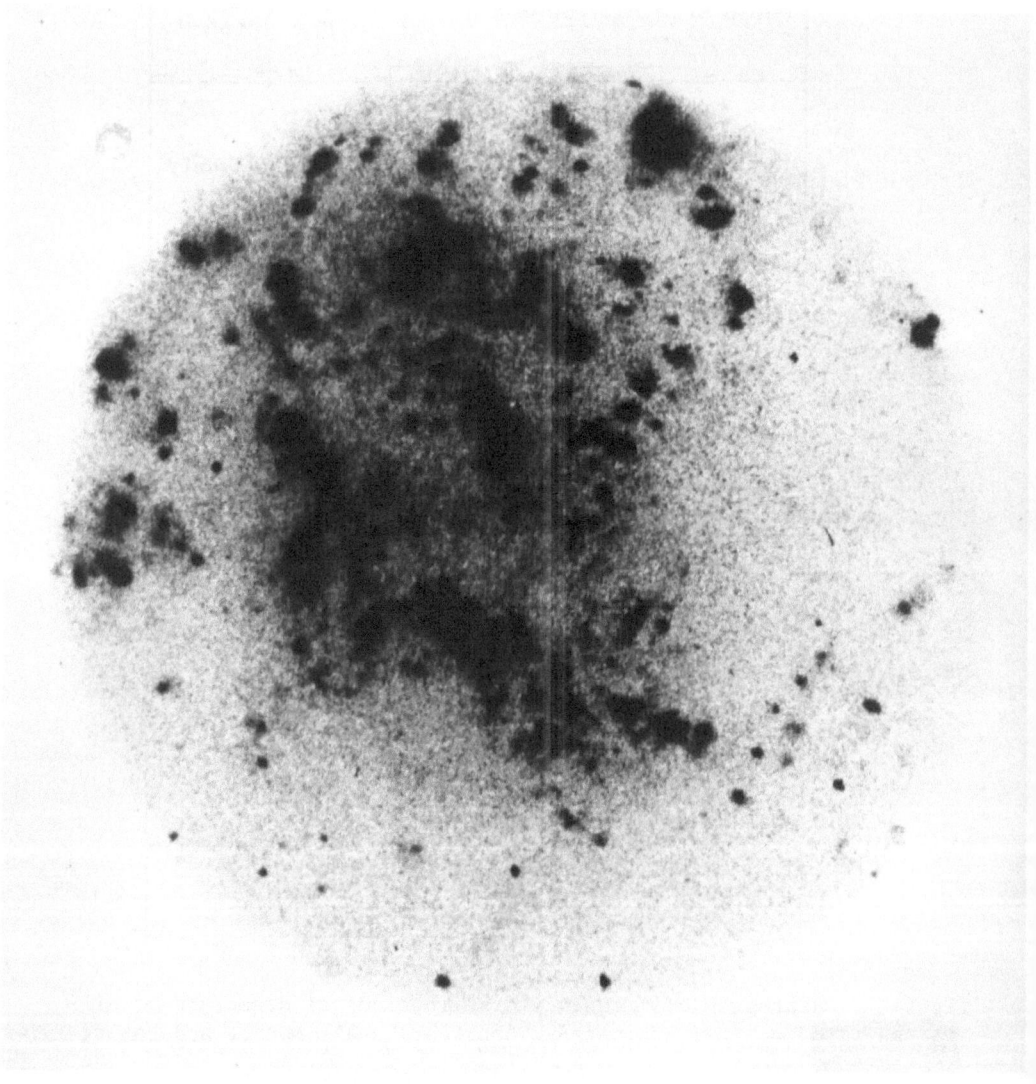

Plate 3

Central part and southern arm of M33 in Hα light.
Focal reducer F/1, with 193-cm telescope of the
Haute Provence Observatory; $\Delta\lambda$ = 4 Å; RCA image tube.

From Boulesteix, Courtès, Laval, Monnet and Petit, 1974.

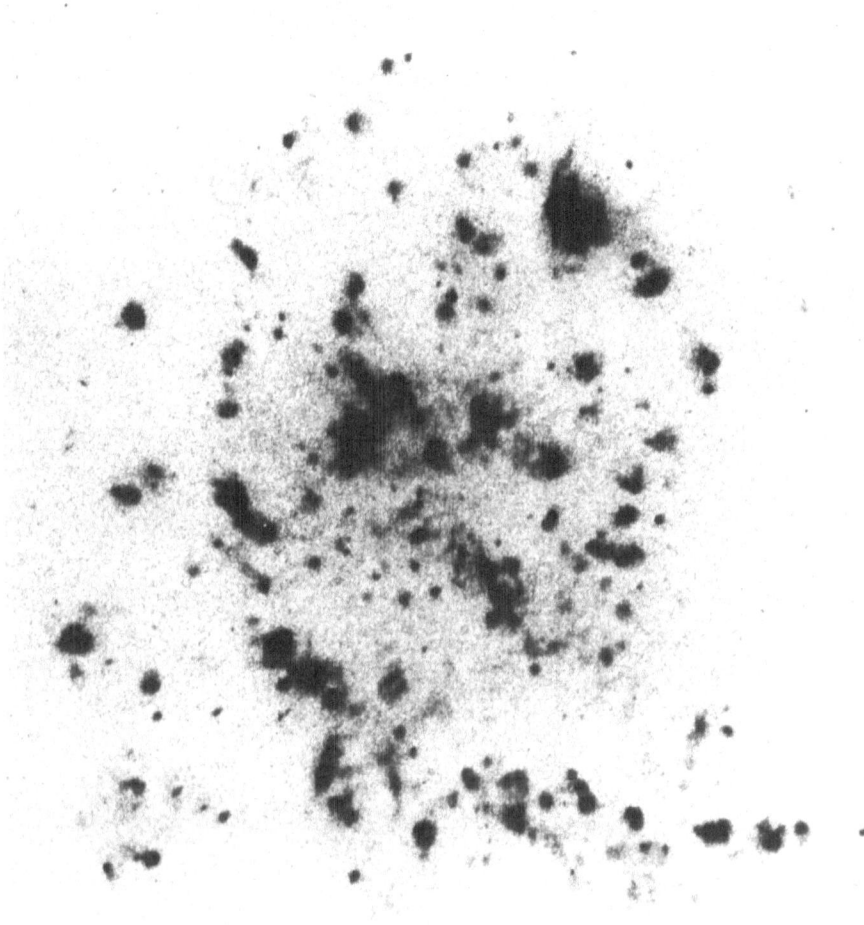

Plate 4

Field slightly different from Plate 3, showing the
relatively constant size of the HII regions in the
southern arm and several filamentary structures.

Plate obtained by Cruvellier and Donas.

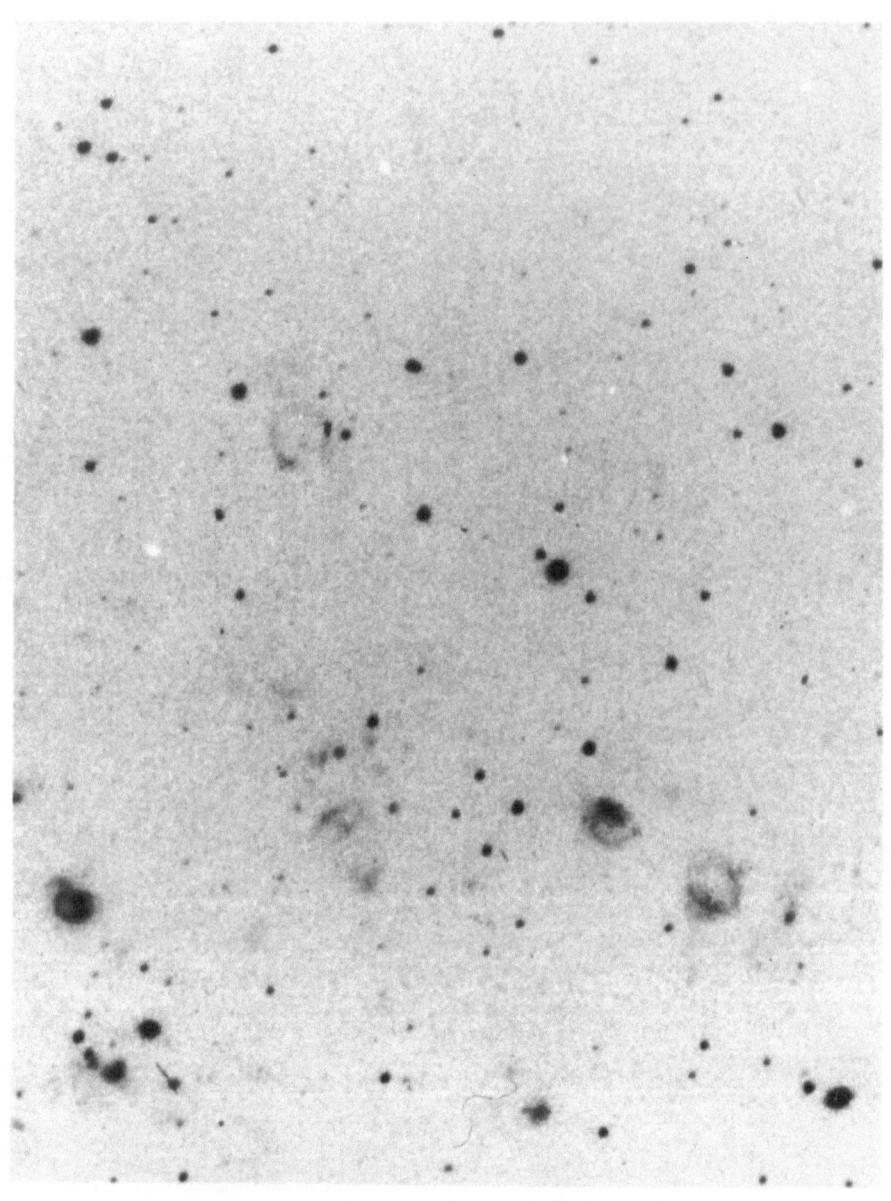

Plate 5

Hα photograph of the northern area in M33, showing
faint ring-like HII regions at 30' from the nucleus.

From Boulesteix, Courtès, Laval, Monnet and Petit, 1974.

## Evidence of star formation

An attempt to interpret this pure, narrow front in terms of a density wave and star formation finds all conditions by chance fulfilled (Courtès and Dubout-Crillon, 1971; Dubout-Crillon, 1976):

1) Individual star clusters and associations observable.

2) Evidence of strong and regular discontinuity in HII and HI across the spiral structure. Chain of HII regions within a small range of diameters (30 pc<d<50 pc) and with similar intensities of about 1600 $cm^{-6}$ pc, suggesting a common age (Plates 3 and 4).

3) Pitch angle large enough to obtain a clear segregation by age after the passage of the density front.

4) Morphological evidence for relative motion of wave and matter, because of the asymmetry in the distributions of stars and HII regions across the arm. A pure expansion of stellar associations would simply widen the arm both on the inner and the outer sides.

Thus, we have evidence for star formation by a sudden gas compression through propagation of a density wave (Fujimoto, 1968a,b; Roberts, 1969; Shu, 1973; Contopoulos, 1972). After the passage of the shock front initiating the formation of stars, the wave and the stars separate, owing to the difference between the pattern velocity and the circular motion of the material which is known from Fabry-Pérot interference measures (Carranza et al., 1968). The stars begin to evolve in brightness and colour and the associations expand. In principle, the rate of expansion of 10 km $s^{-1}$ (van den Bergh, 1964 in M31) and the spectrophotometric evaluation of the age[*] permit to compute by two different ways the time elapsed since the passage of the density wave (Courtès and Dubout-Crillon, 1971) and the velocity of this wave in the material (Fig. 5). Precise UBV photometry and counting of the stars has been made thanks to the Lallemand-Duchesne Electronic Camera by Dubout-Crillon (1976). The same paper gives a detailed interpretation of the spectrophotometric stellar data of Dixon (1971) and Lee (1973).

Between 4.5 and 7.6 kpc, kinematic and dynamic studies (Boulesteix, 1976) have shown that a corotation zone is reached, with similar velocities for the gas and the wave. No geometrical age differentiation can be made in this section of the spiral arm.

---

[*] Associations in the outer parts of the southern arm are older than the associations situated at the inner edge. The velocity of propagation of the density wave is slower than the rotation velocity of the gas (Dubout-Crillon, 1976).

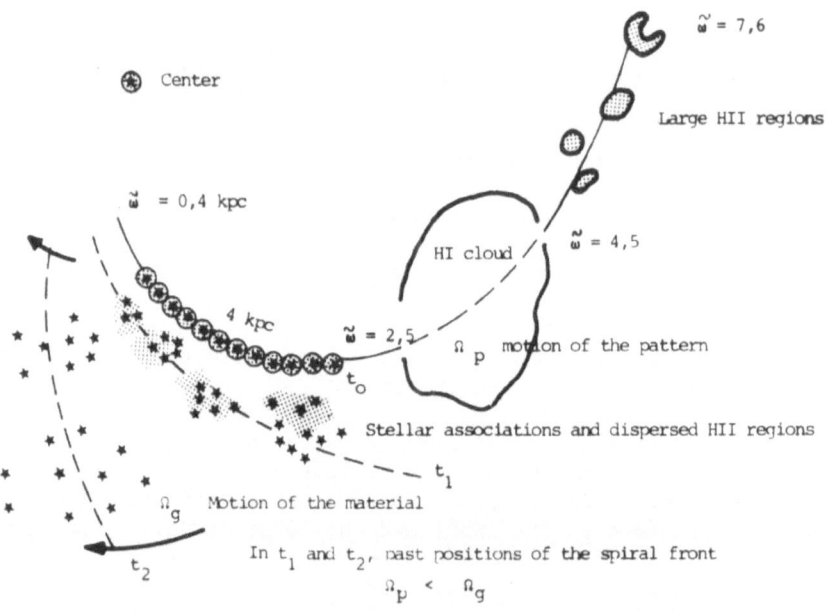

Figure 5.   Structure and motion of the southern spiral arm in M33. From
Courtès and Dubout-Crillon (1971) and Dubout-Crillon (1976).


One concludes (Dubout-Crillon, 1976) that in the section of arm at
1.3 to 2.2 kpc from the centre, both the rate of star formation and the
quantity $\tilde{\omega}$ $(\Omega_g-\Omega_p)$ reach a maximum, in agreement with predictions by
Roberts. The rotational velocity is nearly constant along this section
at 56 to 60 km s$^{-1}$, as measured by the Marseille group from the radial
velocities of the HII regions and the diffuse arm emission. The limiting
values for $\Omega_p$ are:   7 km s$^{-1}$ < $\Omega_p$ < 25 km s$^{-1}$.

Star formation may also appear, with a less clear morphology, near
the regions of strong variation in the HI densities.


## Physical conditions of excitation of the gas in M33

New spectroscopic studies of line-intensity ratios give some possible interpretations of these different kinds of HII regions; Burbidge
and Burbidge (1975) have recently made a synthesis of such studies. The
intensity ratio of emission lines is a difficult observational problem,
especially in the faint, newly discovered, emission of the disk. If we
limit our discussion to the interpretation of the excitation, we can
conclude after Peimbert (1975), Benvenuti et al. (1973), Comte and Monnet
(1974) to the following results, recently clarified by Dubout et al.
(1976) and summarized in Figure 6:

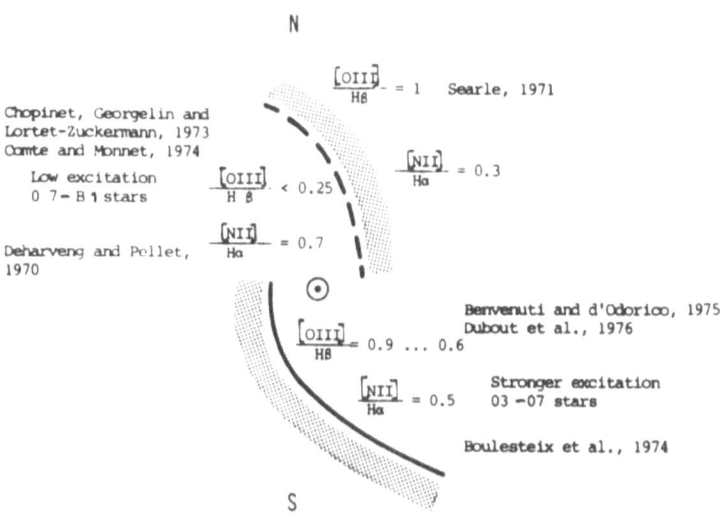

Figure 6.   Condition of excitation in the northern and southern parts
of the disk of M33.

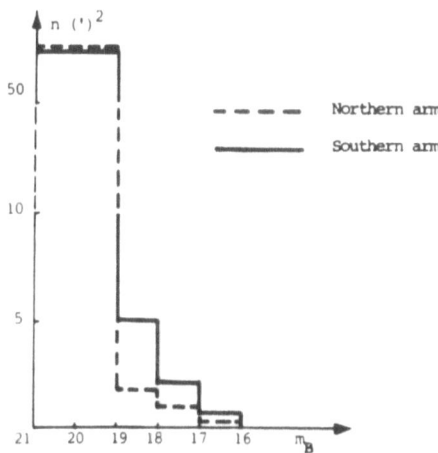

Figure 7.   Histogram of the B magnitudes of stars in the southern and
northern arms of M33. The ordinates given are number of stars per
square minute. Data from Walker (120-inch Lick telescope) and from
Lallemand-Duchesne electronic camera on the 193-cm telescope of the
Haute Provence Observatory.


Figures 6 and 7.    After Dubout, Laval, Maucherat, Monnet, Petit and
                    Simien (1976).

Conventional HII regions and diffuse arm emission (2' to 6' from the centre):

[NII] 6584 / Hα        = 0.3        (Comte and Monnet, 1974), similar to most
                                     of the [NII]/Hα ratios in the arms of our
                                     Galaxy (Courtès, 1960)

[NII] 6584 / Σ[SII] = 1 in the South, 1.2 in the North

[OIII] / Hβ            = 1          (Searle, 1971)

Thus, normal excitation by UV stellar radiation.

Disk emission:

$$[NII] / Hα \quad = \begin{cases} \text{southern disk 0.5} & \text{(Boulesteix et al., 1974;} \\ & \text{Benvenuti et al., 1976)} \\ \text{northern disk 0.7} & \text{(Deharveng and Pellet, 1970)} \end{cases}$$

$$[OIII] / Hβ \quad = \begin{cases} \text{southern disk 0.9} & \text{(Benvenuti et al., 1976)} \\ \qquad\qquad\quad 0.6 & \text{(Dubout et al., 1976)} \quad * \\ \text{northern disk 0.25} & \text{(Comte and Monnet, 1974;} \\ & \text{Chopinet et al., 1973)} \end{cases}$$

[NII] 6584 / Σ[SII] = 1 (North and South).

The different ratios found in the northern and southern parts of the disk can be related to differences in the stellar populations (Dubout et al., 1976): low excitation in the North, which is poor in supergiants 03 to 07 (apparent B magnitude 16 to 19). The surface density of 07 to B1 main-sequence stars (apparent magnitude 19 to 21) is the same in North and South. The southern arm is much richer in very bright stars of types 03 to 07, as shown by a histogram (Dubout et al., 1976) derived from 120" Lick telescope plates, taken and kindly made available by Walker (1976), and from electronic-camera plates taken at the 193-cm telescope of the Observatoire de Haute Provence (Fig. 7).

In the disk, local excitation remains possible by late blue stars, normal B stars, dwarf blue (Hills, 1972; Canochan et al., 1975) or run-away stars (Benvenuti et al., 1976). All these stars are beyond the limiting magnitude, thus undetected. The faintness of [OIII] excludes other exotic sources of excitation, but the higher value of [OIII]/Hβ in the South can be explained by the rich O-star population, if the UV radiation comes from the arms (Comte and Monnet, 1974; Benvenuti et al., 1976).

Estimates of the emission measures E for a temperature similar to the HII regions in the Galaxy (7500 K) are:

---

* Recent measures (November 1976) by Boulesteix confirm this value.

- classical regions      $E = 10^3$ cm$^{-6}$ pc
- diffuse arm emission   $E = 10^2$ cm$^{-6}$ pc
- diffuse disk emission:

$\left[\begin{array}{l}\text{southern half} \quad E = 172 \text{ cm}^{-6} \text{ pc}; \ 150 \text{ cm}^{-6} \text{ pc}\\ \qquad\qquad\qquad\qquad \text{(Dubout et al., 1976; Benvenuti et al., 1976)}\\ \text{northern half} \quad E = 110 \text{ cm}^{-6} \text{ pc (Dubout et al., 1976)}\end{array}\right.$

It is interesting to compare in general the detection performance of radio and optical observations:

|  | Galaxy | Extragalactic |
|---|---|---|
| Radio continuum | 10 cm$^{-6}$ pc | 100 cm$^{-6}$ pc |
| H recombination lines (radio) | 200 cm$^{-6}$ pc | may be of the order of 2000 cm$^{-6}$ pc |
| H$\alpha$ optical emission with large telescope, focal reducers, Fabry-Pérot interferometer and image tube or photon counting | 5 cm$^{-6}$ pc | 20 cm$^{-6}$ pc |

Thus there is a real advantage for optical studies of extragalactic objects as shown by Israel (1975) who has discussed the best radio observations of extragalactic HII regions. In addition, we note with Peimbert (1975) that element abundances in galaxies can be obtained only from optical spectroscopic studies of HII regions. The increase of [NII] toward the centre has been observed (Burbidge and Burbidge, 1962) and explained (Peimbert 1968) as a gradient of the N/H abundance ratio. See also Searle (1971) (in M33), Rubin et al. (1972) (M31), and Comte (1975) (M33, M101, M51), as well as the communication of Webster (1977) in this book. Figure 8 summarizes the observations in M33 by Comte (1975).

The excitation of the central parts for $\overset{\sim}{\omega} < 2' = 400$ pc is easily explained by the large number of blue supergiant stars; the nucleus itself consists of relatively young stars of recent formation (van den Bergh, 1975).

THE LARGE SCALE GALACTIC HII PROBLEMS

Very extensive surveys of the Galaxy have been made in H$\alpha$ light; a complete study of the whole Milky Way has been obtained thanks to $60^\circ$ wide-angle camera techniques (Courtès, Saisse, Sivan, 1976) by Sivan [*] (1974), and one can be certain that almost all HII regions with $E > 30$ cm$^{-6}$

[*] These observations have been made at the European Southern Observatory (ESO) and Haute Provence CNRS Observatory (OHP).

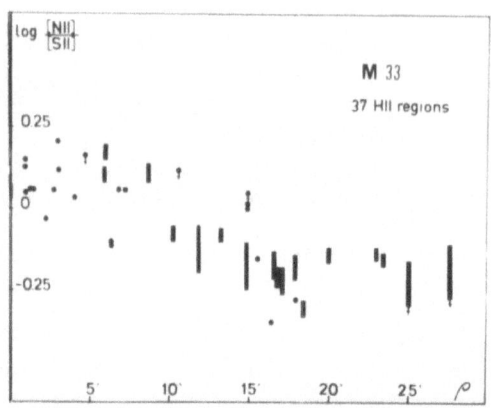

Figure 8. A plot of the observed [NII]/[SII] intensity ratio versus ρ, distance to the centre of M33 (in the plane of the Galaxy). The black bars indicate uncertainty introduced by the contamination of λ 6584 of [NII] with λ 6577 of OH. Spectrographic observation by Comte (1975) with the nebular spectrograph (Baranne et al., 1974) of the 193-cm telescope of the Observatoire de Haute Provence.

pc have been detected by this optical means. Other powerful methods are those with the high-resolution Fabry-Pérot interferometer (Reynolds et al., 1973) and with a photon-counting Hα photometer in the French satellite D2A (Levasseur and Blamont, 1971; Levasseur, 1976).

Some twelve new, very extended regions have been discovered (Meaburn, 1972; Sivan, 1974) and confirmation has been obtained for a diffuse general emission (Courtès, 1951; Reynolds et al., 1974) very similar to that discovered in the M33 galaxy (Carranza et al., 1968) and in some of the nearest spirals (Monnet, 1974). Two general HII region catalogues of the Northern Milky Way (Dubout-Crillon, 1976) and a photographic atlas of the Southern Hemisphere (Y.P. and Y.M. Georgelin, 1970)[*] have been published. A synthetic study has been given by Marshalkova (1974).

Another problem of detection was to try some deep probes in the galactic plane, especially as an optical identification of the hydrogen recombination lines of remote HII regions (Mezger, 1970; Wilson et al., 1970; Reifenstein et al., 1970).

The comparison of sensitivities given above allows us to evaluate the chance to obtain an optical Hα signal from HII regions discovered by radio observations. In the absence of absorption, the optical methods are about a hundred times more sensitive than the radio method (Y.M. and Y.P. Georgelin, 1976). It is always possible in the state of the art to

[*] These observations have been made at the European Southern Observatory (ESO) and Haute Provence CNRS Observatory (OHP).

make an optical identification in spite of an absorption of 5 magnitudes.
Thanks to such observations, it has been possible to reach distances of
9.8 kpc with the interference Fabry-Pérot method applied by Y.M. and Y.P.
Georgelin (1976) to the Sagittarius-Carina arm prolongation (Bok et al.,
1970).

Y.M. and Y.P. Georgelin have optically identified numerous radio
HII regions, comparing 109α and Hα radial velocities and spectrophotometric distances of exciting stars. A new structure of the whole Galaxy
in a four-arm spiral pattern of 12° pitch angle (Y.M. and Y.P. Georgelin,
1976; Crampton and Georgelin, 1975) has been obtained, supporting parts
of the Weaver (1970) model of our Galaxy (Figure 9). Comparative statistics (Churchwell, 1975) of HII region diameters in M33 and the Galaxy
lead to classification of our Galaxy as an Sc-Sb (Georgelin, 1971).

## New comparison of HII rings in M31, M33 and the Galaxy

These new distance determinations permit Sivan (1976) to measure
from his galactic survey the absolute diameter of the largest HII regions,
which are often ring-shaped; he found diameters in good agreement with
previous evaluations in the Galaxy (Georgelin, 1971) and in M33 and M31
(Boulesteix et al., 1974). An unpublished comparison (Maucherat et al.,
1976) of the spherical and ring-like large (> 50 pc) HII regions in M31
and M33 shows diameters of the same order: between 75 to 220 pc, but in
M33 systematically increasing with distance from the nucleus (Boulesteix
et al., 1974). The HII regions of 200 pc diameter appear at 5 kpc in M33
and about 14 kpc in M31. If one plots distances not in absolute measure
but as a function of the HI diameter of each galaxy, the large HII
regions are situated at about the same position with respect to the
galactic structure. The size of the evolved HII regions might be mainly
related to the HI density distribution, rather than to the absolute diameter of the galaxy itself (Figures 10 and 11).

Special mention may be made of the peculiar case of the Gum Nebula.
Studies of expansion (Reynolds, 1976) or nonexpansion (Hippelein and
Weinberger, 1975; Georgelin, 1976) concern mainly the general diffuse
material at normal temperature, 8000 to 11000 K, the only one observable
at the relatively low angular definition of the instrumentation used.
The main support for the supernova-remnant interpretation comes, in fact,
from the filamentary structures; these have not yet been studied at high
spatial and spectrographic resolution, for obvious reasons: a very large
number of big-telescope nights would be required for nebular structures
of such wide angular diameter (Elliott et al., 1976).

As practically not one space programme has been directly devoted to
optical study of the interstellar gas in emission, spectrophotometric UV
observations are very rare. One may note the 1600 Å, 2200, 2550 and 3500
Å survey made with the French satellite D2B-AURA* (Cruvellier et al.,

---

* AURA: Astronomical Ultraviolet Radiation Analysis
      Analyse Ultraviolette du Rayonnement des Astres

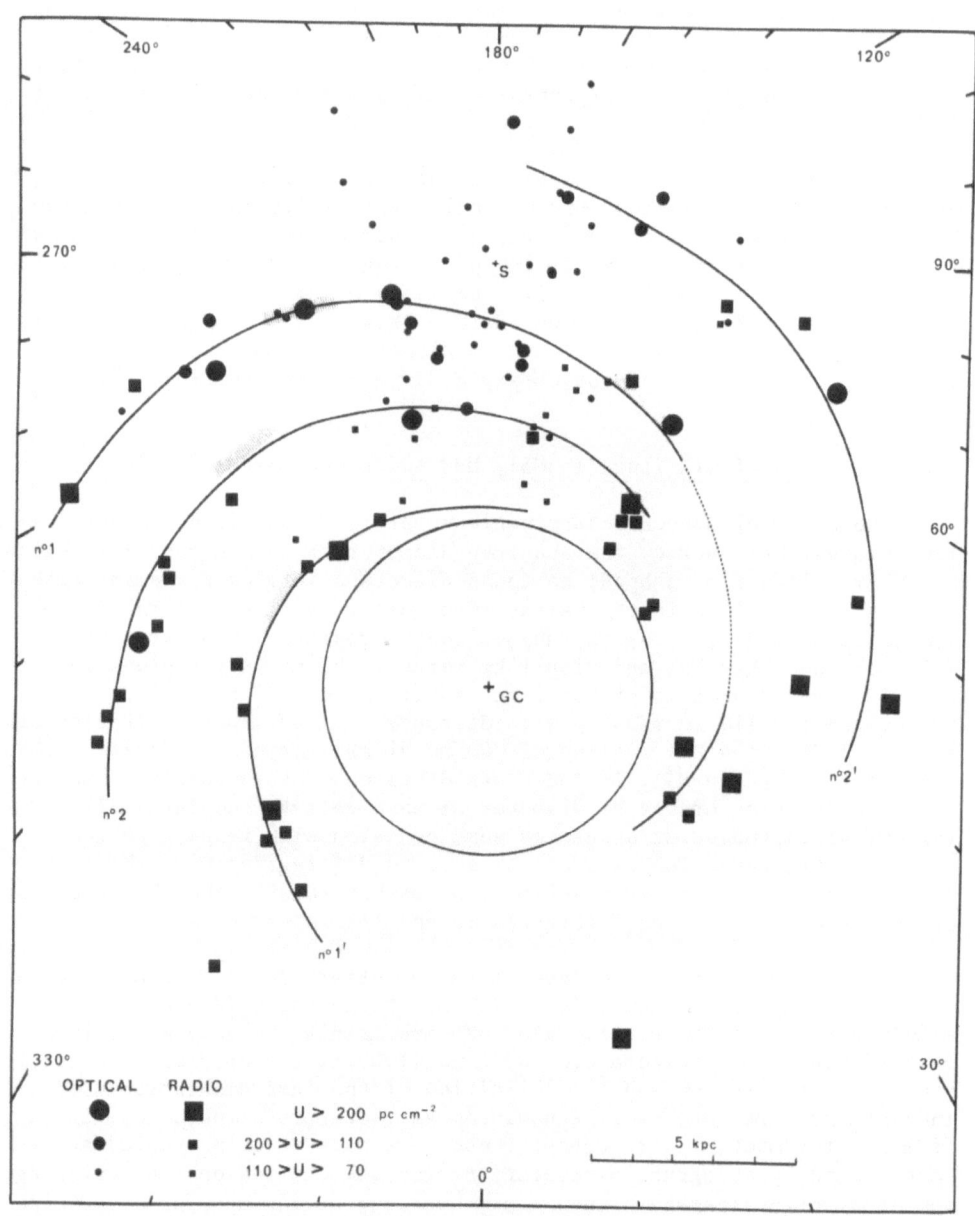

Figure 9

Spiral model of the Galaxy obtained by Y.M. Georgelin and Y.P. Georgelin
(1976) from HII regions of high excitation parameter ($U > 70$ pc cm$^{-2}$).
The resulting spiral pattern has two symmetrical pairs of arms (i.e.
four arms altogether). No. 1, Major arm: Sagittarius-Carina Arm; No. 2,
Intermediate arm: Scutum-Crux Arm; No. 1', Internal arm: Norma Arm;
No. 2', External arm: Perseus Arm.

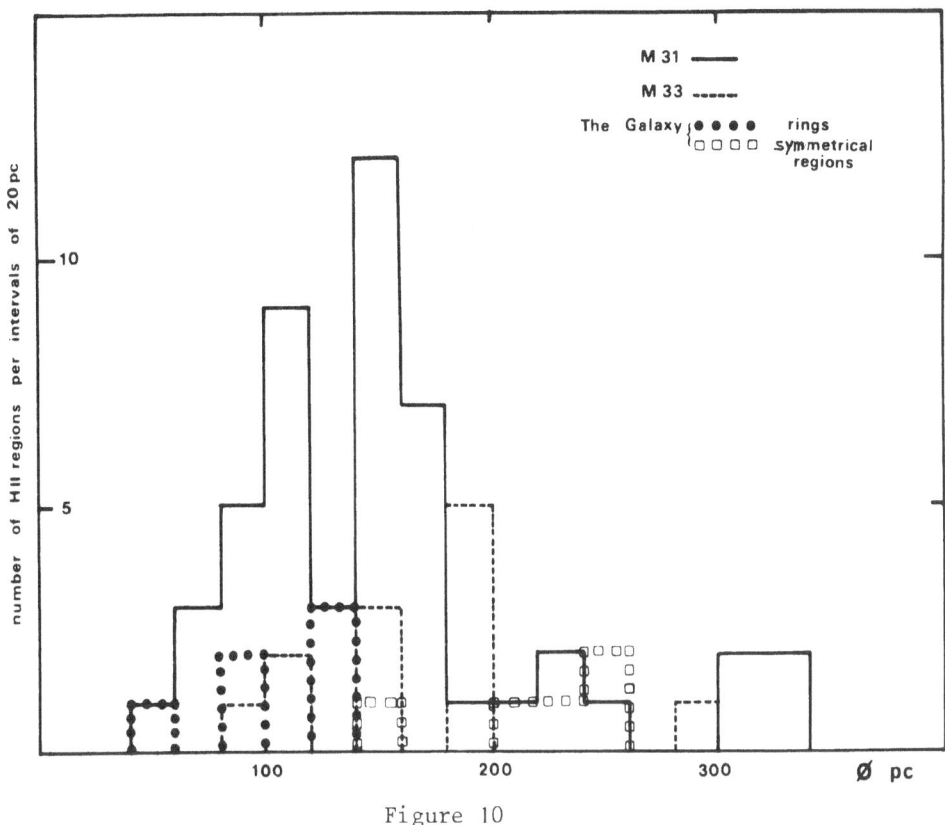

Figure 10

Histogram of the diameters of Large HII regions in M31, M33 and the
Galaxy.   After Georgelin (1971), Boulesteix et al.(1974), Sivan (1976)
and Maucherat et al. (1976).

1976). Preliminary results show a general 1600 Å radiation over the full
diameter of the Gum Nebula. This result may be interpreted, as are the
photographs of the Cygnus Loop by Carruthers (1976), in terms of the
predictions of emission lines given by Osterbrock (1963). However, much
of the Gum Nebula is under low excitation ([OIII] is faint), such as in
conventional HII regions (Reynolds, 1976). The Gum Nebula is probably
excited by the far-UV flux of $\zeta$ Pup and $\gamma^2$ Vel (Reynolds, 1976). Dust
scattering of this radiation on the inner side of the shell is not excluded.

Caption Figure 9, continued:

Hatched areas correspond to intensity maxima in the radio continuum and
in neutral hydrogen. The extreme limit of the Sagittarius-Carina Arm as
seen optically is an HII region situated at 9.8 kpc from the Sun.
Observations made at the Observatoire de Haute Provence and the European
Southern Observatory in Chili.

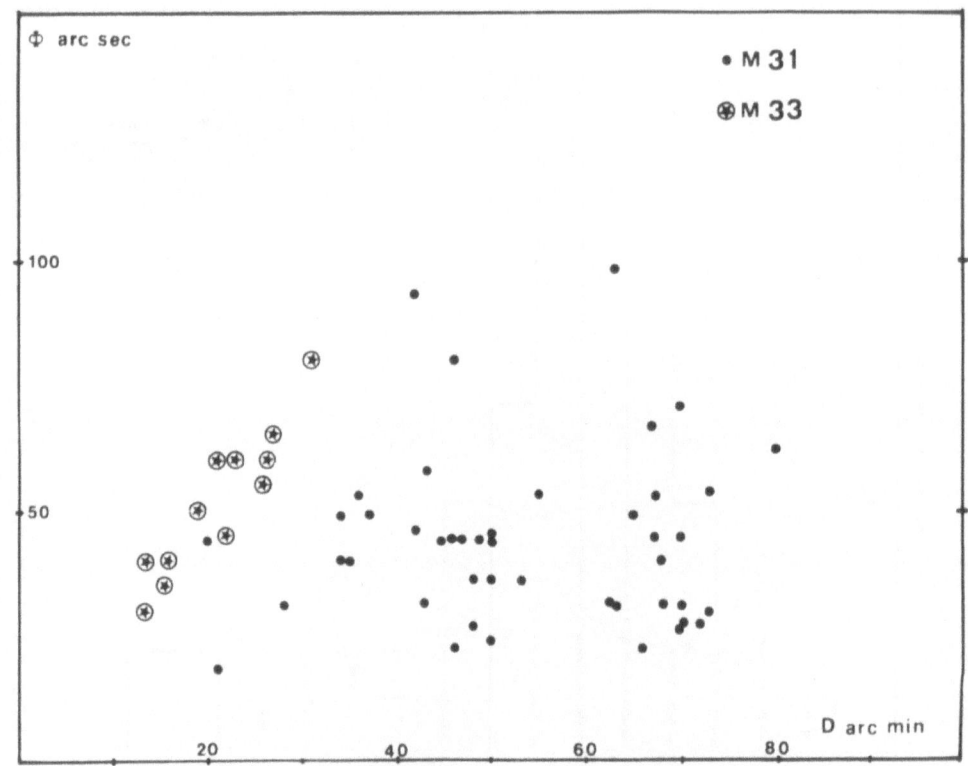

Figure 11

Apparent diameter of ring-like HII regions, as a function of distance
from the nucleus in M31 (black dots) and in M33 (stars in circles).
After Boulesteix et al. (1974) and Maucherat et al. (1976).
Note the regular increase with distance of the diameters in M33, with a
very small spread. The relation diameter-distance is not so obvious in
M31. The average diameter of the large symmetrical HII regions is of the
same order in M31 and M33.


## Spectrographic wide-angle surveys

A general spectrographic study of the Milky Way (HII regions < 40
pc, $n_e$ < 1000 $cm^{-3}$), improving the pioneer work of Johnson (1953), has
been undertaken by Sivan (1976) (Plate 6). It shows the first observation
of a nitrogen to sulphur abundance gradient across the disk of the Galaxy
between 8 to 14 kpc, as was already observed in other galaxies (Benvenuti
et al., 1973; Peimbert et al., 1974; Comte, 1975; Smith, 1975; Maucherat
and Sivan, 1976; Burbidge and Burbidge, 1975). This gradient is similar
to the one observed in M33. This wide-field survey uses a slit length of
$5°30'$ on the sky; the dispersion is about 50 $Å$ $mm^{-1}$, in order to obtain
the best compromise for monochromatic detection (see above) (Courtès and
Cruvellier, 1965). The highest spectral resolution, distinguishing

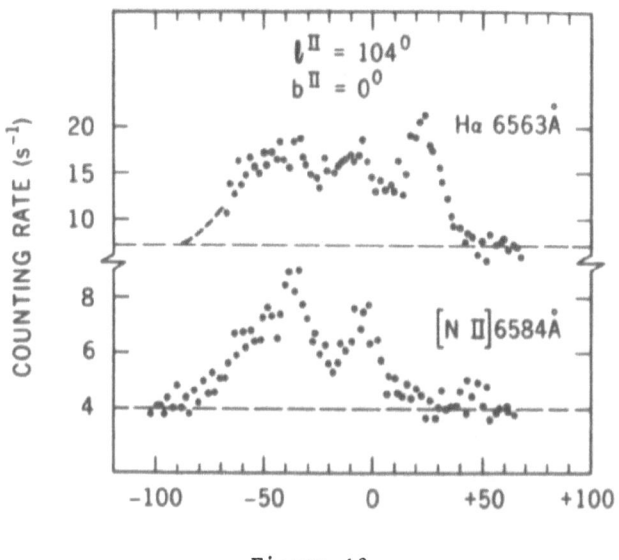

Figure 12

Fabry-Pérot observation of diffuse galactic Hα and [NII] emission:
Scans of Hα and [NII] emission at $\ell = 104°$, $b = 0°$. The photomultiplier
counting rate is plotted as a function of radial velocity with respect
to the local standard of rest. The horizontal dashed line indicates the
background counting rate due to photomultiplier dark pulses, scattered
terrestrial light, and background starlight. In the top scan, three
separate emission lines are clearly resolved: the Hα line from the Earth's
geocorona at $V = + 26$ km s$^{-1}$, the local galactic line at $V = - 7$ km s$^{-1}$
and the galactic line at $V = - 45$ km s$^{-1}$ originating in the Perseus Arm.
(From Reynolds, R.J., Roesler, F.L., Scherb, F.: 1973, Astrophys. J. 185,
869).


different components in velocity and distance, has been obtained by
Reynolds et al. (1973), see Figure 12.


THE IRREGULAR GALAXIES

The Magellanic Clouds

    The last surveys of emission nebulosities (Aller and Czyzak, 1975;
Davies et al., 1976) after the well-known catalogue of Henize (1956)
lead to some comparisons of the HII regions in the LMC and SMC.  Israel
('975) has pointed out a remarkable similarity in the HII distributions
in M33 and the LMC, in spite of the very different masses (a factor of
10) and different morphological types (Sc and Ir).

    The general study of Davies et al. (1976) shows the following
results. The largest bright complexes, of diameter ⩾ 30', occur only in

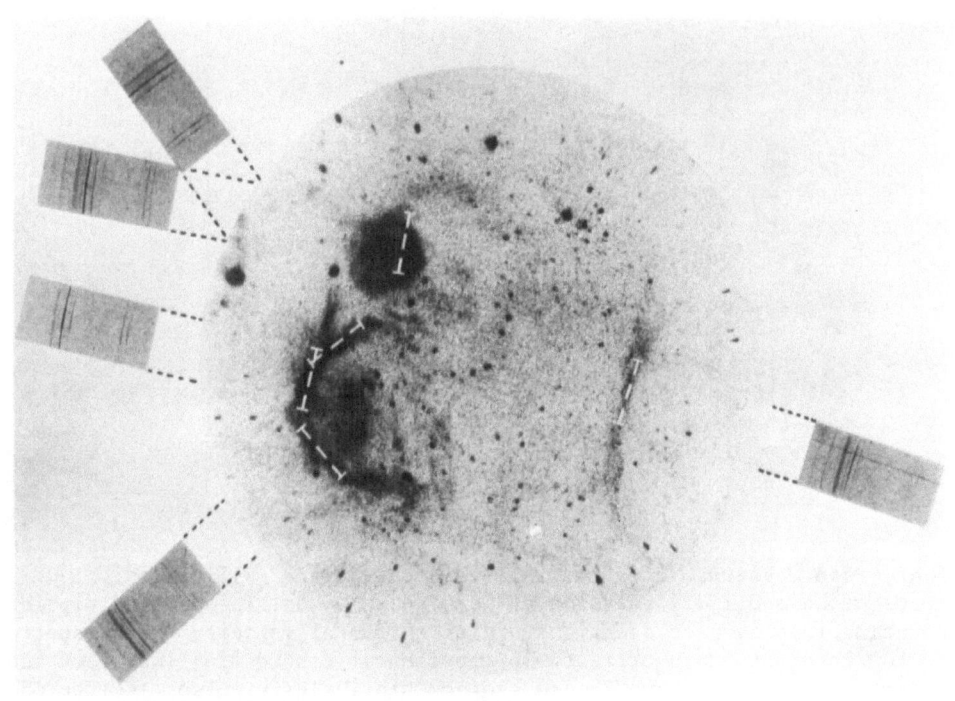

Plate 6

Hα photograph of the Orion-Cetus region; F/0.7; 60° field. One recognizes
the Barnard Loop nebula, the spherical λ Ori HII region (on the left),
and the large-diameter Hα loop in Cetus (on the right). This plate reveals unexpected filamentary extensions to these nebulae (Sival, 1974).
Around the field are reproduced some spectra from a large-scale spectrographic survey of the galactic emission regions (Sivan, 1976). For each
spectrum, the projected slit of the spectrograph (330 x 4 arc min) is
indicated by a dashed line. Hα, [NII] λλ6548, 6584 and [SII] λλ6717,
6731 lines are recorded with a mean dispersion of about 50 Å mm$^{-1}$.


the LMC, and are similar to the giant complexes in distant galaxies (the
30 Doradus complex extends to 80' ≈ 1.3 kpc). Two other complexes measure
400 pc. The SMC shows a maximum HII region diameter of 14' (No.103) and
falls immediately to 6' for the next one (No.32).

The main difference is in their morphology: the SMC HII regions are
mostly diffuse; the LMC HII regions have great variety from diffuse to
complex. The majority of brighter objects show fine structure.  SMC and
LMC are similar at 21 cm. Davies et al. think that the tidal interaction
between the Galaxy and the Clouds should be more disruptive for the SMC.

Perigalactic passage of the Magellanic Clouds occurred $2\text{-}3 \times 10^8$ years ago. That could affect the production of HII regions.

In some cases, the OB stars are situated at the boundaries of HII regions; but there exists a general scattering of OB stars. As was expected, space UV photographs of the Clouds show a very good coincidence with the HII region distribution (Carruthers and Page, 1972) (Apollo 16) (Courtès et al., 1975) (S 183 Skylab). HII regions correspond in most cases to HI surface densities $> 10^{21}$ at.cm$^{-2}$. There is a better HII-HI correlation in the SMC than in the LMC (Davies et al., 1976); the HII regions are situated at the maximum gradient of the HI distribution (Martin et al., 1976).

From McGee and Milton (1964), Hindman (1967), and recent radio results (Davies et al., 1976), it was found that in both Clouds there seems to be a minimum surface density of HI capable of star formation (critical density $10^{21}$ at.cm$^{-2}$). If the thickness of the HI is 200 pc, the volume density would be 1.6 cm$^{-3}$. In the LMC (Davies et al., 1976) there are 46 possible supernova remnants, with diameters from 17 to 400 pc. The most surprising are 16 objects with linear diameters greater than 160 pc, i.e. bigger than the SNRs in our Galaxy (Milne, 1971). In the SMC, two SNRs have been identified. Ten shells are possible SNRs, some with sharp filaments. There is a need for a 1-meter wavelength radio survey and spectrographic studies for verification. The Doradus complex extends filaments up to 1 kpc (Davies et al., 1976).


## NGC 6822

This was classified IB(s)m (de Vaucouleurs and de Vaucouleurs, 1964) after being observed first by Hubble (1925), then by Gum and de Vaucouleurs (1953). Hodge (1969) published a chart of HII regions in this magellanic irregular galaxy. A recent unpublished paper by Hodge (1976) reviews the main observational information on NGC 6822 and in particular its HII regions. Peimbert and Spinrad (1970) made a spectrographic analysis; Sandage and Tammann (1974) made diameter measurements for distance-scale calibration.

Hodge compares the size distributions of HII regions in NGC 6822 and in M33 (Boulesteix et al., 1974) and found quite different distributions. In particular, the steep cut-off in M33 predicted in general by Sandage and Tammann (1974) does not happen clearly in NGC 6822. A closer examination of size and number of HII regions is needed for a better comparison. It is interesting to note that of 16 HII regions measured by Hodge (1976), the average diameter is 40 pc, and the largest five others have an average diameter of 120 pc, not so far from the diameters in M33. There is a large asymmetry in the spatial distribution of the HII regions, mostly situated on one side of the main body, as is common in irregular galaxies (Hodge, 1969).

<u>IC 1613</u>

Classified IBm (de Vaucouleurs, 1959), this magellanic irregular galaxy shows more large HII regions than NGC 6822; their size is of the same order as of those in M33. A subsequent paper, in preparation, from Hodge will give more details on this irregular galaxy. One notes a strong asymmetry with most of the largest ring-like HII regions associated with blue supergiants situated to the northeast of a well-resolved bar (Ables, 1971).

The largest HII region (Sandage, 1962; Sandage, 1971) has a diameter of 167 pc, the next one 127 pc, the others of the 19 catalogued are much smaller.

CONCLUSION

In recent years, the study of HII regions in the Local Group has developed considerably as a powerful tool for understanding spiral structure, star formation problems, and the physics of the interstellar medium. The investigation of HII regions as distance indicators may depend on fundamental uncertainties, with however a relatively more favourable case for the large, ring-shaped HII regions. The use of the 2.4-m Space Telescope will extend the reach of these distance indicators to galaxies ten times more distant. Some very promising results with large, ground-based telescopes may be expected from detailed morphological, spectrographic, and kinematic observations of the sharpest spiral fronts in which we have, perhaps, direct evidence of star formation and evolution of newborn stars.

REFERENCES

Ables, H.D.: 1971, US Naval Obs. Publ. Vol. XX, Part IV, 98
Aller, L.H. and Czysak, S.J.: 1975, Bull. Amer. Astr. Sce. Vol. 7, 255
Arp, H. and Brueckel, F.: 1973, Astrophys. J. 179, 445-451
Baade, W. and Mayall, N.V.: 1950, Problems of Cosmical Aerodynamics
     USAF, Central Document Office, Dayton (Ohio)
Baldwin, J.E.: 1976, private communication
Baranne, A., Carozzi, N., Comte, G., Courtès, G., Deharveng, J.M.,
     Duflot, R., Monnet, G., Pellet, A.: 1975, ESO/CERN/SRC Conference
     on Research Program for new large Telescopes, Geneva 27 May 1974
Benvenuti, P., Odorico, S. d' and Peimbert, M.: 1973, Astron. Astrophys.
     28, 447
Benvenuti, P., Odorico, S. d' and Peimbert, M.: 1976, Rev. Mexicana
     Astron. Astrof.    (in press)
Bergh, S. van den: 1964, Astrophys. J. Suppl. Ser. 9, 65
Bergh, S. van den: 1968, Journal Royal Astr. Soc. Canada, Vol. 62 no.4,
     145
Bergh, S. van den: 1975, Annual Review, Vol. 13, 217
Berkhuijsen, E.M.: 1976, Survey of M31, radio and optical, Preprint

Boulesteix, J., Courtès, G., Laval, A., Monnet, G. and Petit, H.: 1974,
    Astron. Astrophys. 37, 33-48
Boulesteix, J., Courtès, G., Laval, A., Monnet, G. and Petit, H.: 1975,
    ESO/SRC/CERN Conference on programs for large telescopes (A. Reiz,
    editor), 221
Boulesteix, J.: 1976, Tbilissi IAU Proceedings of the third European
    Astron. Meeting, 341
Bok, B.J., Hine, A.A., Miller, E.W.: 1970, IAU Symposium 38, 246
Brand, D.T., Roosen, R.G. and Thompson, J.: 1975, Preprint
Brand, P.W., J.L. and Zealey, W.J.: 1975, Astron. Astrophys. 38, 363
Burbidge, E.M. and Burbidge, G.R.: 1962, Astrophys. J. 135, 694
Burbidge, E.M. and Burbidge, G.R.: 1975, in "HII Regions and Related
    Topics" (Ed. Wilson and Downes), Springer, p. 308
Burton, W.M.: 1974, private communication
Carnochan, D.J., Dworetsky, M.M., Todd, J.J., Willis, A.J. and Wilson,
    R.: 1975, Phil. Trans. R. Soc. Lond. A 279, 479-485
Carranza, G., Courtès, G., Georgelin, Y., Monnet, G. and Pourcelot, A.:
    1968, Ann. Astrophys. 31, 63
Carruthers, G.R. and Page, T.L.: 1972, NASA, SP 315, 13
    Sky and Telescope, Vol. 44 No. 1, July 1972, 6
Carruthers, G.R. and Page, T.L.: 1976, Astrophys. J. 205, 397
Chopinet, M., Georgelin, Y. and Lortet-Zuckermann, M.C.: 1973, Astron.
    Astrophys. 229, 225
Churchwell, E.: 1975, in "HII Regions and Related Topics" (Ed. Wilson
    and Downes), Springer, p. 245
Code, A.D.: 1969, Publ. Astron. Soc. Pacific. 81, 475
Comte, G.: 1975, Astron. Astrophys. Vol. 39, 197
Comte, G. and Monnet, G.: 1974, Astron. Astrophys. Vol. 33, 161
Contopoulos, G.: 1972, The Dynamics of Spiral Structure, Univ. of
    Maryland
Courtès, G.: 1951, Compt. Rendus, Ac. Sc. Paris 232, 1283
Courtès, G.: 1960, Annales d'Astrophys. t 23, 115-217
Courtès, G.: 1964, Astron. J. Vol. 69, Nb 5, 325
Courtès, G. and Cruvellier, P.: 1965, Annales d'Astrophys. t 28,4, 683
Courtès, G. and Dubout-Crillon, R.: 1971, Astron. Astrophys. 11, 468
Courtès, G. and Sivan, J.P.: 1972, Astrophys. Letters 11, 159
Courtès, G.: 1972, Vistas in Astronomy, Vol. 14, 81
Courtès, G., Cruvellier, P., Deharveng, J.M., Maucherat, J., Monnet, G.,
    Pellet, A. and Simien, M.: 1975a, Tbilissi IAU Proceedings of the
    third European Astron. Meeting, 391
Courtès, G., Laget, M., Sivan, J.P., Viton, M., Vuillemin, A. and Atkins,
    H.: 1975b, Phil. Trans. R. Soc. Lond. A. 279, 401-404
Courtès, G., Saisse, M. and Sivan, J.P.: 1976, (in preparation)
Crampton, D. and Georgelin, Y.M.: 1975, Astron. Astrophys. 40, 317
Cruvellier, P., Roussin, A. and Valerio, Y.: 1970, IAU Symp. 36, 130
Cruvellier, P., Maucherat, M., Maucherat, J., Hanus, M. and Courtès, G.:
    1976, IAU Congres Com. 44
Davies, R.D., Elliott, K.H. and Meaburn, J.: 1976, Memoirs of the Royal
    Astron. Society, Vol. 81, 89
Deharveng, J.M. and Pellet, A.: 1970, Astron. Astrophys. 9, 181
Deharveng, J.M. and Pellet, A.: 1975, Astron. Astrophys. 38, 15-28

Deharveng, J.M., Laget, M., Monnet, G. and Vuillemin, A.: 1976,
    Astron. Astrophys. 50, 371-375
Dixon, M.E.: 1971, Astrophys. J. 164, 411
Dubout-Crillon, R.: 1976, Astron. Astrophys. Suppl. 25, 25-54
Dubout-Crillon, R.: 1976, Astron. Astrophys. (in press)
Dubout, R., Laval, A., Maucherat, J., Monnet, G., Petit, M. and Simien,
    F.: 1976, Astrophys. Letters, 141-145
Elliott, K.H., Goudis, C. and Meaburn, J.: 1976, Monthly Not. Royal
    Soc., 175
Emerson, D.T.: 1974, Colloques du C.N.R.S., Ed. L. Weliachew, 243
Emerson, D.T.: 1974, Monthly Notices Royal Astron. Soc. 169, 607
Fujimoto, M.: 1968, IAU Symp. 29, 453; Astrophys. J. 152, 391
Gee, R.X. Mc and Milton, J.A.: 1964, IAU Symp. 20, 291
Georgelin, Y.P. and Georgelin, Y.M.: 1970, Astron. Astrophys. Suppl.
    3, 1
Georgelin, Y.P.: 1971, Astron. Astrophys. 11, 414
Georgelin, Y.P.: 1976, private communication
Georgelin, Y.M. and Georgelin, Y.P.: 1976, Astron. Astrophys. 49, 57
Guibert, J.: 1973, Astron. Astrophys. Suppl. 12, 263
Gum, C.S. and Vaucouleurs, G. de: 1953, The Observatory, 152
Henize, K.G.: 1956, Astrophys. J. Suppl. 2, 315
Hills, J.G.: 1972, Astron. Astrophys. 17, 155
Hindman, J.V.: 1967, Austr. Journal of Physics 20, 147
Hippelein, H.H. and Weinberger, R.: 1975, Preprint, Kinematics of the
    Gum Nebula
Hodge, P.W.: 1966, Atlas and catalogue of HII regions in galaxies,
    Seattle University of Washington
Hodge, P.W.: 1969, Astrophys. J. 156, 847; Astrophys. J. Suppl. Vol.
    18, 73
Hodge, P.W.: 1976, Preprint, The structure and content of NGC 6822
Hodge, P.W.: 1976, in preparation, Atlas of HII regions in M31
Hubble, E.P.: 1925, Astrophys. J. 62, 409
Israel, F.P.: 1975, in "HII Regions and Related Topics" (Ed. Wilson and
    Downes), Springer, p. 288
Johnson, H.M.: 1953, Astrophys. J. 118, 370
Johnson, H.M. and Hauna, M.M.: 1972, Astrophys. J. 174, L 71
Lee, V.: 1973, Thesis Univ. of Indiana
Levasseur, A.C. and Blamont, J.E.: 1971, The Gum Nebula and related
    problems, Ed. S.P. Mazan, J.C. Brandt, T.P. Stecher, NASA Report
    SP-332, 203
Levasseur, A.C.: 1976, Thesis Univ. of Paris
Marshalkova, P.: 1974, Astrophysics and Space Sciences 27, 3
Martin, N., Prevot, L., Rebeirot, E. and Rousseau, J.: 1976, Astron.
    Astrophys. 51, 31-50
Maucherat, J. et al.: 1976, in preparation
Maucherat, J.: 1976, in preparation
Maucherat, J. and Sivan, J.P.: 1976, in preparation
Mayall, N.U. and Aller, L.: 1939, Publ. Astron. Soc. Pacific 51, 112
Meaburn, J.: 1972, Astrophys. Space Sciences 17, 499
Mezger, P.G.: 1970, IAU Symp. 38, 107

Milne, D.K.: 1971, The Crab Nebula, Ed. R.D. Davies and F.G. Smith,
    D. Reidel, Dordrecht/Holland, 248
Monnet, G.: 1974, Galactic Radio-Astronomy (F.J. Kerr and S.C. Simonson,
    editors), IAU Symp. 60, 249.
Münch, G.: 1960, Astrophys. Journal 131, 250
Osterbrock, D.E.: 1963, Planet. Space Sci. 11, 621
Peimbert, M.: 1968, Astrophys. J. 154, 33
Peimbert, M. and Spinrad, H.: 1970, Astron. Astrophys. 7, 311
Peimbert, M., Rodriguez, L. and Torres-Peimbert, S.: 1974, Rev. Mex.
    Astron. Astrof. 1, 129
Peimbert, M.: 1975, Annual Review Astron. Astrophys., Vol. 13, 113
Reifenstein, E.C., Wilson, T.L., Burke, B.F., Mezger, P.G., Altenhoff,
    W.F.: 1970, Astron. Astrophys. 4, 357
Reynolds, R.J., Roesler, F.L. and Scnerb, F.: 1973, Astrophys. J. 179,651
Reynolds, R.J., Roesler, F.L. and Scherb, F.: 1974, Astrophys. J.
    Letters 192, L 53
Reynolds, R.J.: 1976, Astrophys. J. 203, 151; Astrophys. J. 206, 679
Reynolds, R.J.: 1976, The Gum Neb. Old SNR ionized by Zeta Puppis and
    Gamma Velorum, Astrophys. J. (in press)
Roberts, W.W.: 1969, Astrophys. J. 158, 123
Rubin, V.C. and Ford, W.K.: 1970, Astrophys. J. 159, 379
Rubin, V.C. and Ford, W.K.: 1971, Astrophys. J. 170, 25
Rubin, V.C., Kumar, C.K. and Ford, W.K.: 1972, Astrophys.J. Vol.177,
    31-44
Sandage, A.: 1962, IAU Symp. No. 15 (Ed. MacMillian, New York)
Sandage, A.: 1971, Astrophys. J. 166, 13
Sandage, A. and Tammann, G.A.: 1974, Astrophys. J. 194, 559
Searle, L.: 1971, Astrophys. J. 168, 327
Shu, F.H.: 1973, IAU Symp. 52, 257
Sivan, J.P.: 1974, Astron. Astrophys. Suppl. 16, 163
Sivan, J.P.: 1976, in preparation
Sivan, J.P.: 1976, Astron. Astrophys. 49, 173
Smith, H.E.: 1975, Astrophys. J. 199, 591
Ulrich, M.H.: 1975, in "HII Regions and Related Topics" (Ed. Wilson and
    Downes), Springer, p. 472
Vaucouleurs, G. de: 1959, Handbuch der Physik 53 (Berlin, Springer
    Verlag), 275
Vaucouleurs, G. de, and Vaucouleurs, A. de: 1964, Reference catalogue of
    bright galaxies (Univ. Texas Press, Austin)
Walker, M.: 1976, private communication
Warner, J.W.: 1973, Astrophys. J. 186, 21
Weaver, H.: 1970, IAU Symp. 38, 126
Webster, L.: 1976, see paper in this book
Wesselius, P.R.: 1975, Tbilissi IAU Proceedings of the third European
    Astron. Meeting, 67
Wilson, T.L., Mezger, P.G., Gardner, F.F., Milne, D.K.: 1970, Astron.
    Astrophys. 6, 364
Wilson, T.L. and Downes, D. (editors): 1975, "HII Regions and Related
    Topics", Lecture Notes in Physics Vol. 42, Springer
Wright, M.C.H., Warner, P.J. and Baldwin, J.E.: 1972, Monthly Notices
    Roy. Astron. Soc. 155, 337

## ACKNOWLEDGEMENTS

I am grateful to all giving preprint in advance of publication, to Dr. J. Caplan for revision of the manuscript, to Mrs. E. Reforzo for its preparation, to J.P. Goudal, A. Bauthéas, J. Ray and Mrs. M. Leclerc for photographs and graphs.

INTERSTELLAR ABUNDANCES IN EXTERNAL GALAXIES

B. Louise Webster
Anglo-Australian Observatory, P.O. Box 296,
Epping, N.S.W. 2121, Australia.

This field has been excellently reviewed recently by Peimbert
(1975), Burbidge and Burbidge (1975) and Searle (1975).   The present
discussion will cover work since these and in particular the evidence
for composition gradients across galaxies and the systematics with
galactic type.

Evidence has been accumulating for years that, quite distinct from
the classical population separation, there exist large scale inhomogeneities in heavy element abundance over some galaxies and from one
galaxy to another.   In our own Galaxy such evidence comes for example
from the period-frequency distribution of cepheids, spectra of G and K
stars and planetary nebulae (Janes and McClure 1972; D'Odorico et al.
1976), radio recombination lines (e.g. Churchwell et al. 1974; but see
Brown and Lockman 1975), H II regions (Sivan 1976), but no completely
self-consistent picture has emerged.   The colours and stellar absorption features in elliptical galaxies and the centres of spirals also
suggest abundance trends.   The relative ease of measuring emission
lines from ionized gas has made this galactic constituent a most suitable one for systematizing abundance variations in the wide range of
galaxies displaying such lines.

The ionized gas on which the observational work is based belongs
to three distinct regimes, the volume within about 50 pc of the centre
of a galaxy, the general low density disc emission including the interarm region, and bright nebulae such as occur in the spiral arms.   The
well known emission-line anomalies in the very central regions form a
separate problem from that of the more widespread variation in line
strengths (Warner 1973; Searle 1975) so will not be discussed further
here.   In the interarm regions the radiation is so dilute that the
ionization level is low irrespective of the temperature of the exciting
stars (Peimbert et al. 1974; my calculations).   The relative intensities of lines such as [N II] 6584 and [S II] 6717-31 are insensitive
to electron temperature also, so an accurate measure of N/S can be
readily obtained from them.   Observations of the interarm regions of
M 33 (Benvenuti et al. 1973; Comte and Monnet 1974) thus provide

probably the least controversial means for estimating the nitrogen abundance relative to the other elements, but do not yet give all the line intensities required over a large enough radial distance to demonstrate a N/S gradient.

The major observational effort has gone into the giant H II regions in external galaxies, in which the line intensities depend strongly on the stellar radiation field and the electron temperature in addition to abundance, and on the elements O, N, S.   (The helium problem is discussed separately by Peimbert at this conference.)   The two fundamental papers in this field are those by Aller (1942) who discovered a relation between the ionization level of nebulae in M 31 and their distance from the centre of that galaxy, and by Searle (1971) who interpreted these and further observations of his own as consequences of a composition gradient in the sense that the inner regions are richer in heavy elements.   Shields (1974) was able to compute model H II regions confirming the general predictions from Searle's analytical arguments. The existence of composition gradients was put beyond doubt when Smith (1975) measured the electron temperatures from forbidden line ratios and was thus able to measure the abundances of some elements directly. His work demonstrated that the outermost H II regions in M 33 and M 101, at least, are heavy element deficient relative to those at intermediate galactocentric distances.

The observational points most critical to the interpretation are (1) the He 5876/Hβ ratio in the nebulae does not vary much with radial distance from the centre of a galaxy; (2) [O III] 5007 is weaker at small radial distances; (3) [O II] is stronger at small radial distances, as are the [N II] and [S II] lines, so the electron temperature must be high enough to excite the [O II] lines; (4) the Hβ equivalent width increases with radial distance, though the continuum comes primarily from cooler stars associated either with the nebula or with the background; (5) the electron temperature inferred from the [O III] lines increases and therefore the O/H ratio decreases with radius as described above; (6) there are some bright low-excitation nebulae in the inner regions of M 83 with densities high enough to significantly de-excite the main infrared coolant lines.

There are three pieces of work that together summarize the remaining uncertainty concerning the interpretation of these observational facts.

Balick and Sneden (1976) have computed models that are self-consistent in that the effect of heavy element opacities on the energy distributions of the exciting stars is taken into account.   (Rodriguez et al. (1974) had pointed out the importance of this effect with particular reference to helium.)   Balick and Sneden find that while the total emergent flux from the star is insensitive to metal abundance, the emergent flux of photons more energetic than 35 or 40 eV is strongly dependent on abundance which has a drastic effect on the ionization structure in the surrounding nebula.   Using their models it is possible to explain most of the observational points listed above, though the difficulties with the strength of [O II] (point 2) found by Shields remain.

Shields and Tinsley (1976) have argued that point 4 implies a gradient in the temperature of the hottest exciting star. This would be expected in the presence of a composition gradient if Kahn's (1974) mechanism for star formation were applicable, since radiation pressure on the grains of an opaque cocoon would limit the accretion onto the newly-formed star at a point depending on the dust and thus the metal abundance.

Sarazin (1976) has also examined the evidence and decided that a variation in the ionizing radiation field with radial distance is needed to explain the observations. He considers that there is strong reason to believe that dust must be included in the calculations. With a dust to gas ratio proportional to heavy element content and assumptions about the relative abundances of He, N, and the other heavy elements, he has computed a series of models that agree rather well with all the important observed quantities.

The general conclusions are that gradients in the heavy element abundance undoubtedly exist in some spiral galaxies. There is also a gradient in the ionizing radiation field that is related to the abundance gradient, partly through the effect of heavy element opacity on the emergent flux, but possibly also through extinction by dust in the nebula and/or a variation in the initial mass function. Until these mechanisms are better understood the magnitude of the abundance gradient remains somewhat uncertain.

H II regions in a fairly large number of normal galaxies have now been observed (Aller 1942; Searle 1971; Rubin et al. 1972; Searle and Sargent 1972; Comte and Monnet 1974; Smith 1975; Comte 1975; Alloin et al. 1976; my AAT observations). Of course extensive data exist on both gaseous and stellar components of each Magellanic Cloud. It seems well established that heavy element abundances in the H II regions (Peimbert and Torres-Peimbert 1974, 1976; Dufour 1975; Aller et al. 1974; Pagel 1976), supernova remnants (Dopita et al. 1976) and less certainly the planetary nebulae (Sanduleak et al. 1972; Osmer 1976; Webster 1976) decrease in the order of our Galaxy to the LMC to the SMC.

Some general statements can be made about trends in abundances. There is a very tight correlation between the morphological type of a galaxy and the excitation properties of its H II regions. Early-type supergiant galaxies (M 51, M 83) have high abundances with relatively little radial variation; later-type spirals (M 33, NGC 300, M 101) have pronounced composition gradients; some irregulars (LMC, NGC 55, NGC 6822 etc) are underabundant relative to Orion by a factor of perhaps two, with little variation. Finally there is a loose class of dwarf galaxy, often rich in H I, that is everywhere more than 5 or 10 times underabundant relative to Orion. Such galaxies include the SMC, the compact gassy dwarfs (Searle and Sargent 1972; Alloin et al. 1976) and extremely low surface brightness galaxies discovered on the SRC Schmidt deep sky survey (Goss et al. 1976).

Smith (1975) has emphasized that morphological type, rather than total mass, determines the abundance properties, but that there is a relation also between heavy element abundance and the ratio of gaseous to total mass.   There have been several discussions of how these results relate to the general problem of star formation in different galaxies (Searle and Sargent 1972; Talbot and Arnett 1973) and a summary of these is given by Audouze and Tinsley (1976).   It is critical to these to determine the magnitude of the O/H gradient and the behaviour of N/O in a sample of external galaxies.

As a final example of metal abundance in a 'spiral' galaxy consider the Cartwheel galaxy studied by Fosbury and Hawarden (1976). They present a convincing case that the galaxy was formed by the passage of one galaxy along the polar axis of another, a spiral galaxy, (Lynds and Toomre 1976) and that the gas on the ring is a mixture of all gas out to a radius much greater than is normally observable. An abundance analysis of H II regions on the ring shows that the heavy elements are an order of magnitude less abundant than in Orion.   This may be the best estimate of the average heavy element content of a spiral galaxy.

REFERENCES

Aller, L.H., 1942, Astrophys. J. 95, 52.
Aller, L.H., Czyzak, S.J., Keyes, C.D. and Boeshaar, G., 1974, Proc.
     Nat. Acad. Sci. 71, 4496.
Alloin, D., Bergeron, J. and Pelat, D., 1976, Astron. Astrophys.,
     in press.
Audouze, J. and Tinsley, B.M., 1976, Ann. Rev. Astron. Astrophys.
     14, 43.
Balick, B. and Sneden, C., 1976, Astrophys. J. 208, 336.
Benvenuti, P., D'Odorico, S. and Peimbert, M., 1973, Astron. Astrophys.
     28, 447.
Brown, R.L. and Lockman, F.J., 1975, Astrophys. J. Lett. 200, L155.
Burbidge, E.M. and Burbidge, G.R., 1975, H II Regions and Related
     Topics, T.L. Wilson and D. Downes (ed), p.304.
Comte, E., 1975, Astron. Astrophys. 39, 197.
Comte, E. and Monnet, G., 1974, Astron. Astrophys. 33, 161.
Churchwell, E., Mezger, P.G. and Huchtmeier, W., 1974, Astron. Astro-
     phys. 32, 283.
D'Odorico, S., Peimbert, M. and Sabbadin, F., 1976, Astron. Astrophys.
     47, 341.
Dopita, M.A., Mathewson, D.S. and Ford, V.L., 1976, Astrophys. J.,
     in press.
Dufour, R.J., 1975, Astrophys. J. 195, 315.
Fosbury, R.A.E. and Hawarden, T.G., 1976, Mon. Not. Roy. Astron. Soc.,
     in press.

Goss, W.H., Hawarden, T.G., Longmore, A.J., Mebold, U. and
    Webster, B.L., 1976, in preparation.
Janes, K.A. and McClure, R.D., 1972, IAU Coll. No.17, p.28.
Kahn, F.D., 1974, Astron. Astrophys. 37, 149.
Lynds, R. and Toomre, A., 1976, Astrophys. J. 209, 382.
Osmer, P.S., 1976, Astrophys. J. 203, 352.
Pagel, B.E.J., 1976, Observatory 96, 229.
Peimbert, M., 1975, Ann. Rev. Astron. Astrophys. 13, 113.
Peimbert, M., Rodriguez, L.F. and Torres-Peimbert, S., 1974, Rev. Mex.
    Astron. Astrof. 1, 129.
Peimbert, M. and Torres-Peimbert, S., 1974, Astrophys. J. 193, 327.
Peimbert, M. and Torres-Peimbert, S., 1976, Astrophys. J. 203, 581.
Rodriguez, L.F., Torres-Peimbert, S. and Peimbert, M., 1974, Rev. Mex.
    Astron. Astrof. 1, 161.
Rubin, V.C., Kumar, C.K. and Ford, W.K., 1972, Astrophys. J. 177, 31.
Sanduleak, N., MacConnell, D.J. and Hoover, P.S., 1972, Nature 237, 28.
Sarazin, C.L., 1976, Astrophys. J. 208, 323.
Searle, L., 1971, Astrophys. J. 168, 327.
Searle, L., 1975, Royal Greenwich Observatory Tercentenary Symposium,
    in press.
Searle, L. and Sargent, W.L.W., 1972, Astrophys. J. 173, 25.
Shields, G.A., 1974, Astrophys. J. 193, 335.
Shields, G.A. and Tinsley, B.M., 1976, Astrophys. J. 203, 66.
Sivan, J.P., 1976, Astron. Astrophys. 49, 173.
Smith, H.E., 1975, Astrophys. J. 199, 591.
Talbot, R.J. and Arnett, W.D., 1973, Astrophys. J. 186, 51.
Warner, J.W., 1973, Astrophys. J. 186, 21.
Webster, B.L., 1976, Mon. Not. Roy. Astron. Soc. 174, 513.

# THE HELIUM PROBLEM

MANUEL PEIMBERT
Instituto de Astronomía, Universidad Nacional Autónoma de México

Abstract. Helium abundance determinations based on observations of
interstellar matter are reviewed.  Some of the conditions that these
results impose on stellar models and cosmological models are discussed.

## 1. THE GALAXY AND THE MAGELLANIC CLOUDS

### 1.1 Pregalactic helium abundance

The pregalactic helium abundance can be obtained from extremely metal
poor H II regions which presumably have not been appreciably enriched
by the products of stellar evolution.  The best candidates seem to be
the bright H II regions in the SMC due to their low metal abundance,
low reddening, high emission measure and high ionization degree.
Peimbert and Torres-Peimbert (1976) find for 3 H II regions in the SMC
that the heavy element content by mass, Z, is 0.003 and that $N(He)/N(H) =$
$0.078 \pm 0.005$; this result corresponds to a helium abundance by mass, Y,
equal to 0.237.  Dufour and Killen (1977) find $N(He)/N(H) = 0.077 \pm 0.008$
for IC 1644, a small H II region of relatively high ionization degree
in the SMC, in excellent agreement with the previous result.

By assuming that the helium enrichment is proportional to the
heavy element enrichment and by comparing the chemical composition of
the H II regions in the SMC with that of the Orion Nebula (Peimbert
and Torres-Peimbert, 1976,1977) it is found that for Z= 0 the pregalactic
helium abundance is given by $Y_p = 0.228 \pm 0.014$ and $N(He)/N(H) = 0.074 \pm 0.006$.
In this computation it was assumed that for the SMC H II regions the
mass fraction of heavy elements embedded in dust grains is negligible.
This assumption is based on: a) the dust-to-gas ratio of the SMC is
smaller than that of the solar neighborhood (van den Bergh 1968,1974),
b) in the Orion Nebula $\leq$ 25% of the most abundant heavy elements are
embedded in dust grains (Peimbert and Torres-Peimbert, 1977), and c)
the reddening of the SMC H II regions is small (Peimbert and Torres-
Peimbert, 1976).

*Hugo van Woerden (ed.), Topics in Interstellar Matter, 249-254. All Rights Reserved.*
*Copyright © 1977 by D. Reidel Publishing Company, Dordrecht-Holland.*

The He/H abundance determinations based on optical data of main
sequence OB stars, H II regions and planetary nebulae in the Galaxy do
not indicate a significantly different pregalactic helium abundance.

A few years ago there were two pieces of evidence that pointed to
a smaller pregalactic helium abundance: 1) The very low photospheric
helium abundances derived from Bw and sdB stars; these abundances do
not correspond to the stellar internal composition and cannot be used
to obtain information about the pre-stellar He/H ratio (Baschek et al.,
1972).  2) The galactic center observations of three H II regions
within 150 pc of the center where $N(He^+ + He^{++})/N(H^+) < 0.03$
(Churchwell and Mezger,1973;Huchtmeier and Batchelor, 1973); these
determinations are model dependent and do not include the amount of
neutral helium.  Recently Brown and Lockman (1975) detected $H76\alpha$ and
$He76\alpha$ in Sgr B2 and found that $N(He^+)/N(H^+)=0.085\pm0.015$; Mezger and
Smith have also reported that $N(He^+)/N(H^+)=0.095$ for Sgr A; however
additional work on this source by Mezger's group seem to indicate that
the feature attributed to $He^+$ might be due to a high velocity hydrogen
cloud (Shaver 1976).  Moreover the relatively high $Ne^+/H^+$ ratio of
Sgr A West (Aitken et al.,1976) and the large infrared excesses of H II
regions close to the center of the Galaxy probably imply the presence
of large amounts of neutral helium and that the total He/H ratio in
the center of the Galaxy could very well be considerably larger than
the pregalactic value (Mezger et al.,1974;Churchwell and Mezger, 1974;
Mezger and Smith, 1976).

The increase of helium abundance, $\Delta Y$, for different objects in the
Galaxy can be obtained from the pregalactic helium abundance.  It can
also be used to choose appropriate models of stellar evolution for
metal poor objects; for example Peimbert and Torres-Peimbert (1976)
find an age of $10.4\times10^9$ years for the globular cluster M15.

1.2 Helium to heavy elements enrichment ratio, $\Delta Y/\Delta Z$

The $\Delta Y/\Delta Z$ ratio provides us with an additional restriction for models
of the chemical evolution of the Galaxy.  In Table I we present values
of this ratio derived from different objects.  The H II regions'
result is based on the SMC H II regions and the Orion Nebula. The
planetary nebulae values were derived from objects in the solar
neighborhood; however their nebular shells might have been contaminated
by stellar evolution products.  There are stellar evolution computations
that predict a small to moderate increase in the helium abundance
(Torres-Peimbert and Peimbert, 1971;Demarque, 1975;Gingold, 1976).
Consequently the helium present in the shells might not correspond to
the pre-stellar helium abundance.  The entry in Table I by Chiosi and
Nasi (1974) was derived by comparing the theoretical frequency
distribution of massive stars in different parts of the HR diagram with
the observed one in the Galaxy and the Magellanic Clouds.  The value
by Perrin et al., (1976) was derived by comparing, on the HR diagram,
stellar evolution predictions with observations of stars near the

TABLE I

Helium to heavy elements enrichment ratio derived from observations

| Object | $\Delta Y/\Delta Z$ | Reference |
|---|---|---|
| H II Regions | 3.3±0.6 | Peimbert and Torres-Peimbert 1977 |
| Planetary Nebulae | 2.4±1 | Torres-Peimbert and Peimbert 1977 |
| Planetary Nebulae | 2.9±1 | D'Odorico et al., 1976 |
| Supergiant Stars | ∿4.5 | Chiosi and Nasi 1974 |
| Main Sequence FGK Stars | 5. ±3 | Perrin et al., 1976 |
| Main Sequence FGK Stars | ∿3.5 | Faulkner 1967 |

zero age main sequence. The value by Faulkner (1967) was derived by comparing quasi-homology relations for stellar models with main sequence field stars observed by Eggen and Sandage (1962).

To compute the expected $\Delta Y/\Delta Z$ ratio in order to compare it with observations it is necessary to adopt: a) a stellar initial mass function, IMF, b) a model for the chemical evolution of the solar neighborhood, and c) the amount and chemical composition of the ejected material for all stellar masses during the objects' lifetime.

Hacyan et al.,(1976) and Gingold (1976) obtain $\Delta Y/\Delta Z \sim 0.4$ by adopting a power law IMF with $\alpha = 1.55$ (where the number of stars between M and M+dM is proportional to $M^{-\alpha-1}$), the Simple model for galactic chemical evolution and the stellar evolution predictions by Talbot and Arnett (1974). This value is about a factor of eight smaller than the values presented in Table I. Stellar evolution models by Gingold in the 1-4$M_\odot$ range increase the theoretical value by about 50%, while the adoption of other IMFs and other models for the chemical evolution of the solar neighborhood reduce the discrepancy to a factor of three (Hacyan et al.). These results suggest that objects more massive than 4$M_\odot$ eject more helium to the interstellar medium than predicted by the models of Talbot and Arnett. In particular models with continuous mass loss and helium mixing to the surface seem to be needed to explain the large $\Delta Y/\Delta Z$ observed values.

1.3 $\Delta Y+\Delta Z$

$\Delta Y+\Delta Z$ is a measure of the energy produced by the stars after galaxies formed. In the interstellar medium of the solar neighborhood it has been found that $\Delta Y/\Delta Z \sim 0.1$ (Peimbert and Torres-Peimbert, 1976, 1977). This value coupled with a model of the chemical evolution of the solar neighborhood can be used to estimate the total energy emitted by the Galaxy in its lifetime.

## 2. COSMOLOGY

### 2.1 Evidence for a universal pregalactic helium abundance

There are observations of normal H II regions in about 25 galaxies the farthest one located at about 100 Mpc, which have $N(He)/N(H) \sim 0.10$. Moreover there is some evidence that the metal poorer H II regions have a smaller helium to hydrogen abundance ratio than the metal richer ones, in very good agreement with the results for the Galaxy and the Magellanic Clouds (Shields, 1974b; Peimbert, 1975; Smith, 1975).

Shields (1974a) has calculated models of the Seyfert galaxy 3C 120 ($z = 0.033$) which indicate that $0.09 < N(He)/N(H) < 0.23$.

It is well known that some quasars exhibit very faint helium emission lines that might imply an intrinsically low helium abundance (Osterbrock and Parker, 1966; Wampler, 1967,1968; Bahcall and Kozlovsky, 1969a,b; Peimbert and Spinrad, 1970; Bahcall and Oke, 1971). However the derived He/H abundance ratios are strongly model dependent, so these abundance determinations are not reliable at present (Burbidge et al., 1966; Williams, 1971; MacAlpine, 1972; Jura, 1973; Chan and Burbidge, 1975). Recently Baldwin (1975) has studied a set of quasars with $z < 0.3$, under the assumption that the emission lines are produced only by recombination and that $N(He)/N(H) = N(He^+ + He^{++})/N(H^+)$; he found an average value for seven objects of $N(He)/N(H) = 0.14$.

From this discussion it follows that there is a general process, like the big-bang, which is responsible for a $N(He)/N(H) \sim 0.1$ pregalactic production.

### 2.2 Big-bang models

Up to this point all the results mentioned in this review are independent of the cosmological model adopted for the Universe.

The pregalactic helium abundance of $Y_p = 0.228 \pm 0.014$ coupled with the standard big-bang model by Wagoner (1973) implies an open Universe for values of $H_0$ larger than 17 km s$^{-1}$ Mpc$^{-1}$ (Peimbert and Torres-Peimbert, 1974,1976). For this model the helium and deuterium restrictions are not independent and a similar result is derived from the deuterium abundance (Rogerson and York, 1973; Gott et al., 1974).

Under the adoption of a big-bang model with nonzero lepton numbers two more degrees of freedom are introduced. For this case the pregalactic Y value provides us with a restriction which is independent of the deuterium restriction (Yahil and Beaudet, 1976; Beaudet and Goret, 1976). Therefore in addition to the helium and deuterium pregalactic values another independent restriction is needed to specify the parameters of the Universe.

## References

Aitken, D.K., Griffiths, J., and Jones, B.: 1976, Monthly Notices Roy. Astron. Soc. 176, 73 p.

Bahcall, J.N. and Kozlovsky, B.: 1969a, Astrophys. J. 155, 1077.

Bahcall, J.N. and Kozlovsky, B.: 1969b, Astrophys. J. 158, 529.

Bahcall, J.N. and Oke, J.B.: 1971, Astrophys. J. 163, 235.

Baldwin, J.A.: 1975, Astrophys. J. 201, 26.

Baschek, B., Sargent, W.L.W., and Searle, L.: 1972, Astrophys. J. 173, 611.

Beaudet, G. and Goret, P.: 1976, Astron. Astrophys. 49, 415.

Brown, R.L. and Lockman, F.J.: 1975, Astrophys. J. 200, L155.

Burbidge, G.R., Burbidge, E.M., Hoyle, F., and Lynds, C.R.: 1966, Nature 210, 774.

Chan, Y.T. and Burbidge, E.M.: 1975, Astrophys. J. 198, 45.

Chiosi, C. and Nasi, E.: 1974, Astron. Astrophys. 35, 81.

Churchwell, E. and Mezger, P.G.: 1973, Nature 242, 319.

Churchwell, E. and Mezger, P.G.: 1974, Astron. Astrophys. 32, 283.

Demarque, P.: 1975, private communication.

D'Odorico, S., Peimbert, M., and Sabbadin, F.: 1976, Astron. Astrophys. 47, 389.

Dufour, J.R. and Killen, R.M.: 1977, Astrophys. J.,in press.

Eggen, O.J. and Sandage, A.R.: 1962, Astrophys. J. 136, 735.

Faulkner, J.: 1967, Astrophys. J. 147, 617.

Gingold, R.A.: 1976, preprint.

Gott, R.J., III, Gunn, J.E., Schramm, D.N., and Tinsley, B.M.: 1974, Astrophys. J. 194, 543.

Hacyan, S., Dultzin-Hacyan, D., Torres-Peimbert, S., and Peimbert, M.: 1976, Rev. Mexicana Astron. Astrof. 1, 355.

Huchtmeier, W.K. and Batchelor, R.A.: 1973, Nature 243, 155.

Jura, M.: 1973, Astrophys. J. 181, 627.

MacAlpine, G.M.: 1972, Astrophys. J. 175, 11.

Mezger, P.G. and Smith, L.F.: 1976, Astron. Astrophys. 47, 143.

Mezger, P.G., Smith, L.F., and Churchwell, E.: 1974, Astron. Astrophys. 32, 269.

Osterbrock, D.E. and Parker, R.A.R.: 1966, Astrophys. J. 143, 268.

Peimbert, M.: 1975, Ann. Rev. Astron. Astrophys. 13, 113.

Peimbert, M. and Spinrad, H.: 1970, Astrophys. J. 159, 809.

Peimbert, M. and Torres-Peimbert, S.: 1974, Astrophys. J. 193, 327.

Peimbert, M. and Torres-Peimbert, S.: 1976, Astrophys. J. 203, 581.

Peimbert, M. and Torres-Peimbert, S.: 1977, Monthly Notices Roy. Astron. Soc.,in press.

Perrin, M.N., Hejlesen, P.M., Cayrel de Strobel, G., and Cayrel R.: 1976, preprint.

Rogerson, J.B., Jr. and York, D.G.: 1973, Astrophys. J. 186, L95.

Shaver, P.A.: 1976, private communication.

Shields, G.A.: 1974a, Astrophys. J. 191, 309.

Shields, G.A.: 1974b, Astrophys. J. 193, 335.

Smith, H.E.: 1975, Astrophys. J. 199, 591.

Talbot, R.J., Jr. and Arnett, W.D.: 1974, Astrophys. J. 190, 605.

Torres-Peimbert, S. and Peimbert, M.: 1971, Bol. Obs. Tonantzintla y
  Tacubaya 6, 101.
Torres-Peimbert, S. and Peimbert, M.: 1977, Rev. Mexicana Astron. Astrof.
  in press.
van den Bergh, S.: 1968, The Galaxies of the Local Group, David Dunlap
  Observatory Communications, No. 195.
van den Bergh, S.: 1974, Astrophys. J. 193, 63.
Wagoner, R.V.: 1973, Astrophys. J. 179, 343.
Wampler, E.J.: 1967, Publ. Astron. Soc. Pacific 79, 210.
Wampler, E.J.: 1968, Astrophys. J. 153, 19.
Williams, R.E.: 1971 Astrophys. J. 167, L27.
Yahil, A. and Beaudet, G.: 1976, Astrophys. J. 206, 26.

# GALACTIC WARPS: OBSERVATIONS

R. Sancisi
Kapteyn Astronomical Institute, University of Groningen

## Abstract

Recent observations of neutral hydrogen in nearby galaxies have revealed a significant bending of their outer gas layers. The most striking warp is found in NGC 5907. Most of these galaxies have no bright close companions.

It has long been known from the tilted absorption lane of NGC 5866 - first noticed by Pease (1917) - and from later photographs of systems like NGC 3190 (cf. Arp 1966) and NGC 4762 (cf. Sandage 1961) that the disks of galaxies are sometimes quite warped rather than flat. About two decades ago a similar large-scale bending of the HI layer in the outer parts of our Galaxy was discovered and first reported by Burke (1957) and Kerr (1957). At that stage, such a warp was still considered an exceptional feature.

In recent years, high-resolution 21-cm line studies of external galaxies have indicated that quite a few other systems may possess large-scale warps of at least their layers of neutral hydrogen. The two most impressive examples of this sort are probably M33 (Gordon 1971, Wright Warner and Baldwin 1972, Rogstad Wright and Lockhart 1976) and M83 (Lewis 1968, Rogstad Lockhart and Wright 1974).

A sample of nearby edge-on galaxies have been recently observed in the 21-cm line with the Westerbork Synthesis Radio Telescope to study the gas distribution in the direction perpendicular to the plane. A significant bending of the outer gas layers has been found in four out of five systems. Three of the galaxies with warps (NGC 5907, 4565 and 4244) are fairly isolated in the sky. The most pronounced and unambiguous bending is found in NGC 5907 and is shown in Figure 1. The deviation of the HI layer from the optical plane is about 20 percent of the radius. The warp is not entirely restricted to the two extreme ends of the galaxy, perpendicular to the line of sight (i.e., in the line of nodes of galactic plane on sky plane), as the picture in Figure 1 might

*Hugo van Woerden (ed.), Topics in Interstellar Matter, 255-259. All Rights Reserved.*
*Copyright © 1977 by D. Reidel Publishing Company, Dordrecht-Holland.*

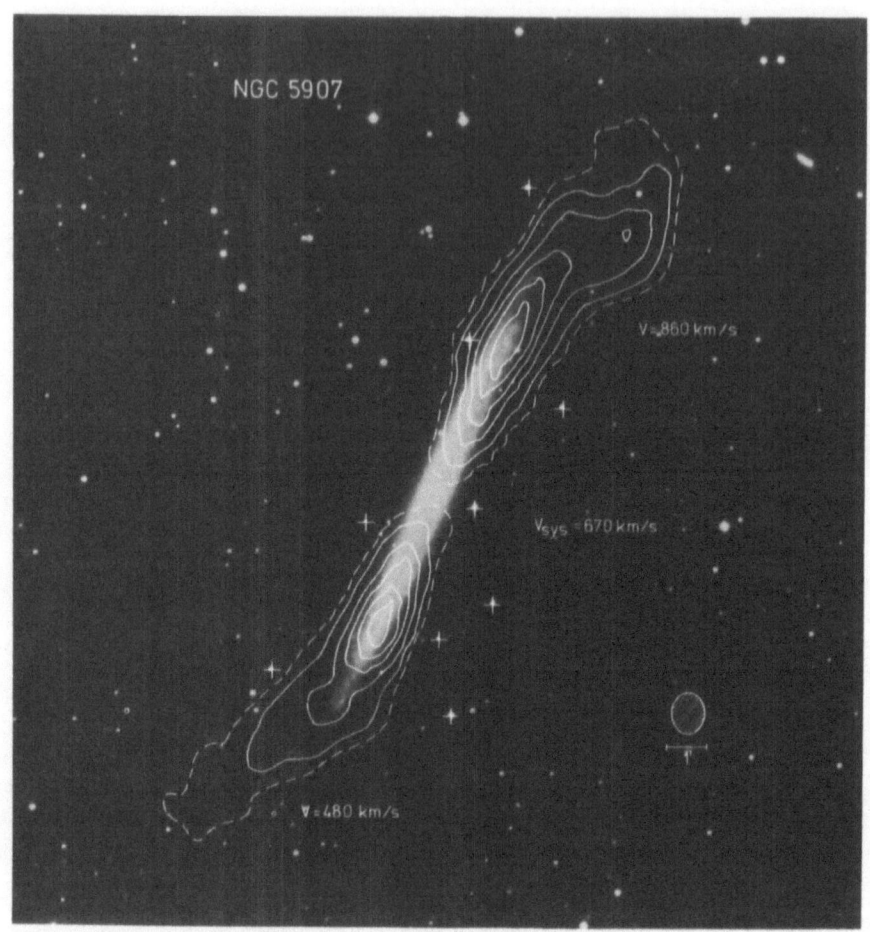

Figure 1. Two channel maps of neutral hydrogen receding and
approaching with almost maximum rotational speed, superposed
on a photograph of NGC 5907. The photograph is from two stacked
IIIa-J plates taken by Van der Kruit and Bosma with the 48-
inch telescope of the Hale Observatories. The contours show
the distribution of beam-averaged brightness temperature at
2 ($\simeq 3\sigma$, dashed), 4, 8, 12, 16, 20 K. The velocity channels at
and near the systemic velocity are not shown; this explains the
gap in the radio map towards the centre. The velocities are
heliocentric. The beamwidth at half-power (hatched ellipse)
is 51" x 61". North is at the top, east at left.

suggest. It continues through azimuth angles (in the plane of the galaxy)
of at least 60° from the line of nodes. This is shown by the diagram in
Figure 2, where the main ridge lines of HI in the individual velocity
channel maps (cf. Sancisi 1976, Fig. 2) are drawn. Some radiation remains
present also on the major axis. Less striking is the bending found in

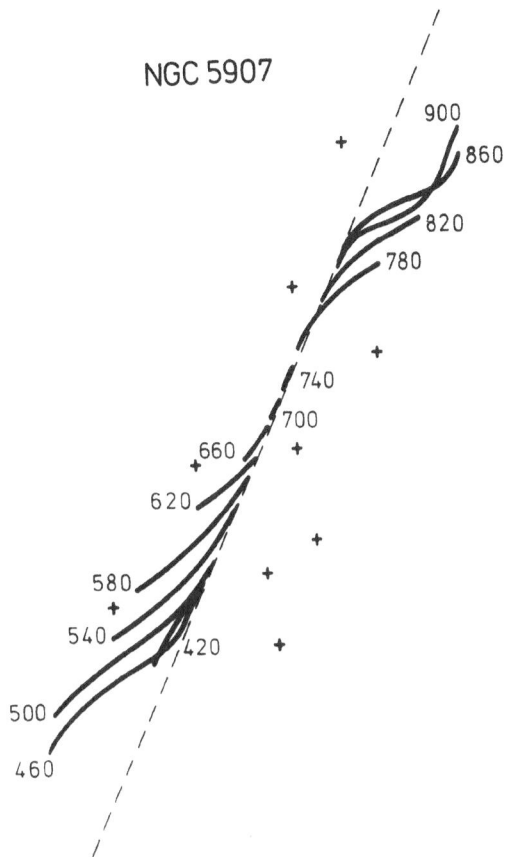

NGC 5907

Figure 2. Diagram showing the main-ridge line of HI in each
individual velocity channel map of NGC 5907. The radial velocity (km/s, heliocentric) is indicated for each channel. The
crosses mark the positions of the stars as in Figure 1. The
thin dashed line shows the position of the optical major axis
of NGC 5907. North is at the top, east at left.

NGC 4565 and in NGC 4244, where the deviations are only of order 10 and
5 percent.

   The other two edge-on galaxies observed at Westerbork are NGC 4631
and 891. The first has a close, bright and much disturbed companion,
NGC 4656/57, the second is well isolated in the sky. The HI layer of
NGC 4631 shows a strong bending (Figure 3). However, the picture of the
hydrogen distribution (integrated over all velocities) for this galaxy
is more complicated and seems to be dominated by other conspicuous HI

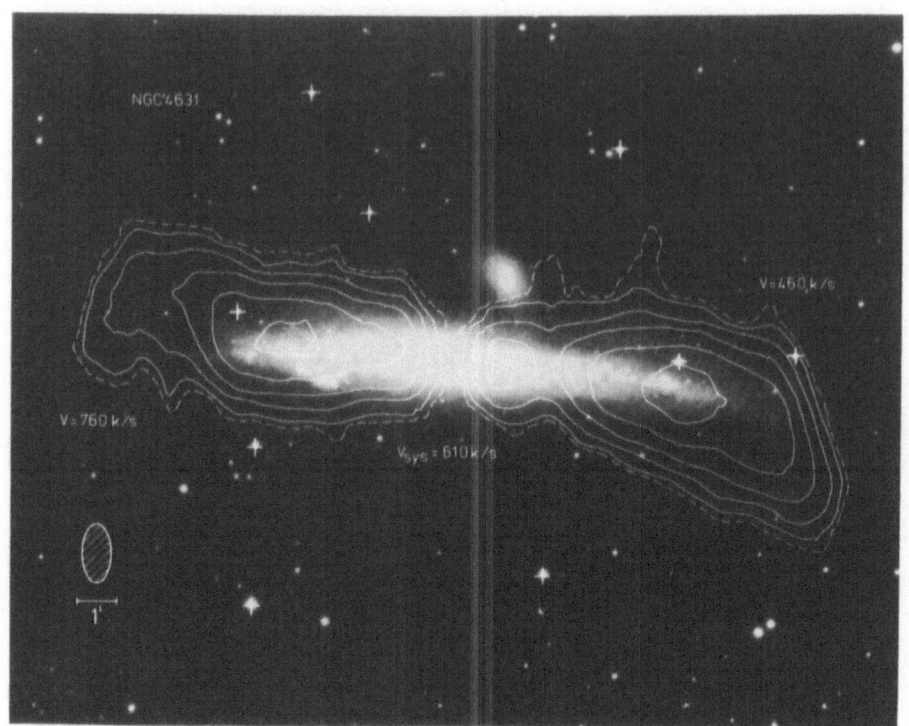

Figure 3. Two channel maps of neutral hydrogen emission
superposed on a photograph of NGC 4631 (reproduced from a
IIIa-J plate taken with the Hale Observatories 48-inch
Schmidt telescope and kindly made available by Arp). The
contour values of beam-averaged brightness temperature are
0.6 ($\sim$ 3$\sigma$, dashed), 1, 2, 4, 8, 16 K. The half-power beamwidth is 48" x 89". North is at the top, east at left.

features stretching out of the HI disk up to large projected distances
from the plane (Weliachew, Sancisi and Guélin, in preparation). In
NGC 891 the HI layer is essentially flat and coincident with the optical disk (cf. Sancisi, Allen and Van Albada, 1974).

All these observations strongly suggest that warps are a common,
rather than an exceptional, phenomenon among disk galaxies.

I am grateful to H. van Woerden for helpful discussions.

References

Arp, H. 1966, Atlas of Peculiar Galaxies, Calif. Inst. of Technology,
    Pasadena, Calif.
Burke, B.F. 1957, Astron.J. 62, 90.
Gordon, K.J. 1971, Astrophys.J. 169, 235.

Kerr, F.J. 1957, Astron.J. 62, 93.

Lewis, B.M. 1968, Proc. Astron. Soc. Australia 1, 104.

Pease, F.G. 1917, Astrophys.J. 46, 24.

Rogstad, D.H., Lockhart, I.A., Wright, M.C.H. 1974, Astrophys.J. 193, 309.

Rogstad, D.H., Wright, M.C.H., Lockhart, I.A. 1976, Astrophys.J. 204, 703.

Sancisi, R. 1976, Astron.Astrophys. 53, 159.

Sancisi, R., Allen, R.J. and Van Albada, T.S. 1974, Colloques Interna-
    tionaux du C.N.R.S. No. 241, p. 295, La Dynamique des Galaxies
    Spirales, Bûres-sur-Yvette, Ed. L. Weliachew.

Sandage, A. 1961, The Hubble Atlas of Galaxies, Carnegie Institution of
    Washington.

Wright, M.C.H., Warner, P.J., Baldwin, J.E. 1972, M.N.R.A.S. 155, 337.

WARPED HYDROGEN LAYERS:   A theoretical comment, by Alar Toomre.

The warped neutral-hydrogen layers of the seemingly isolated edge-on spirals NGC 5907, 4244 and 4565 reported by Sancisi constitute a fascinating dynamical puzzle. To this theorist, who with Hunter (1969, Ap.J. 155, 747) concluded that such integral-sign shapes could not long endure, they are indeed a source of embarrassment: Unlike the warps of our Galaxy or those of NGC 3190 and 3628, they cannot even hypothetically be attributed to any visible companions - nor, unlike with the already serious analogous claims made for M31, M33 and M83 by Roberts, Lewis, Rogstad and others, can one remotely claim these data to be geometrically ambiguous.

I have no new answers to offer. I may recall the recognition already by Kahn and Woltjer (1959, Ap.J. 130, 705) that bent shapes are very prone to distort and ultimately destroy themselves via differential precession, or the counterclaim by Lynden-Bell (1965, M.N. 129, 299) that at least in highly-flattened Maclaurin spheroids and some closely related (if perhaps artificial) models one can find genuine modes of bending which precess indefinitely without any deformation. I may also touch on the pros and cons of Kahn and Woltjer's old suggestion of distortion due to intergalactic winds, and the suggestion by Rogstad et al. (1976, Ap.J. 204, 703) that the blame might instead lie in some fairly recent but massive infall of gas into those galaxies. Most likely, this reviewer feels, Hunter and Toomre were simply mistaken in concluding that long-lived, discrete modes of bending are impossible in realistic disks - but he certainly remains at a loss to know just where or why such a mistake may have occurred!

# THE GAS CONTENT OF EARLY-TYPE GALAXIES

Hugo van Woerden
Kapteyn Astronomical Institute, University of Groningen,
Netherlands
and
Division of Radiophysics, CSIRO, Sydney, Australia.

ABSTRACT

This paper reviews recent observations of neutral hydrogen in elliptical and lenticular galaxies.

Of about 50 ellipticals observed, most have hydrogen masses $M_H < 0.5 \times 10^9 \, M_\odot$ and hydrogen/luminosity ratios $M_H/L_B < 0.05$; for the giant N 4472, these limits are a factor 10 lower. In NGC 4278, several independent observers report a hydrogen mass of about $0.7 \times 10^9 \, M_\odot$, or $M_H/L_B \sim 0.05$.

Among some 150 lenticulars observed, more than 30 have been reliably detected in tne hydrogen line. Values of $M_H$ and $M_H/L_B$ both range over at least two orders of magnitude. Some S0 galaxies are as rich in gas as late spirals and magellanic irregulars. There is no clear cut correlation of gas-richness with any optical property, although blue colours and peculiar morphology both are indicators of gas-richness.

In most early-type galaxies, the amount of gas observed is much smaller than expected from mass loss by cooling stars; galactic winds, intergalactic ram pressure, or formation of new stars may be responsible for this.

## 1. INTRODUCTION

Ever since the first reviews of neutral hydrogen in extragalactic systems (Volders and Van de Hulst, 1959; Heidmann, 1961; Epstein, 1964), it has been clear that the gas content of galaxies varies systematically with morphological type. Figure 1, from a recent compilation (Roberts, 1976) of data on over 100 galaxies, shows that the ratios of neutral hydrogen mass $M_H$ to both total mass $M_T$ and photographic luminosity $L_{pg}$ increase along the Hubble sequence, from early to late spirals and irregulars. More recent surveys (e.g. Shostak, 1977) confirm the trend, though with reduced scatter and slope.

Until 1976, no convincing detection of neutral hydrogen in any elliptical galaxy had been reported. Upper limits on $M_H$ ranged from $10^8$

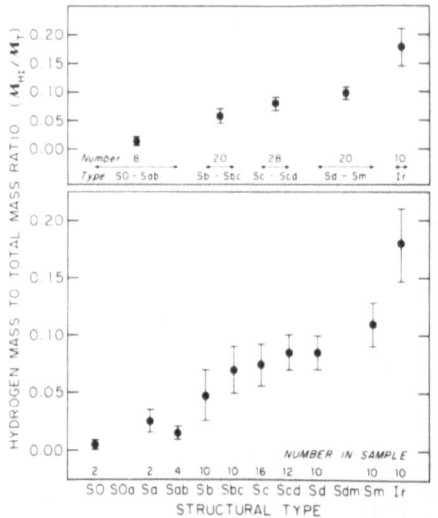
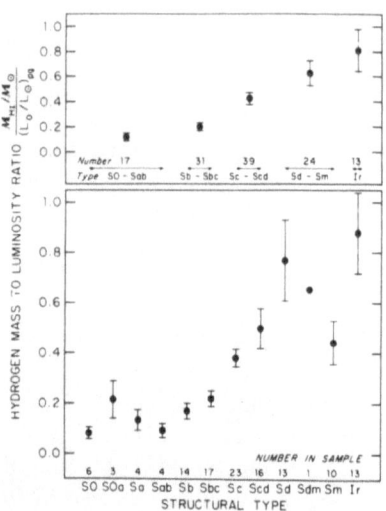

Figure 1.   Gas content as a function of morphological type.

Left: The ratio of hydrogen mass $M_{HI}$ to total mass $M_T$ for
86 galaxies (Roberts 1976, figure 13).
Right: The ratio of hydrogen mass $M_{HI}$ to photographic
luminosity $L_{pg}$ for 124 galaxies (Roberts 1976, figure 16).
The lower panels show average values for individual
structural types; the upper panels give averages for
broader type intervals. The error bars represent the mean
errors of the averages; the spread of individual values
around these averages is a factor $N^{\frac{1}{2}}$ greater.

to $10^9$ $M_\odot$; on $M_H/L_{pg}$ they were of order 0.01. Ionized gas had been found
in a considerable number of elliptical galaxies (Mayall, 1958), but the
amounts appeared to be of order $10^4$-$10^6$ $M_\odot$ only (Osterbrock, 1960;
Spinrad and Peimbert, 1976, Table 11), that is, negligible compared to
the HI limits. It appeared, then, that less than 1/1000 of the matter in
ellipticals is in gaseous form, consistent with the notion (Sandage,
Freeman and Stokes, 1970) that star formation was essentially complete
in the first $10^9$ years of an elliptical's life.

Sandage et al. (1970) also suggested that the morphological types
of spiral and SO systems are essentially defined at the time of the
formation of old disk stars, probably by the amount of gas left over.
The suggestion that lenticular (or, equivalently, SO) systems are
essentially gas-less disks appeared to be supported by the absence of
spiral arms, young stars and HII regions. The first surveys of neutral
hydrogen in lenticular galaxies (Gouguenheim, 1969; Bottinelli et al.,

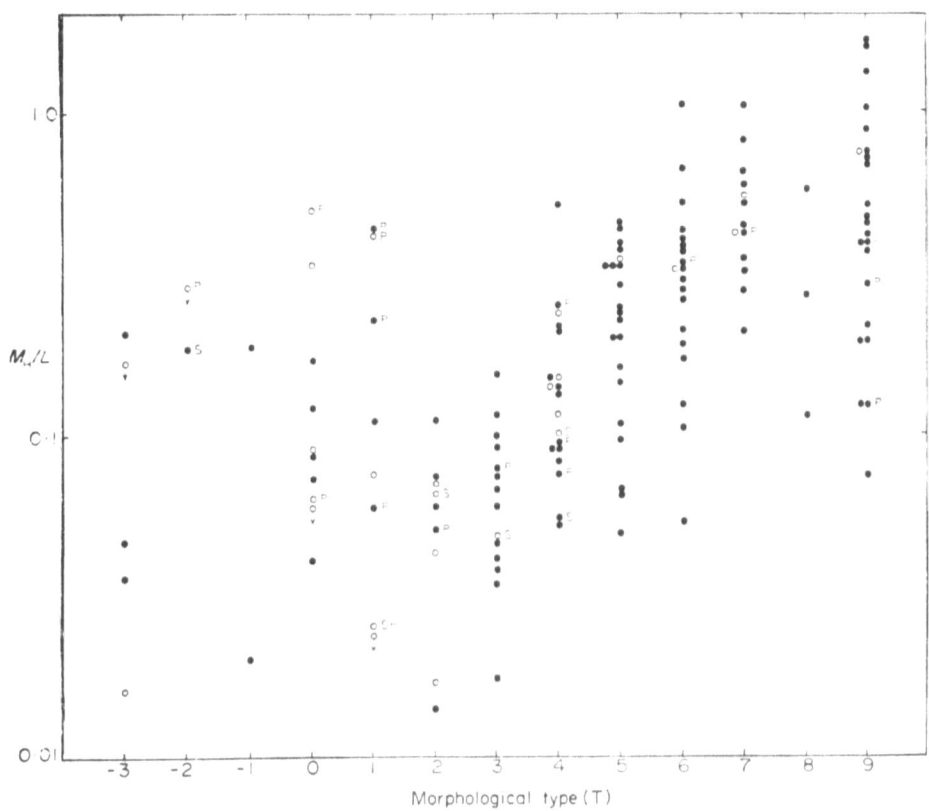

Figure 2. Hydrogen/luminosity ratio versus type (Lewis and Davies, 1973).

The morphological types (from the Reference Catalogue, De Vaucouleurs and De Vaucouleurs, 1964) are coded: T = -3, -1, 1, 3, 5, 7, 9 for SO-, SO+, Sa, Sb, Sc, Sd, Sm; IO and SO/a galaxies are both plotted at T = 0. Open circles: Jodrell Bank observations by Lewis and Davies (1973); filled circles: other observations. S = Seyfert galaxy, P = peculiar galaxy. Individual galaxies are plotted, hence the scatter is much greater than in figure 1. Note that SO galaxies are richer in gas than Sab (T = 2).

1970; Balkowski et al., 1972; Lewis and Davies, 1973) brought a considerable number of detections (cf. Table 6), some of them with $M_H/L$ ratios
comparable to those of Sb and Sc spirals (Figure 2). However, many of
these detections were marginal and some were not confirmed by later
surveys of higher sensitivity. After critical discussion, Gallagher,
Faber and Balick (1975) concluded that both hydrogen-rich and hydrogen-
poor S0's exist, comparable in gas content to Sb's and ellipticals,
respectively, but that the S0 class still appeared to form a true transition between the HI properties of spirals and ellipticals; peculiar S0's
and Irr II (= I0) systems tended to be gas-rich.

Faber and Gallagher (1976) have stressed that the low upper limits
on neutral hydrogen content found in many elliptical and lenticular
galaxies disagree with predictions from stellar evolution. The amounts
of gas returned to the interstellar medium by dying stars would exceed a
limit of 0.01 on $M_H/L_B$ already in $10^9$ years. Faber and Gallagher show
that unseen storage of this gas (in extended HII regions of $10^4$K, as hot
gas of $10^6$K, in dense clouds, or as cold gas at 3K) is unlikely. Hence,
it must be used up in new generations of stars, swept out by intergalactic
ram pressure, or driven out by hot galactic winds. Faber and Gallagher
consider the latter mechanism the most likely.

New observations, carried out in 1975 and 1976 but unpublished at
the time of the Assembly (August 1976), have thrown new light on these
problems. The first convincing detection of HI in an elliptical galaxy,
by Gallagher, Knapp, Faber and Balick, was reported at the Assembly.
Major new surveys of hydrogen in S0 galaxies, at Green Bank, Parkes and
Arecibo, have brought many firm detections, including some with very high
hydrogen-to-luminosity ratios. This paper reviews these new developments,
updated to early 1977. Section 2 summarizes the observations of HI in
elliptical galaxies, and Section 3 those of S0 galaxies. Section 4
discusses a survey of neutral hydrogen in a well-defined sample of
southern lenticular systems. Section 5 reviews the results and outlines
some needs for further investigation.


2. NEUTRAL HYDROGEN IN ELLIPTICAL GALAXIES

a) Searches, limits and detections(?)

Table 1 summarizes all searches of which I am aware; $N_E$ is the
number of elliptical galaxies observed. The observations give directly
the quantity $M_H \Delta^{-2}$, where $\Delta$ = distance; the derived hydrogen mass $M_H$
depends on assumed distance, but its ratio to blue luminosity $L_B$ is
distance-independent. (Since the derivation of $M_H/M_T$ depends on distance
and on knowledge of internal motions, which is often lacking, I prefer
$M_H/L_B$ as measure of the gas content.)

Virtually all searches sofar have only produced upper limits. With
low-noise receivers on the major radio telescopes (43-100m diameter),
these limits are of order $1 \times 10^6$ $M_\odot$ $Mpc^{-2}$ for $M_H \Delta^{-2}$, or $M_H < 10^8-10^9$ $M_\odot$

TABLE 1

Searches for neutral hydrogen in elliptical galaxies

| References | $N_E$ | $M_H \Delta^{-2}$ $(10^6 \, M_\odot \, Mpc^{-2})$ | $M_H$ $(10^9 M_\odot)$ | $M_H/L_B$ $(M_\odot/L_\odot)$ |
|---|---|---|---|---|
| Heeschen (1957) | 1 | see Table 3 | | |
| Wentzel, van Woerden (1959) | 1 | see Table 3 | | |
| Davies et al. (1963) | 2 | see Table 3 | | |
| Robinson, Koehler (1965) | 4 | $\leq$ 2 - 5 | $\leq$0.3 - 0.7 | $\leq$0.02 |
| Gallagher (1972) | 4 | $\leq$ 1 - 2 | $\leq$0.3 - 0.7 | <0.008-0.04 |
| Bottinelli et al. (1973) | 6 | < 1.4 - 5 | <0.2 - 0.7 | <0.005-0.07 |
| Davies, Lewis (1973) | 3 | 6 - < 17 | 1.3 - <4 | 0.017-<0.11 |
| Guibert (1973) | 1 | see Table 3 | | |
| Knapp, Kerr (1974) | 1 | < 0.6 | < 0.08 | <0.0024 |
| Huchtmeier et al. (1975) | 9 | < 0.2 - 0.5 | <0.1 - 0.4 | <0.002-0.04 |
| Shostak et al. (1975) | 15 | $\leq$ 1 | $\leq$0.05 - 1.5 | $\leq$0.01 -0.10 |
| Gallagher et al. (1975) | 9 | < 1.6 - 2.8 | <0.05 - 0.9 | <0.01 -0.05 |
| Huchtmeier et al. (1976) | 6 | < 0.7 - 2.5 | <0.3 - 1.6 | <0.01 -0.08 |
| Rood, Dickel (1976) | 8 | < 4 | <0.3 - 1.0 | |
| Krumm, Salpeter (1976) | 7 | ? | | |
| Balick et al. (1976) | 2 | $\leq$ 3 | $\leq$1.8, 13 | $\leq$0.09, 0.18 |
| Bottinelli, Gouguenheim (1977) | 1 | 1.8 ± 0.6 | 0.17 | 0.041 |
| Knapp et al. (1977a) | 4 | < 0.7 - 3.5 | $\leq$0.3 - 2 | $\leq$0.05 -0.10 |
| Huchtmeier et al. (1977) | 5 | < 0.2 - 0.7 ✻ | <0.03 - 0.6 | <0.003-0.027 |
| Gallagher et al. (1977) | 1 | 2.8 ± 0.4 | 0.72 | 0.05 |
| Bieging, Biermann (1977) | 3 | $\leq$ 0.1 - 0.2 | $\leq$0.01 - 1.3 | $\leq$0.001-0.03 |
| Knapp et al. (1977b) | ∿40 | unpublished | | |

✻ For detections see Table 2

TABLE 2

Reported detections of neutral hydrogen in elliptical galaxies

| Object NGC | Type, $D_{25}$ | References | $M_H \, \Delta^{-2}$ $(10^6 \, M_\odot \, Mpc^{-2})$ | $M_H/L_B$ $(M_\odot/L_\odot)$ |
|---|---|---|---|---|
| 221 | cE2, 7!6 | Four observations (see Table 3); no reliable detection | | |
| 4105 | E3, 2!4 | Huchtmeier et al. (1977) | $0.7 \pm 0.2$ | 0.027 |
| 4278 | E1+, 3!6 | Five observations (see Table 4); detection well confirmed | | |
| 4374 | E1, 5!0 | Davies, Lewis (1973) | $6.5 \pm 1.2$ | 0.045 |
| | | Bottinelli et al. (1973) | <4.7 | <0.041 |
| | | Huchtmeier et al. (1977) | <0.5 | <0.007 |
| 4472 | E2, 9!0 | Seven observations (see Table 5); no reliable detection | | |
| 4636 | E0, 6!2 | Robinson, Koehler (1965) | <2 | <0.02 |
| | | Huchtmeier et al. (1975) | <0.3 | <0.005 |
| | | Gallagher et al. (1975) | <2.8 | <0.03 |
| | | Huchtmeier et al. (1976) | <2.2 | <0.04 |
| 5846 | E0+, 3!4 | Bottinelli et al. (1973) | <1.6 | <0.031 |
| | | Huchtmeier et al. (1975) | <0.22 | <0.008 |
| | | Huchtmeier et al. (1977) | $0.5 \pm 0.2$ | 0.015 |

for distances of 10-30 Mpc; and 0.01-0.05 in $M_H/L_B$. Long integrations with the Effelsberg 100-m dish (Huchtmeier et al., 1975/1977) and, especially, observations with the Arecibo 300-m telescope (Bieging and Biermann, 1977b; Knapp, Kerr and Williams, 1977b) reach even lower limits. Results for about 50 galaxies have been published so far.

Table 2 lists the ellipticals for which detections have been reported. The numerous observations of NGC 221, 4278 and 4472 are summarized separately, in Tables 3, 4 and 5. We discuss the detection claims one by one.

The small ellipticals in the Local Group, NGC 205 and 221 (Table 3), with their large angular size, have been attempted early. Observations are difficult, because of the proximity of the large spiral, M31. At any velocity, radiation from the Andromeda Nebula may be picked up in the near side lobes or even the main beam of the antenna pattern; the signal measured by Heeschen (1957) was due to this. The careful, high-sensitivity study by Guibert (1973) at Nançay has set fairly low limits. Improve-

TABLE 3

Neutral hydrogen in elliptical galaxies in the Local Group

| Name<br><br>Type, $D_{25}$, $B_T^o$ | NGC 205<br>E5p, 17', 8.44 | | NGC 221<br>cE2, 7!6, 8.75 | |
|---|---|---|---|---|
| | $M_H \, \Delta^{-2}$ | $M_H/L_B$ | $M_H \, \Delta^{-2}$ | $M_H/L_B$ |
| References | $(10^6 \, M_\odot \, Mpc^{-2})$ | $(M_\odot/L_\odot)$ | $(10^6 \, M_\odot \, Mpc^{-2})$ | $(M_\odot/L_\odot)$ |
| Heeschen | | | $\sim 100$ | |
| Wentzel and Van Woerden (1959) | | | $\leq 60$ | $\leq 0.12$ |
| Davies et al. (1963) | $\leq 120$ | $\leq 0.18$ | $\leq 35$ | $\leq 0.05$ |
| Guibert (1973) | $< 10$ | $< 0.015$ | $< 16$ | $< 0.033$ |

Notes: Optical data (type, angular diameter $D_{25}$, total blue magnitude $B_T^o$) taken from Second Reference Catalogue (De Vaucouleurs et al. 1976).
For an assumed distance $\Delta = 0.7$ Mpc, the hydrogen mass limits are $M_H < 5$ and $< 8 \times 10^6 \, M_\odot$, respectively.

Blue luminosity $L_B$ (face-on, extinction-corrected) from:

$$\log L_B = 12.19 - 0.4 \, B_T^o + 2 \log \Delta.$$

ment will be difficult, even with aperture-synthesis techniques, since Guibert's limits show that true brightness temperatures are below 1 K.

NGC 4105 has a companion, NGC 4106 (type LBS+, diameter 2!0), at 1.1 arcmin SE. The optical velocities differ by 300 km/s only (De Vaucouleurs et al., 1976). Although the narrow profile (half-width ∿140 km/s) given for NGC 4105 may seem to preclude confusion with its companion, this reviewer considers the source of the measured line in doubt. Huchtmeier et al. (1977) have applied a fourth-order baseline correction and state that the profile structure is uncertain. Without recourse to the original records, it is impossible to decide what part of the profile might be due to NGC 4106.

NGC 4278 is discussed separately in Section 2b.

NGC 4374. The detection announced by Davies and Lewis (1973) was not confirmed by Bottinelli et al. (1973) and now appears refuted by the more sensitive observation of Huchtmeier et al. (1977). The profile of Davies and Lewis looked suggestive, but the $M_H \, \Delta^{-2}$ value derived from the profile integral does not meet the detection criterion for Jodrell

TABLE 4

Neutral hydrogen in NGC 4278

| Reference | $M_H \Delta^{-2}$ $(10^6 \ M_\odot \ Mpc^{-2})$ | $M_H/L_B$ $(M_\odot/L_\odot)$ | V (km/s) | W (km/s) |
|---|---|---|---|---|
| Gallagher (1972 | $\leq 2$ | $\leq 0.043$ | ($\sim$550) | ($\sim$500) |
| Huchtmeier et al. (1975) | $\leq 0.3$ | $\leq 0.01$ | $\sim$750 | ($\sim$200) |
| Gallagher et al. (1975) | $< 2.7$ | $< 0.05$ | | |
| Bottinelli and Gouguenheim (1977) | $1.8 \pm 0.6$ | 0.041 | 680±40 | 510±100 |
| Gallagher et al. (1977)<br>91m<br>43m | <br>$2.8 \pm 0.4$<br>2 | <br>0.05 | <br>670±17<br>610 | <br>470± 40<br>510 |

Notes: Values in brackets determined by present author from published
profiles.
For distance $\Delta = 16$ Mpc, the 91-m result of Gallagher et al.
corresponds to $M_H = 0.7 \times 10^9 \ M_\odot$.

TABLE 5

Neutral hydrogen in NGC 4472

| References | $M_H \Delta^{-2}$ $(10^6 \ M_\odot \ Mpc^{-2})$ | $M_H/L_B$ $(M_\odot/L_\odot)$ |
|---|---|---|
| Robinson and Koehler (1965) | 2 $\pm$ 1 | 0.007 |
| Gallagher (1972) | $\leq 2$ | $\leq 0.0084$ |
| Bottinelli et al. (1973) | $< 1.4$ | $< 0.005$ |
| Davies, Lewis (1973) | 6 $\pm$ 3 | 0.017 |
| Knapp, Kerr (1974) | $< 0.6$ | $< 0.0024$ |
| Huchtmeier et al. (1975) | $< 0.24$ | $< 0.002$ |
| Huchtmeier et al. (1976) | $< 1.8$ | $< 0.01$ |

Note: With distance $\Delta = 15$ Mpc, the lowest limit corresponds
to $M_H < 54 \times 10^6 \ M_\odot$.

Bank measurements established by Gallagher et al. (1975) through inter-comparisons of several observers.

NGC 4472, the biggest giant elliptical in the northern sky, has been most intensively observed (Table 5). The early detections by Robinson and Koehler (1965) and Davies and Lewis (1973) are not confirmed by later, more sensitive measurements. The fraction of mass in neutral hydrogen, $M_H/M_T$, is below $10^{-4}$, if $M_T/L_B$ is not smaller than 20. The amount of neutral hydrogen in this supergiant of $M_B \sim -22$ is probably smaller than in a dwarf irregular galaxy of $M_B \sim -15$.

For NGC 4636, Huchtmeier et al. (1975) reported a "marginal line"; in their 1977 paper they call it a detection. In view of the weakness of the signal, and the involved procedure for removal of baseline ripple, this detection remains uncertain until confirmed independently.

The detection of NGC 5846 reported by Huchtmeier et al. (1977) is marginally consistent with their earlier (1975) upper limit. The published profile, obtained with first-order baseline correction only, looks good. Confusion by NGC 5850 (SBrb, 4.3 arcmin diameter, $10\!\cdot\!3$ separation, velocity difference 500 km/s) appears very unlikely. Yet, confirmation remains desirable.

The conclusion of this section is that, apart from NGC 4278, no detection of HI in any elliptical is firmly established. Confirmation of the detections recently announced is urgently required, as are further highly sensitive searches. A further important point is that, with the bandwidths of 10 MHz $\approx$ 2000 km/s currently employed, detection of faint signals broader than about 500 km/s is virtually impossible because of baseline ripple. Support of a mass of $10^{12}$ $M_\odot$ confined within 25 kpc radius requires motions of order 400 km/s, hence a profile width $\sim 800$ km/s. Detection of such a broad hydrogen line may require a bandwidth broader than 10 MHz.

b) NGC 4278: the only elliptical with detected hydrogen.

NGC 4278 is a structurally normal elliptical galaxy with normal colours, but strong nuclear emission lines (Osterbrock, 1960) and with an intense (0.5 Jy) non-thermal radio source in its nucleus (Heeschen, 1968). The first HI observations by Gallagher (1972) showed a marginal line, of "doubtful reality". Huchtmeier et al. (1975) also found a "rather definite" "marginal line", but with rather different parameters (Table 4). The observations by Gallagher et al. (1975) agreed with those of Gallagher (1972) but were also inconclusive.

Gallagher, Knapp, Faber and Balick (1977; private communication, August 1976), from 8 nights' observations at the NRAO 91-m telescope in May 1976, have now obtained a firm detection (Figure 3). This result agreed with earlier observations at the NRAO 43-m telescope, and was confirmed (Knapp, Kerr and Williams, private communication, August 1976)

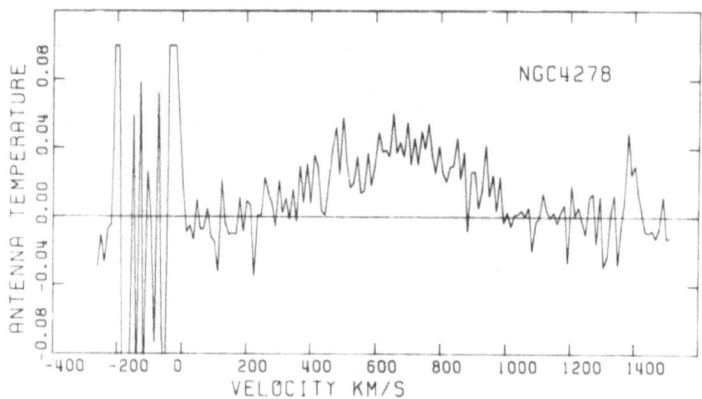

Figure 3.   Neutral hydrogen in the elliptical galaxy,
            NGC 4278 (Gallagher, Knapp, Faber and Balick,
            1977).

The profile is the average of 3 nights of observations
at the NRAO 91-m telescope. Data are unsmoothed, and only
a linear baseline has been removed. The galaxy contains
$7 \times 10^8$ solar masses of hydrogen, if at 16 Mpc distance.

at Arecibo in June 1976. Mrs. Knapp kindly allowed me to show the profile
and announce the detection at the Assembly. The detection has been confirmed by Bottinelli and Gouguenheim (1977), and the earlier marginal
detections further corroborate it.

    The possibility of confusion by neighbouring galaxies has been
discarded after careful checks by Gallagher et al. (1977). The distribution and dynamics of the gas are obviously of great importance in connection with its origin and evolution, but the data are far from complete.
The profile shape allows both a turbulent disk and a quiescent, uniform
disk in solid-body rotation; the extent of the gas is probably not greater than 10' in diameter. The Arecibo observations (Knapp et al., 1977b)
suggest a disk with major axis East-West, roughly perpendicular to the
optical major axis and to the gradient of motions in the nucleus. This
result, and the unique strength of the hydrogen line in this elliptical
(without evidence for dust, HII regions or young stars), suggest that
the gas may be due to accretion. Clearly, however, detailed interferometric measurements of the hydrogen distribution and velocity field are
urgently required.


## 3. NEUTRAL HYDROGEN IN LENTICULAR (S0) GALAXIES

    Table 6 summarizes observations of neutral hydrogen in lenticular
(or, equivalently, S0) galaxies. $N_L$ is the number of L, S0 or S0/a

TABLE 6

Observations of neutral hydrogen in lenticular galaxies

| References | $N_L$ | $M_H \Delta^{-2}$ $(10^6\ M_\odot\ \mathrm{Mpc}^{-2})$ | $M_H$ $(10^9\ M_\odot)$ | $M_H/L_B$ $(M_\odot/L_\odot)$ |
|---|---|---|---|---|
| Dieter (1962) | 1 | <300 | <7 | |
| Robinson, Koehler (1965) | 2 | $\leq$ 3 | $\leq$0.4 | <0.02 |
| Gouguenheim (1969) | 2 | $\sim$ 6 | 0.3 −0.8 | 0.02, 0.04 |
| Bottinelli et al. (1970) | 6 | 2.3 −24 | 0.4 −1.9 | 0.04 −0.14 |
| Roberts (1970) | 1 | Absorption Cen A | | |
| Lewis (1970) | 1 | > 17 | | |
| Whiteoak, Gardner (1971) | 1 | Abs. + em. Cen A | | |
| Balkowski et al. (1972) | 9 | 1.6 −15 | 0.18 −2.0 | 0.03 −0.42 |
| Lewis, Davies (1973) | 6 | 5 −26 | 0.10 −2.3 | 0.02 −0.09 |
| Davies, Lewis (1973) | 3 | 3 − 7 | 0.7 −1.6 | 0.02, 0.04 |
| Peterson, Shostak (1974) | 5 | 1 −15 | 0.25 −3.4 | 0.01 −0.22 |
| Huchtmeier, Bohnenstengel (1975) | 1 | 2.6 | 1.0 | |
| Gallagher et al. (1975) | 17 | < 2 − 6<br>2 Det. 6 , 17 | <0.05 −1.3<br>0.95, 0.33 | <0.01 −0.09<br>0.22, 0.12 |
| Huchtmeier et al. (1976) | 4 | $\leq$ 0.8 − 1.7 | $\leq$0.3 −0.7 | $\leq$0.004−0.07 |
| Rood, Dickel (1976) | 8 | $\leq$ 2 − 4 | | |
| Gardner, Whiteoak (1976) | 1 | > 62 ✳ | >3.0 ✳ | >0.036 ✳ |
| Van Woerden et al. (1976, 1977)    + | 54 | 12 Det. 3 −55<br>Limits 2 −5 | 0.6 −53 | 0.05 −1.0<br><0.02 −0.3 |
| Krumm, Salpeter (1976) | 36 | 5 Det. 0.3 −1.6 | 0.1 −0.5 | 0.006−0.12 |
| Balick et al. (1976) | 14 | (6) 3 −18<br>Lim. 2 − 4 | 0.5 −17<br><0.3 −13 | 0.10 −0.54<br><0.05 −0.21 |
| Knapp et al. (1977a) | 28 | (3) 1.5 −4.8<br>Lim. 0.3 −1.2 | 0.5 −0.9<br><0.15 −4.0 | 0.12 −0.31<br><0.02 −0.25 |
| Bieging, Biermann (1977) | 35 | (12) 0.2 −2.6<br>Lim. 0.05−0.4 | 1.3 −28<br><0.005−3.6 | 0.004−0.16<br><0.001−0.04 |

✳ Values determined by present author from emission in Cen A profile published by Gardner and Whiteoak (1976).

+ Values given here are from Van Woerden et al. (1977); the values for $M_H/L_B$ are based on magnitudes $B_T^0$ from the Second Reference Catalogue (De Vaucouleurs et al., 1976), while those in Van Woerden et al. (1976) were based on B(o) from the First Reference Catalogue (De Vaucouleurs and De Vaucouleurs, 1964). Consequently the new $M_H/L_B$ values tend to be a factor $\sim$2 lower than in the 1976 paper. A similar difference applies with respect to the other surveys, except that of Bieging and Biermann (1977).

galaxies observed; classifications are usually by Sandage or from the Reference Catalogue. Columns 3, 4 and 5 indicate the ranges of gas content (or upper limit thereof) measured in each survey.

As indicated in the Introduction, the early surveys by Bottinelli et al. (1970), Balkowski et al. (1972), and Lewis and Davies (1973) yielded many detections with considerable values of $M_H/L_B$, but some of these results were not confirmed by later observations of higher quality. Gallagher et al. (1975), while admitting that some morphologically normal SO's are as rich in hydrogen as Sb spirals, still found that the majority of normal lenticulars have no significant amount of neutral hydrogen; peculiar objects tend to be richer in gas.

Since 1975, major surveys of hydrogen in lenticular galaxies have been carried out at Parkes (Van Woerden et al., 1976, 1977), Green Bank (Balick et al., 1976; Knapp et al., 1977a) and Arecibo (Krumm and Salpeter, 1976; Bieging and Biermann, 1977). These surveys have brought dozens of reliable detections. Both at Parkes and at Green Bank very high values of $M_H/L_B$, comparable to those normal for late-type spirals, have been found. The Arecibo surveys, with the great collecting power of the 305-m dish, have lowered the detection limit on $M_H/L_B$ by an order of magnitude; nevertheless, even in these surveys large numbers of lenticulars have remained undetected.

The Parkes survey is discussed in detail in Section 4. Figure 4 shows the profiles detected by Balick et al. (1976) with the Green Bank 43-m telescope (beamwidth 21'). The profile of NGC 0936 may be confused with that of the nearby, smaller and fainter, spiral NGC 0941 (separation 13'), whose velocity is unknown. Similar situations occur for NGC 6340 (companion IC 1254, at 7') and NGC 7679 (NGC 7682 at 5'), although in these cases attribution of the hydrogen profile to the fainter companion would imply unlikely high values for $M_H/L_B$. NGC 4670 is in a clear field. The detections of NGC 1291, 1326 and 6902 ($M_H/L_B$ values 0.10, 0.12 and 0.54) have been confirmed at Parkes (cf. Section 4). In fact, the profile of NGC 1326 includes contributions from its two companions, NGC 1326A (V ∿ +1800 km/s) and 1326B (V ∿ +1000 km/s); the Parkes observations have clearly separated these (Van Woerden et al., 1977; see also Mebold et al., 1977). The discussion of this paragraph emphasizes that measurement of the hydrogen content of a galaxy requires not only the detection of a signal, but also demonstration that it cannot be due to another galaxy in the field.

Figure 5 shows profiles of 2 lenticulars in the Virgo Cluster measured by Krumm and Salpeter (1976). The $M_H/L_B$ ratios for these objects are modest: 0.12 for N 4324, 0.04 for N 4694. Comparison of profiles obtained away from the centre suggests that the gas in N 4324 is distributed in a ring rotating at ∿ 150 km/s, while in N 4694 it is concentrated close to the centre and there is no measurable rotation. The Arecibo dish has obvious merits for the determination of the distribution and internal motions of the gas! Also, with its 3.5 beam, it can more easily discriminate between neighbouring galaxies.

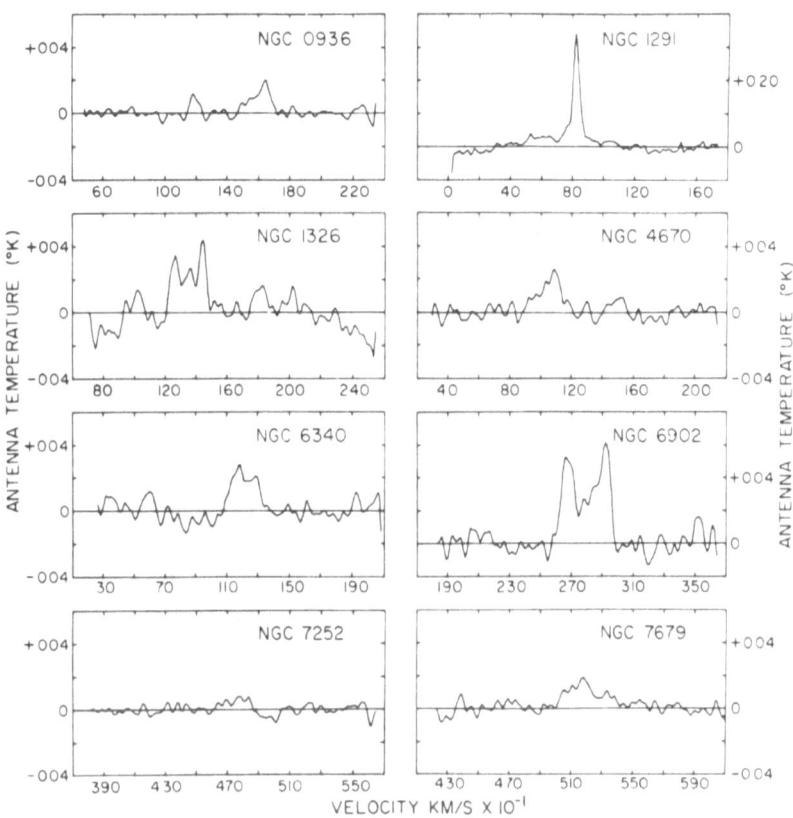

Figure 4.  Lenticular galaxies detected by Balick, Faber and Gallagher
          (1976).
          The profiles of NGC 0936, 6340, and 7679 may be confused with
          those of nearby spirals — see text. The detections of NGC
          1291, 1326 and 6902 have been confirmed at Parkes (cf. Section
          4 and Tables 7 and 8). NGC 1326 has companions at velocities
          of 1000 and 1800 km/s.


     Table 8 lists the lenticulars and S0/a galaxies for which reliable
detections are now available.

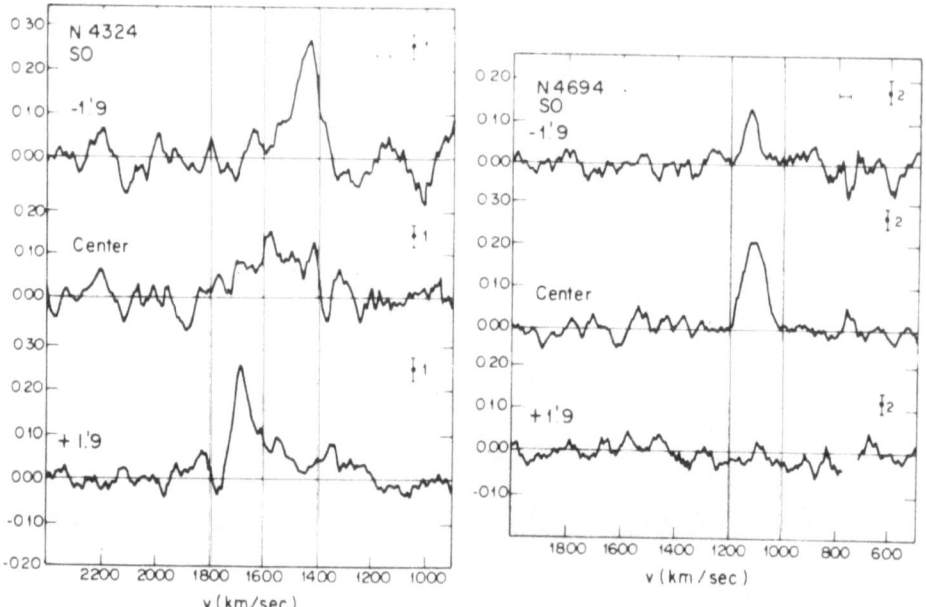

Figure 5. Distribution and motions of neutral hydrogen within two S0
         galaxies in the Virgo Cluster (Krumm and Salpeter, 1976).

         Profiles at positions offset from the centre show that the
         gas in N 4314 is in a disk or ring rotating at 150 km/s;
         in N 4694, the gas is concentrated at the centre, and rotation
         is negligible.

## 4. A SURVEY OF HYDROGEN IN SOUTHERN LENTICULAR GALAXIES

In the course of a survey of southern galaxies ($\delta < -18^{\circ}$) with the
Parkes telescope, Van Woerden, Goss, Mebold, Siegman and Hawarden (1976)
measured the neutral-hydrogen content of all 37 lenticulars with diameter
$D > 2'$, and all 17 S0/a galaxies of any size, in the Reference Catalogue.
This survey constitutes a complete sample in terms of Reference Catalogue
parameters. For the bigger and/or stronger objects, the hydrogen distribution was mapped; mapping also served to clarify cases of possible
confusion with neighbouring galaxies. The thirteen strong, unconfused
detections are listed in Table 7. Among these detected objects, there is
no correlation whatsoever of hydrogen content (in terms of $M_H/L_B$) with
colour or luminosity.

In view of the surprising number of strong detections, Van Woerden
et al. (1976, 1977) inspected the deep IIIa-J plates taken at Siding
Spring for the SRC Schmidt survey, to look for optical peculiarities and
possible classification errors. Several of the detected galaxies show

TABLE 7

Thirteen southern early-type galaxies rich in neutral hydrogen

| Name | Type (RC1) | $M_H \Delta^{-2}$ ($10^6 M_\odot$ Mpc$^{-2}$) | $M_H/L_B$ ($M_\odot/L_\odot$) | $M_H$ ($10^9 M_\odot$) | $L_B$ ($10^9 L_\odot$) | $(B-V)^0_T$ | Modified type (U.K. Schmidt) | Notes |
|---|---|---|---|---|---|---|---|---|
| N 1079 | RSAB(rs)0/ap | 7.7 | 0.28 | 5.8 | 20 | 0.83 | R'SAB(rs)a | |
| N 1291 | RS B( s)0/a | 17.1 | 0.050 | 3.3 | 65 | 0.86 | R'S B(r's)0+ | |
| N 1302 | RS B(r )0/a | 3.4 | 0.06: | 3.5 | 58: | | | T |
| N 1326 | RS B(r )0+ | 9.4 | 0.15 | 5.6 | 37 | 0.74 | R SAB(r )0+ | |
| N 1512 | S B(r )0+ | 54.8 | 1.0 | 10.9 | 10.7 | 0.76 | S B(r )aP | T ✱ |
| N 1533 | S B  0- | 20.6 | 0.57 | 2.7 | 4.9 | 0.89 | S B(r s)0+ | |
| N 1808 | RSAB( s)0/a | 13.5 | 0.11 | 3.4 | 32 | 0.69 | R'SA:( s)aP | T |
| N 2217 | RS B(rs)0+ | 6.7 | 0.10 | 5.1 | 51 | 0.90 | R S B(r s)0+ | |
| N 5084 | S  0Sp | 22.9 | 0.55: | 22.1 | 40: | | SA  PSp | |
| N 5101 | RSAB(rs)0/a | 8.9 | 0.17: | 10.0 | 58: | | R'S B(r s)a | T |
| N 5102 | SA  0- | 21.0 | 0.12 | 0.59 | 5.0 | 0.58 | SA  0- | |
| N 6902 | S B(r )0/a | 17.3 | 0.67 | 53 | 79 | ✱ | SAB(r )b | T ✱ |
| IC 5267 | SA ( s)0/a | 13.6 | 0.24 | 15.6 | 66 | 0.84 | ?R'S?A( ?s)a | |

Legend to Table 7

Types in column 2 are from the 1964 Reference Catalogue; modified types in column 8 are from U.K. Schmidt survey plates (T.G. Hawarden). Types are decoded; S0 ≡ L. Distances Δ have been derived from velocities with respect to the Local Group, using a Hubble constant of 50 km s⁻¹ Mpc⁻¹. Blue luminosities $L_B$ were derived from Δ and magnitudes $B_T$ from the Second Reference Catalogue (De Vaucouleurs et al. 1976), or magnitudes $m_c$ with appropriate corrections.

Notes

T = Type changed in Second Reference Catalogue: N1302, RSB(s)0/a; N1512, SB(r)a; N1808, RSAB(s)a; N5101, RSB(rs)0/a; N6902, R'SB(r)a.

Other notes (✱): N1512: Hawarden et al. (1977), using photometry by Disney et al. (1977), find $L_B = 7.7 \pm$ 1.4 x 10$^9$ $L_\odot$, $M_H/L_B = 1.4 \pm 0.3$.

N6902: Balick et al. (1976) give $C_0' = 0.59$, roughly equivalent to $(B-V)^0_T$.

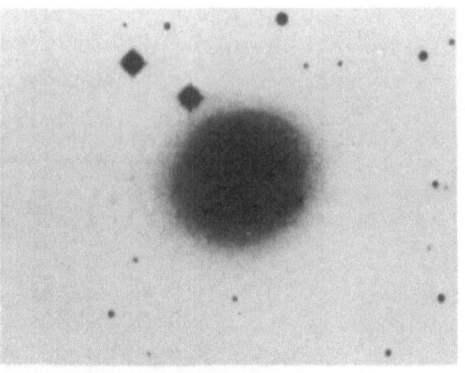

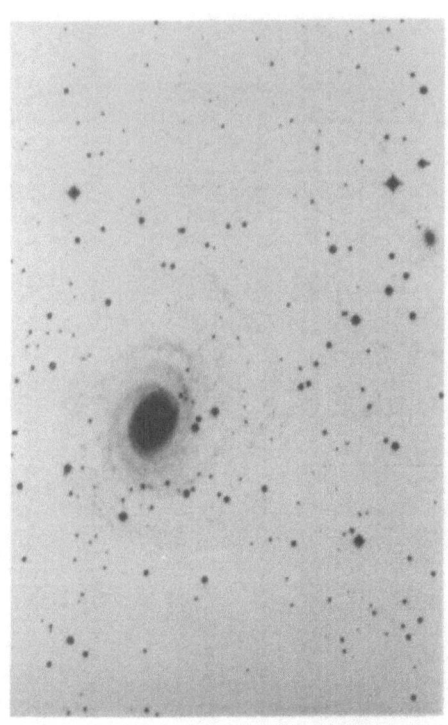

Figure 6 a-b.
NGC 6902 (left) and NGC 1533
(right) from Siding Spring Schmidt
IIIa-J plates. NGC 6902 has well-
developed spiral structure extending over 8 arcmin, several times
De Vaucouleurs' diameter of 2.2
arcmin; revised classification is
SB(r)b. NGC 1533 is an almost-
pure lenticular (type SB(s)0+),
with hardly any sign of spirality.
The $M_H/L_B$ ratios in both systems
are typical for magellanic spirals.

extended -though faint- spiral structure, not recorded in the Reference
Catalogue. An extreme case is NGC 6902 (Figure 6a), which has well-
developed spiral structure extending over 8 arcmin diameter, while the
De Vaucouleurs diameter $D_{25}$ is only 2.2 arcmin. This galaxy is very rich
in hydrogen: $M_H/L_B$ = 0.67, a value typical for magellanic spirals and
irregulars. On the other hand, NGC 1533 (Figure 6b), with a similar $M_H/L_B$,
has hardly any sign of spirality. Its hydrogen distribution extends
asymmetrically over 19 × 12 arcminutes, while $D_{25}$ = 4 arcmin.

NGC 5084 is an edge-on system with $D_{25}$ = 4.8 arcmin. The deep
Schmidt plate shows a warped disk of 15 arcmin diameter; the hydrogen
extends over at least 9 arcmin. The edge-on orientation of this system
makes reliable classification difficult.

The one peculiar object is NGC 1512 (Figure 7), a barred ring
galaxy with compact blue elliptical companion NGC 1510 (Disney et al.,
1977). The Siding Spring Schmidt photo shows extensive, filamentary arms
strongly suggestive of tidal interaction (Hawarden et al., 1977). The
great amounts of gas in this system are spread over a region 24 × 15
arcmin, and Hawarden et al. argue that the companion may well have
accreted enough gas from this distribution to form a new generation of
young stars. Thus the elliptical may be rejuvenated, rather than newly

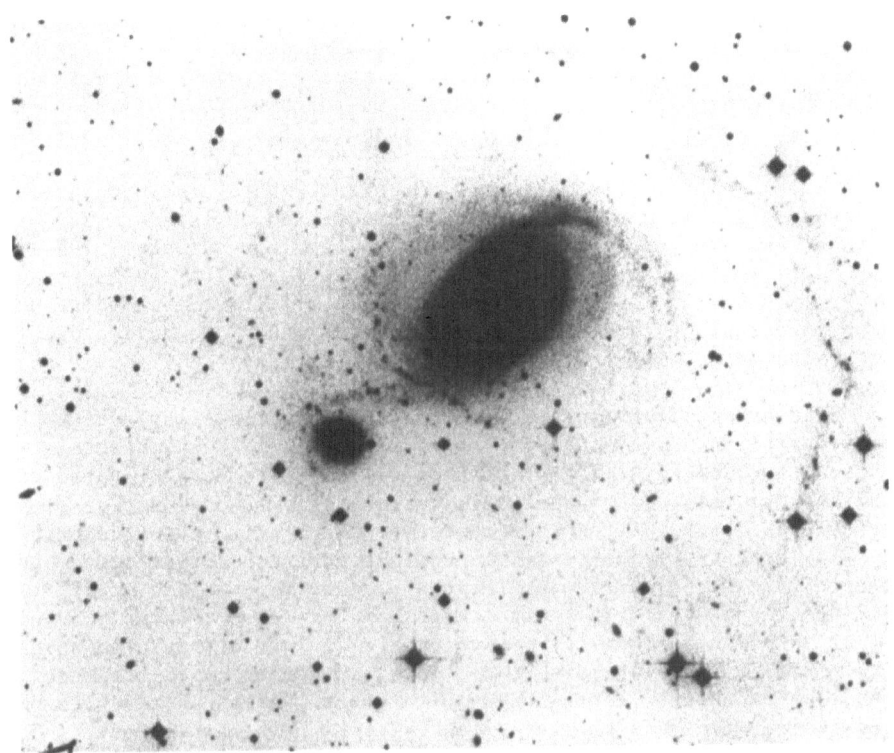

Figure 7. NGC 1512 and 1510 from a Siding Spring Schmidt IIIa-J plate.
         The larger galaxy has extensive, filamentary spiral arms
         which are probably due to tidal interaction with the elliptic-
         al companion at 5.0 arcmin SW. The hydrogen extends over a
         region 24 × 15 arcmin in size. The elliptical may owe its blue
         colour to young stars formed from gas accreted recently
         (Hawarden et al, 1977).

formed as suggested by Disney et al.

    The last column of Table 7 gives classifications based on the
Siding Spring Schmidt plates. In several cases these modified types,
which are intended to be on the De Vaucouleurs (1959) system, are somewhat later than those in the Reference Catalogue, but the difference is
never great enough to explain the high values of $M_H/L_B$, and no correlation of this ratio with morphology is apparent.

    In the Parkes survey of southern lenticulars, many galaxies remained
undetected. Obviously the intrinsic spread in gas content among S0 galaxies must be very great.

## 5. DISCUSSION

From Sections 2-4 it is clear that substantial amounts of gas are present in some non-spiral galaxies. In Section 2 we have concluded that NGC 4278 is the only elliptical galaxy in which neutral hydrogen has been demonstrated beyond doubt; other recent detections remain to be confirmed. The six major surveys of hydrogen in lenticular galaxies mentioned in Table 6 are generally in very good agreement. There are 12 galaxies observed in two or more of these six surveys and detected in at least one of them; only one of these 12 (NGC 4762; Krumm and Salpeter, 1976) fails to be confirmed. Hence, the many other detections reported in these six surveys may also be considered as generally reliable, even where no second observation is available yet. In total then, we have probable detections of neutral hydrogen in 37 galaxies (Table 8).

The most striking feature of the results reviewed appears to be the great variation in gas content, both among elliptical galaxies and among lenticulars. Clearly, the gas content (as described by the ratio of neutral hydrogen mass $M_H$ to blue luminosity $L_B$) is not a function of morphological type alone. Attempts to correlate $M_H/L_B$ with other quantities have met with little success. The correlation with colour found by Balick et al. (1976) is clearly weak (Figure 8), and Van Woerden et al. (1977) conclude that among detected galaxies $M_H/L_B$ is uncorrelated both with colour and with luminosity. No systematic analyses of relationships between $M_H/L_B$ and bulge/disk ratio (a major morphological feature!) have been reported. Also, correlations with group membership and with presence of emission lines and/or nuclear activity might be attempted.

The question where the gas shed by evolving stars has gone in those galaxies having low limits on hydrogen content remains (cf. Faber and Gallagher, 1976). Are galactic winds, intergalactic ram pressure, or formation of new stars responsible? Bieging and Biermann (1977) find that the distribution in the $(M_H/L_B$, B-V) plane is consistent with the occurrence of weak bursts of star formation. However, no "blue" object (B-V < 0.7) is known in which the gas has been used up. A search for HII regions in true S0's with detected HI would be of great importance. Van den Bergh (1976b) reports "a sprinkling of faint (V $\sim$ 22) stars across the disk" in the normal lenticular, NGC 5102, and "two prominent groupings of young stars embedded in diffuse HII regions". (In the same galaxy, Danks et al. (1977) have found a pronounced dust lane, close to the nucleus.) Spectral analysis of such HII regions might yield their chemical composition, thus possibly indicating whether the gas is internal or external in origin.

Of vital importance, too, would be determinations of the distribution and velocity field of the neutral hydrogen. The 300-meter Arecibo dish has already contributed to this problem (cf. Sections 2b and 3). One should hope that synthesis observations at Westerbork will follow. Such observations may indicate whether gas removal occurs in the inner or outer parts of a galaxy, and whether the gas distribution is related to group membership.

TABLE 8

Probable detections of neutral hydrogen in lenticular galaxies

| Galaxy NGC | Values of $M_H \Delta^{-2}$ measured | | | | | Other detections; notes |
|---|---|---|---|---|---|---|
| | BFG 76 | KS 76 | KGFB 77 | BB 77 | WMHGS 77 | |
| 262 | | | | 1.9 | | |
| 1023 | | | | | | 4.6(PS74); 6.7(G69) |
| 1079 | | | | | 7.7±0.5 | |
| 1291 | 18 ±2 | | | | 17.1±1.8 | |
| 1302 | | | 4.3±0.6 | | 3.4±0.4 | |
| 1326 | 4.7±1.2 | | | | 9.4±1.0 | 6.6(BCGL70); note a) |
| 1512 | | | | | 54.8±4.7 | 35±3(DRP77); note b) |
| 1533 | | | | | 20.6±2.1 | |
| 1808 | | | | | 13.5±1.4 | 15.4(BBGH72) |
| 2217 | | | 6.4±1.2 | | 6.7±0.7 | 2.3?(BCGL70) |
| 2685 | | | 3.6±0.8 | | | 6.7(BBGH72); 9(LD73); 8.2(PS74); 5.8(GFB75) |
| 2859 | | | ≤1.2 | 0.20: | | |
| 2962 | | | | 0.79 | | |
| 3032 | | | ≤1.2 | 0.2: | | |
| 3414 | | | | 0.24 | | |
| 3593 | | | | 1.6 | | 2(RD76); <3.5(BBGH72) |
| 3626 | | | 1.0±0.4 | | | 2(RD76) |
| 3941 | | | | 1.1 | | |
| 4203 | | | 4.8±1.1 | 1.5 | | |
| 4262 | | ≥1.3 | ≤1.9 | 1.0 | | note c) |
| 4324 | | 1.6 | | | | |
| 4385 | | 4.6 | | | | |
| 4438 | | | | | | 2.6(HB75) |
| 4670 | 3.1±0.9 | | | | | 2(RD76) |
| 4694 | | 2.6 | | | | |
| 4866 | | | 3.2±0.7 | | | |
| 4958 | | | 1.5±0.4 | | | |
| 5084 | | | | | 22.9±2.4 | |
| 5101 | | | | | 8.9±0.7 | |
| 5102 | | | | | 21.0±2.1 | 24(BCGL70); 17(GFB75) |
| 5128 | | | | | | >62(GW76; note d) |
| 6340 | 3.6±1.1 | | | | | 2.8(BCGL70) |
| 6902 | 10.6±1.6 | | | | 17.3±1.7 | note e) |
| 7280 | | | | 0.38 | | |
| 7625 | | | | 2.6 | | 4.3(PS74); 3.65(BBGH72) |
| 7679 | 3.4±1.1 | | | | | |
| IC5267 | | | | | 13.6±2.4 | |

Key for references: BFG = Balick, Faber, Gallagher; KS = Krumm and Salpeter; KGFB = Knapp, Gallagher, Faber, Balick; BB = Bieging and Biermann; WMHGS = van Woerden, Mebold, Hawarden, Goss, Siegman; PS = Peterson and Shostak; G = Gouguenheim; BCGL = Bottinelli, Chamaraux,

280

H. VAN WOERDEN

Gouguenheim, Lauqué;  DRP = Disney, Rodgers, Pottasch;  BBGH = Balkowski,
Bottinelli, Gouguenheim, Heidmann;  LD = Lewis and Davies;  GFB =
Gallagher, Faber, Balick;  HB = Huchtmeier and Bohnenstengel;  RD = Rood
and Dickel;  GW = Gardner and Whiteoak.

Notes (to Table 8):

a) The low value of $M_H \Delta^{-2}$ in BFG 76 may be due to baseline error (cf.
   Figure 4).
b) The $M_H \Delta^{-2}$ value given by DRP 77 was measured by Fisher at NRAO in a
   single observation; since the hydrogen distribution is very extended,
   part of the flux was missed.
c) The KS 76 value may be low, since only the centre of the galaxy was
   observed.
d) Measured by Van Woerden on profile published by GW 76; part of the
   flux may have been missed.
e) BFG 76 value may be low due to low elevation.

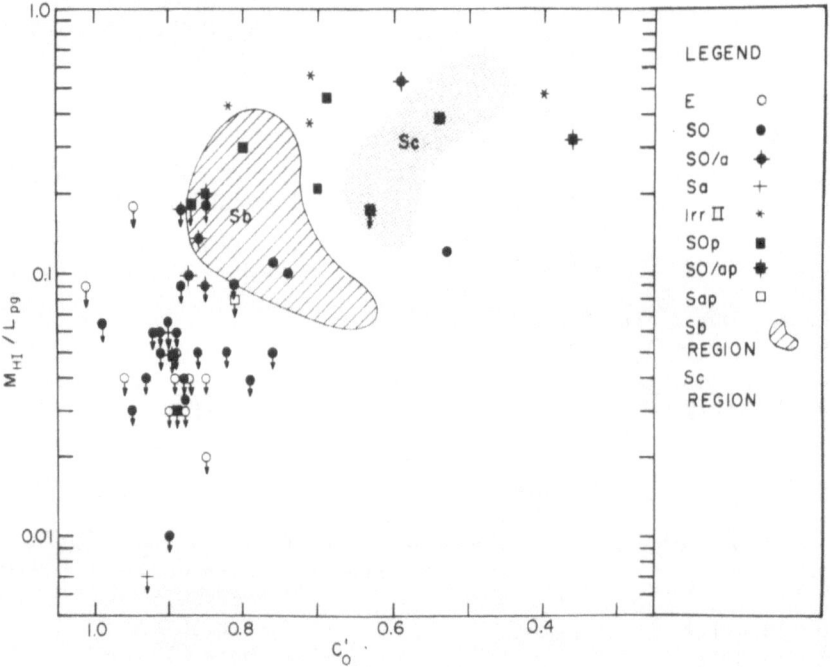

Figure 8. Relationship between gas content and colour? (Balick, Faber
          and Gallagher, 1976)

          The ratio $M_H/L_{pg}$ (hydrogen mass by photographic luminosity)
          plotted versus colour index. Material from Balick et al.
          (1976) and Gallagher et al. (1975). The symbols Sb and Sc
          represent average values for galaxies of those types.

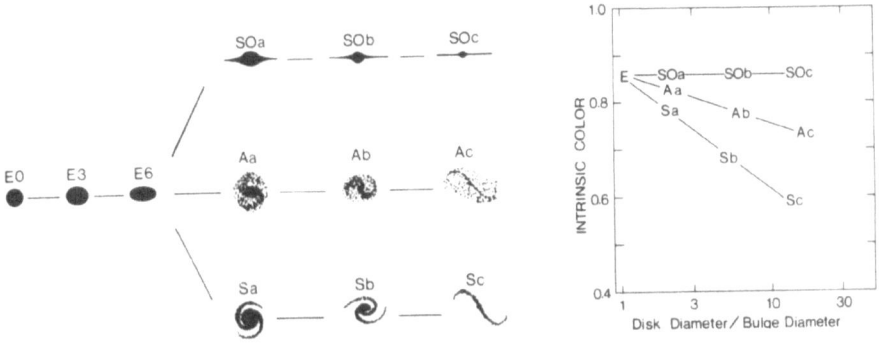

Figure 9. Van den Bergh's (1976a) classification scheme.

Left: The system has three parallel sequences: lenticulars, anemic spirals and normal spirals. Along each sequence, the disk/bulge ratio increases from a to c. The system is further described in the text. Right: Intrinsic colours for Van den Bergh's sequences.


If lenticulars may contain much gas, why do they not show any spiral structure? Clearly, there are objects such as NGC 5102 and 1533 in which no spiral pattern exists. Van den Bergh (1976a) has recently proposed a new classification system (Figure 9a), with parallel sequences of gas-less lenticular galaxies (SOa-b-c), gas-poor "anemic spirals" (Aa-b-c), and gas-rich normal spirals (Sa-b-c). The subdivisions a-b-c are based on the ratio of bulge and disk diameters alone. In this scheme, gas content, rate of star formation, and strength of development of spiral structure are supposed to be related. Van den Bergh also suggests (Figure 9b) that the three sequences differ in intrinsic colour, but no detailed evidence for this is presented. Van den Bergh shows that in the Virgo Cluster SO galaxies and anemic spirals are more frequent than normal spirals, and suggests that this is due to gas depletion by inter-galactic ram pressure in clusters.

This classification scheme is attractive in many respects. Clearly, many of the objects called "lenticulars" in this review (and in the papers on which it is based) may be anemic spirals in Van den Bergh's system (cf. his Table 1). However, we have shown that some of these "lenticulars" are as gas-rich as magellanic spirals and have colours similar to those of Van den Bergh's Sc's. Hence, even with Van den Bergh's new classification there appear to be no simple relationships between gas content, colour and morphological type.

We must conclude that further observations, of both atomic hydrogen and interstellar molecules, and theoretical analyses are required for a full understanding of the gas content of "early-type" galaxies.

ACKNOWLEDGEMENTS

I wish to thank Drs. P. Biermann, L. Bottinelli and L. Gouguenheim, W.K. Huchtmeier, F.J. Kerr, N. Krumm, and especially Mrs. G.R. Knapp for unpublished information. Also, I am grateful to Drs. W.M. Goss and U. Mebold and Mrs. B. Siegman of CSIRO Radiophysics and to Dr. T.G. Hawarden of the U.K. Schmidt Telescope Unit, Siding Spring, for their dedicated cooperation on the Parkes survey. Finally, I acknowledge with gratitude the CSIRO Fellowship, arranged by Drs. J.P. Wild and Dr. B.J. Robinson, during the tenure of which parts of this review were prepared and the Parkes survey was carried out.

REFERENCES

Balick, B., Faber, S.M., Gallagher, J.S.: 1976, Astrophys.J. 209, 710
Balkowski, C., Bottinelli, L., Gouguenheim, L., Heidmann, J.: 1972,
    Astron. Astrophys. 21, 303
Bieging, J.H. and Biermann, P.: 1977, Astron. Astrophys. in press
Bottinelli, L., Chamaraux, P., Gouguenheim, L., Lauqué, R.: 1970,
    Astron. Astrophys. 6, 453
Bottinelli, L. and Gouguenheim, L.: 1977, Astron. Astrophys. 54, 641
Bottinelli, L., Gouguenheim, L., Heidmann, J.: 1973, Astron. Astrophys.
    25, 451
Danks, A.C., Laustsen, S., Van Woerden, H.: 1977, Astron. Astrophys.,
    in preparation
Davies, R.D., Gottesman, S.T., Reddish, V.C., Verschuur, G.L.: 1963,
    Observatory 83, 245
Davies, R.D., Lewis, B.M.: 1973, Monthly Notices Roy. Astron. Soc.
    165, 231
De Vaucouleurs, G.: 1959, Hdb. Physik 53, 275
De Vaucouleurs, G. and De Vaucouleurs, A.: 1964, Reference Catalogue of
    Bright Galaxies (Austin: Univ. Texas Press)
De Vaucouleurs, G., De Vaucouleurs, A., Corwin, H.G.: 1976, "Second
    Reference Catalogue of Bright Galaxies" (Austin: Univ. of Texas
    Press)
Dieter, N.H.: 1962, Astron. J. 67, 222
Disney, M.J., Rodgers, A.W., Pottasch, S.R.: 1977, Astron. Astrophys.,
    in press
Epstein, E.E.: 1964, Astron. J. 69, 490
Faber, S.M. and Gallagher, J.S.: 1976, Astrophys. J. 204, 365
Gallagher, J.S.: 1972, Astron. J. 77, 568
Gallagher, J.S.,Faber, S.M., Balick, B.: 1975, Astrophys. J. 202, 7
Gallagher, J.S., Knapp, G.R., Faber, S.M., Balick, B.: 1977, Astrophys.
    J. 215, 463
Gardner, F.F. and Whiteoak, J.B.: 1976, Proc. Astron. Soc. Australia 3,
    63
Gouguenheim, L.: 1969, Astron. Astrophys. 3, 281
Guibert, J.: 1973, Astron. Astrophys. 29, 335
Hawarden, T.G., Van Woerden, H., Mebold, U., Goss, W.M., Peterson, B.A.:
    1977, Monthly Notices Roy. Astron. Soc., in press

Heeschen, D.S.: 1957, Astrophys. J. 126, 471
Heeschen, D.S.: 1968, Astrophys. J. (Letters) 151, L35
Heidmann, J.: 1961, Bull. Astr. Inst. Netherlands 15, 314
Huchtmeier, W.K. and Bohnenstengel, H.D.: 1975, Astron. Astrophys. 44, 479
Huchtmeier, W.K., Tammann, G.A., Wendker, H.J.: 1975, Astron. Astrophys. 42, 205
Huchtmeier, W.K., Tammann, G.A., Wendker, H.J.: 1976, Astron. Astrophys. 46, 381
Huchtmeier, W.K., Tammann, G.A., Wendker, H.J.: 1977, Astron. Astrophys. 57, 313
Knapp, G.R., Gallagher, J.S., Faber, S.M., Balick, B.: 1977a, Astron.J. 82, 106
Knapp, G.R. and Kerr, F.J.: 1974, Astron. J. 79, 667
Knapp, G.R., Kerr, F.J., Williams, B.A.: 1977b, in preparation
Krumm, N. and Salpeter, E.: 1976, Astrophys. J. (Letters) 208, L7
Lewis, B.M.: 1970, Observatory 90, 264
Lewis, B.M. and Davies, R.D.: 1973, Monthly Notices Roy. Astron. Soc. 165, 213
Mayall, N.U.: 1958, in "Comparison of the Large-Scale Structure of the Galactic System with that of other Stellar Systems" (ed. N.G. Roman), IAU Symp. 5, 23
Mebold, U., Goss, W.M., Van Woerden, H., Hawarden, T.G., Siegman, B.: 1977, Astron. Astrophys., in preparation
Osterbrock, D.E.: 1960, Astrophys. J. 132, 325
Peterson, S.D., Shostak, G.S.: 1974, Astron. J. 79, 767
Roberts, M.S.: 1970, Astrophys. J. (Letters) 161, L9
Roberts, M.S.: 1976, in "Galaxies and the Universe" (eds. A.R. Sandage, M. Sandage and J. Kristian), Stars and Stellar Systems Vol. 9, 309 (Univ. Chicago Press)
Robinson, B.J. and Koehler, J.A.: 1965, Nature 208, 993
Rood, H.J. and Dickel, J.R.: 1976, Astrophys. J. 205, 346
Sandage, A.R., Freeman, K.C., Stokes, N.R.: 1970, Astrophys. J. 160, 831
Shostak, G.S.: 1977, Astron. Astrophys., in preparation
Shostak, G.S., Roberts, M.S., Peterson, S.:1975, Astron. J. 80, 581
Spinrad, H. and Peimbert, M.: 1976, in "Galaxies and the Universe" (eds. A.R. Sandage, M. Sandage and J. Kristian), Stars and Stellar Systems Vol. 9, 37 (Univ. Chicago Press)
Van den Bergh, S.: 1976a, Astrophys. J. 206, 883
Van den Bergh, S.: 1976b, Astron. J. 81, 795 and 895
Van Woerden, H., Goss, W.M., Mebold, U., Siegman, B., Hawarden, T.G.: 1976, Proc. Astron. Soc. Australia 3, 68
Van Woerden, H., Mebold, U., Hawarden, T.G., Goss, W.M., Siegman, B.: 1977, Astron. Astrophys., in preparation
Volders, L.M.J.S. and Van de Hulst, H.C.: 1959, in "Paris Symposium on Radio Astronomy" (ed. R.N. Bracewell), IAU Symp. 9, 423 (Stanford, Univ. Press)
Wentzel, D.G. and Van Woerden, H.: 1959, Bull. Astr. Inst. Netherlands 14, 335
Whiteoak, J.B. and Gardner, F.F.: 1971, Astrophys. Letters 8, 57

Underlined numbers refer to the first page of papers in the present volume contributed by the author in question.

Milgrom, M. 131,133
Milione, V. 205
Miller, E.W. 239
Milman, A.S. 110,112
Milne, D.K. 193,237,241
Milton, J.A. 86,88,237,240
Moffat, A.F.J. 149,153
Monnet, G. 213,219,220,221, 222,224,227,228, 230,238,239,240, 241,243,245,246
Montgomery, J.A. 146
Moore, R.T. 6,16,40,44
Moorwood, A.F.M. 72,77,78,79
Moran, J. 99,103,104
Morgan, D.J. 149,154,162
Morgan, W.W. 211
Mori. T.T. 59,88
Morris, G. 78,121,122
Morris, M. 110,112
Morton, D.C. 8,9,16,35,44, 126,127,128,133
Morton, W. 133
Mufson, S.L. 109,112
Mukai, T. 78
Münch, G. 75,78,88,218,241
Myers, P.C. 54,59,112
Nakagawa, Y. 132
Nandy, K. 61,72,79,155, 157,162
Nasi, E. 250,251,253
Natta, A. 53,59
Neugebauer, G. 59,69,77,102, 103,112
Ney, E.P. 61,75,78
Noguchi, K. 67,78
Norcross, D.W. 153
Nordsieck, K.H. 73,78
O'Dell, C.R. 61,64,65,75,76, 78
O'Donnell, E.J. 129,133,136,147
Ögelman, H. 185
Oka, T. 112
Oke, J.B. 252,253
Okuda, H. 78
Olnon, F.M. 104
O'Neill, A. 153
Onishi, T. 161
Oort, J.H. 201,205
Oppenheimer, M. 132,133,146
Osmer, P.S. 245,247

Osterbrock, D.E. 64,75,79,233,241, 252,253,262,269, 283
Ostriker, J.P. 27,28,30,31,33
Özel, M.A. 185
Page, T.L. 237,239
Pagel, B.E.J. 245,147
Palmer, P. 122,123,160,162, 171,177,181,185
Panagia, N. 53,59,68,79,133
Paresce, F. 33
Parker, R.A.R. 252,253
Patterson, T.A. 153
Paul, J. 179,183,185
Pearson, E.F. 139,146
Pearson, P. 147
Pease, F.G. 259
Pedlar, A. 170,177,187,188, 192,193
Peimbert, M. 47,61,64,76,78,79, 226,229,234,237, 238,241,243,244, 245,246,247,249, 250,251,252,253, 254,262,283
Pelat, D. 246
Pellet, A. 212,213,215,218, 228,238,239
Penman, J.M. 151,153
Pennypacker, C.R. 152,135
Penston, M.V. 102,103,149,153
Penzias, A.A. 88,111,112,122,123
Perek, L. 94
Perrin, M.N. 250,251,253
Persson, S.E. 59,75,78
Peters, W.L. 59,78,164,175, 177,204,205
Peterson, B.A. 271,279,282,283
Petit, H. 214,221,222,224, 227,239,240
Phillips, T.G. 112
Piccinotti, G. 179,185
Pipher, J.L. 67,68,79
Porco, C.C. 154
Pottasch, S.R. 280,282
Pourcelot, A. 239
Prevot, L. 240
Price, R.M. 171,177
Puget, J.L. 177,179, 181,183, 185
Purcell, E.M. 149,150,152,153, 155,161

# ASTROPHYSICS AND SPACE SCIENCE LIBRARY

Edited by

J. E. Blamont, R. L. F. Boyd, L. Goldberg, C. de Jager, Z. Kopal, G. H. Ludwig, R. Lüst,
B. M. McCormac, H. E. Newell, L. I. Sedov, Z. Švestka, and W. de Graaff

24. B. M. McCormac (ed.), *The Radiating Atmosphere. Proceedings of a Symposium Organized by the Summer Advanced Study Institute, held at Queen's University, Kingston, Ontario, August 3–14, 1970.* 1971, XI + 455 pp.
25. G. Fiocco (ed.), *Mesospheric Models and Related Experiments. Proceedings of the 4th ESRIN-ESLAB Symposium, held at Frascati, Italy, July 6–10, 1970.* 1971, VIII + 298 pp.
26. I. Atanasijević, *Selected Exercises in Galactic Astronomy.* 1971, XII + 144 pp.
27. C. J. Macris (ed.), *Physics of the Solar Corona. Proceedings of the NATO Advanced Study Institute on Physics of the Solar Corona, held at Cavouri-Vouliagmeni, Athens, Greece, 6–17 September 1970.* 1971, XII + 345 pp.
28. F. Delobeau, *The Environment of the Earth.* 1971, IX + 113 pp.
29. E. R. Dyer (general ed.), *Solar-Terrestrial Physics/1970. Proceedings of the International Symposium on Solar-Terrestrial Physics, held in Leningrad, U.S.S.R., 12–19 May 1970.* 1972, VIII + 938 pp.
30. V. Manno and J. Ring (eds.), *Infrared Detection Techniques for Space Research. Proceedings of the 5th ESLAB-ESRIN Symposium, held in Noordwijk, The Netherlands, June 8–11, 1971.* 1972, XII + 344 pp.
31. M. Lecar (ed.), *Gravitational N-Body Problem. Proceedings of IAU Colloquium No. 10, held in Cambridge, England, August 12–15, 1970.* 1972, XI + 441 pp.
32. B. M. McCormac (ed.), *Earth's Magnetospheric Processes. Proceedings of a Symposium Organized by the Summer Advanced Study Institute and Ninth ESRO Summer School, held in Cortina, Italy, August 30–September 10, 1971.* 1972, VIII + 417 pp.
33. Antonin Rükl, *Maps of Lunar Hemispheres.* 1972, V + 24 pp.
34. V. Kourganoff, *Introduction to the Physics of Stellar Interiors.* 1973, XI + 115 pp.
35. B. M. McCormac (ed.), *Physics and Chemistry of Upper Atmospheres. Proceedings of a Symposium Organized by the Summer Advanced Study Institute, held at the University of Orléans, France, July 31–August 11, 1972.* 1973, VIII + 389 pp.
36. J. D. Fernie (ed.), *Variable Stars in Globular Clusters and in Related Systems. Proceedings of the IAU Colloquium No. 21, held at the University of Toronto, Toronto, Canada, August 29–31, 1972.* 1973, IX + 234 pp.
37. R. J. L. Grard (ed.), *Photon and Particle Interaction with Surfaces in Space. Proceedings of the 6th ESLAB Symposium, held at Noordwijk, The Netherlands, 26–29 September, 1972.* 1973, XV + 577 pp.
38. Werner Israel (ed.), *Relativity, Astrophysics and Cosmology. Proceedings of the Summer School, held 14–26 August, 1972, at the BANFF Centre, BANFF, Alberta, Canada.* 1973, IX + 323 pp.
39. B. D. Tapley and V. Szebehely (eds.), *Recent Advances in Dynamical Astronomy. Proceedings of the NATO Advanced Study Institute in Dynamical Astronomy, held in Cortina d'Ampezzo, Italy, August 9–12, 1972.* 1973, XIII + 468 pp.
40. A. G. W. Cameron (ed.), *Cosmochemistry. Proceedings of the Symposium on Cosmochemistry, held at the Smithsonian Astrophysical Observatory, Cambridge, Mass., August 14–16, 1972.* 1973, X + 173 pp.
41. M. Golay, *Introduction to Astronomical Photometry.* 1974, IX + 364 pp.
42. D. E. Page (ed.), *Correlated Interplanetary and Magnetospheric Observations. Proceedings of the 7th ESLAB Symposium, held at Saulgau, W. Germany, 22–25 May, 1973.* 1974, XIV + 662 pp.
43. Riccardo Giacconi and Herbert Gursky (eds.), *X-Ray Astronomy.* 1974, X + 450 pp.
44. B. M. McCormac (ed.), *Magnetospheric Physics. Proceedings of the Advanced Summer Institute, held in Sheffield, U.K., August 1973.* 1974, VII + 399 pp.
45. C. B. Cosmovici (ed.), *Supernovae and Supernova Remnants. Proceedings of the International Conference on Supernovae, held in Lecce, Italy, May 7–11, 1973.* 1974, XVII + 387 pp.
46. A. P. Mitra, *Ionospheric Effects of Solar Flares.* 1974, XI + 294 pp.
47. S.-I. Akasofu, *Physics of Magnetospheric Substorms.* 1977, XVIII + 599 pp.
48. H. Gursky and R. Ruffini (eds.), *Neutron Stars, Black Holes and Binary X-Ray Sources.* 1975, XII + 441 pp.
49. Z. Švestka and P. Simon (eds.), *Catalog of Solar Particle Events 1955–1969. Prepared under the Auspices of Working Group 2 of the Inter-Union Commission on Solar-Terrestrial Physics.* 1975, IX + 428 pp.
50. Zdeněk Kopal and Robert W. Carder, *Mapping of the Moon.* 1974, VIII + 237 pp.
51. B. M. McCormac (ed.), *Atmospheres of Earth and the Planets. Proceedings of the Summer Advanced Study Institute, held at the University of Liège, Belgium, July 29–August 8, 1974.* 1975, VII + 454 pp.
52. V. Formisano (ed.), *The Magnetospheres of the Earth and Jupiter. Proceedings of the Neil Brice Memorial Symposium, held in Frascati, May 28–June 1, 1974.* 1975, XI + 485 pp.

53. R. Grant Athay, *The Solar Chromosphere and Corona: Quiet Sun.* 1976, XI + 504 pp.
54. C. de Jager and H. Nieuwenhuijzen (eds.), *Image Processing Techniques in Astronomy. Proceedings of a Conference, held in Utrecht on March 25–27, 1975,* XI + 418 pp.
55. N. C. Wickramasinghe and D. J. Morgan (eds.), *Solid State Astrophysics. Proceedings of a Symposium, held at the University College, Cardiff, Wales, 9–12 July 1974.* 1976, XII + 314 pp.
56. John Meaburn, *Detection and Spectrometry of Faint Light.* 1976, IX + 270 pp.
57. K. Knott and B. Battrick (eds.), *The Scientific Satellite Programme during the International Magnetospheric Study. Proceedings of the 10th ESLAB Symposium, held at Vienna, Austria, 10–13 June 1975.* 1976, XV + 464 pp.
58. B. M. McCormac (ed.), *Magnetospheric Particles and Fields. Proceedings of the Summer Advanced Study School, held in Graz, Austria, August 4–15, 1975.* 1976, VII + 331 pp.
59. B. S. P. Shen and M. Merker (eds.), *Spallation Nuclear Reactions and Their Applications.* 1976, VIII + 235 pp.
60. Walter S. Fitch (ed.), *Multiple Periodic Variable Stars. Proceedings of the International Astronomical Union Colloquium No. 29, Held at Budapest, Hungary, 1–5 September 1975.* 1976, XIV + 348 pp.
61. J. J. Burger, A. Pedersen, and B. Battrick (eds.), *Atmospheric Physics from Spacelab. Proceedings of the 11th ESLAB Symposium, Organized by the Space Science Department of the European Space Agency, held at Frascati, Italy, 11–14 May 1976.* 1976, XX + 409 pp.
62. J. Derral Mulholland (ed.), *Scientific Applications of Lunar Laser Ranging. Proceedings of a Symposium held in Austin, Tex., U.S.A., 8–10 June, 1976.* 1977, XVII + 302 pp.
63. Giovanni G. Fazio (ed.), *Infrared and Submillimeter Astronomy. Proceedings of a Symposium held in Philadelphia, Penn., U.S.A., 8-10 June, 1976.* 1977, X+226 pp.
64. C. Jaschek and G. A. Wilkins (eds.), *Compilation, Critical Evaluation and Distribution of Stellar Data. Proceedings of the International Astronomical Union Colloquium No. 35, held at Strasbourg, France, 19-21 August, 1976.* 1977, XIV+316 pp.
65. M. Friedjung (ed.), *Novae and Related Stars. Proceedings of an International Conference held by the Institut d'Astrophysique, Paris, France, 7-9 September, 1976.* 1977, XIV+228 pp.
66. David N. Schramm (ed.), *Supernovae. Proceedings of a Special IAU Session on Supernovae held in Grenoble, France, 1 September, 1976.* 1977, X+192 pp.
67. Jean Audouze (ed.), *CNO Isotopes in Astrophysics. Proceedings of a Special IAU Session held in Grenoble, France, 30 August, 1976.* 1977, XIII+195 pp.